TELOMERASES

Cover art: Model of telomerase extending telomeric DNA (blue). This model is rendered from available crystal and NMR structures of the telomerase RNA (green; 2K95, 2L3E, 1Z31, and 1OQ0), the telomerase reverse transcriptase (TEN, pink; 2B2A; TRBD, light red; RT, red; and CTE, dark red; 3KYL), the H/ACA snoRNP complex (dyskerin, light blue; Gar1, blue; Nop10, sky blue; and Nhp2, dark blue; 2HVY), and the Pot1-Tpp1 complex (yellow, 1XJV and orange, 2I46; respectively). Image provided by Josh D. Podlevsky and Julian J.-L. Chen (Arizona State University).

TELOMERASES

Chemistry, Biology, and Clinical Applications

Edited by

NEAL F. LUE
Weill Medical College of Cornell University, New York, NY, USA

CHANTAL AUTEXIER
Departments of Anatomy and Cell Biology, and Medicine, McGill University
Bloomfield Centre for Research in Aging, Lady Davis Institute for Medical Research,
Jewish General Hospital, Montreal, Quebec, Canada

WILEY

A JOHN WILEY & SONS, INC., PUBLICATION

Published by John Wiley & Sons, Inc., Hoboken, New Jersey
Published simultaneously in Canada

Library of Congress Cataloging-in-Publication Data:

Telomerases : chemistry, biology, and clinical applications / edited by Neal F. Lue, Chantal Autexier. – 1st. ed.
p. ; cm.
Includes bibliographical references and index.
ISBN 978-0-470-59204-5 (hardback)
I. Lue, F. Neal, 1962- II. Autexier, Chantal, 1963-
[DNLM: 1. Telomerase. QU 56]
572.8′6–dc23 2011047556

Printed in the United States of America

ISBN: 9780470592045

10 9 8 7 6 5 4 3 2 1

CONTENTS

PREFACE

This year marks the 27th anniversary of the discovery of telomerase. In retrospect, even though hints of a special activity needed to maintain linear chromosome ends could be traced to earlier theoretical arguments and experimental observations, it was the exposure of an autoradiogram on Christmas day, 1984 that finally brought the activity into sharp focus and enabled it to be captured, dissected, and manipulated. The fascinating story of the discovery of telomerase has been told elsewhere and will not be repeated here. Our goal for this volume is instead to take stock of what has been learned about this fascinating reverse transcriptase in the ensuing 27 years, in the hope of providing more impetus for the investigation into its chemistry, biology, and clinical applications. If the past 27 years can serve as a guide, than the payoff for the next 27 years of telomerase research would be great indeed.

We have organized this compendium with a view toward offering integrated discussions of the three aspects of telomerase covered by the subtitle. The collection starts with an overview of the telomerase complex, followed by in-depth discussions of the chemistry of its two critical components: TERT and TER. The next two chapters highlight the biological regulatory mechanisms that control the synthesis and assembly of the telomerase complex. Equally significant are the regulations imposed by the nucleoprotein complex at chromosome ends, the topics of the two ensuing chapters. Three more chapters accent studies that bring considerable spotlight to telomerase as a promising target and a useful tool in medical interventions. The collection then concludes with an essay that puts telomerase in evolutionary context and illuminates its place in the extraordinarily diverse family of reverse transcriptases.

Although telomerase research is far from unique in the exploitation of model organisms, it has perhaps uniquely benefited from this approach, as evidenced by the initial discovery of the enzyme in ciliated protozoa, and the demonstration of its importance in chromosome maintenance in budding yeast. The proliferation of model system analysis, while arguably indispensable, also made it difficult even for specialists to keep abreast of all the relevant developments, not to say students and investigators newly attracted to a vibrant research field. A main objective for authors of this volume, then, is not only to gather significant experimental observations, but also to provide an integrated discussion of each significant topic across different model systems. We thank all of the authors for their tremendous efforts in this difficult but admirable endeavor.

This project would not have taken place without the initial suggestion and expert guidance of Anita Lekwani at Wiley. Rebekah Amos and Catherine Odal's help in shepherding the initial drafts into the final texts is greatly appreciated. Finally, we thank our coworkers and colleagues for making the study of telomerase an "endlessly" stimulating and fascinating endeavor.

Neal F. Lue
Chantal Autexier

CONTRIBUTORS

Irina Arkhipova, Josephine Bay Paul Center for Comparative Molecular Biology and Evolution, Marine Biological Laboratory, Woods Hole, MA, USA

Chantal Autexier, Departments of Anatomy and Cell Biology, and Medicine, McGill University; Bloomfield Centre for Research in Aging, Lady Davis Institute for Medical Research, Jewish General Hospital, Montreal, Quebec, Canada

Tara Beattie, Southern Alberta Cancer Research Institute and Departments of Biochemistry and Molecular Biology and Oncology, University of Calgary, Calgary, Alberta, Canada

Pascal Chartrand, Département de Biochimie, Université de Montréal, Montréal, Quebec, Canada

Julian J.-L. Chen, Department of Chemistry and Biochemistry, and School of Life Sciences, Arizona State University, Tempe, AZ, USA

Yu-Sheng Cong, Institute of Aging Research, Hangzhou Normal University School of Medicine, Hangzhou, China

Antonella Farsetti, National Research Council (CNR) and Department of Experimental Oncology, Regina Elena Cancer Institute, Rome, Italy

William Hahn, Department of Medical Oncology, Dana-Farber Cancer Institute and Departments of Medicine, Brigham and Women's Hospital and Harvard Medical School, Boston, MA, USA; Broad Institute of Harvard and MIT, Cambridge, MA, USA

Lea Harrington, Wellcome Trust Centre for Cell Biology, University of Edinburgh, Edinburgh, United Kingdom

Joachim Lingner, Swiss Institute for Experimental Cancer Research (ISREC), School of Life Sciences, *Frontiers in Genetics* National Center of Competence in Research, Ecole Polytechnique Fédérale de Lausanne (EPFL), Lausanne, Switzerland

Yie Liu, Laboratory of Molecular Gerontology, National Institute on Aging, National Institutes of Health Baltimore, MD, USA

Neal F. Lue, Department of Microbiology and Immunology, Weill Medical College of Cornell University, New York, NY, USA

Johanna Mancini, Bloomfield Centre for Research in Aging, Lady Davis Institute for Medical Research, Jewish General Hospital, Montreal, Quebec, Canada

Kenkichi Masutomi, Cancer Stem Cell Project, National Cancer Center Research Institute, Chuo-ku, Tokyo, Japan; PREST, Japan Science and Technology Agency, Saitama, Japan

Jerry W. Shay, Department of Cell Biology, UT Southwestern Medical Center, Dallas, TX, USA

David Shore, Department of Molecular Biology, University of Geneva, *Frontiers in Genetics* National Center of Competence in Research, Geneva, Switzerland

Emmanuel Skordalakes, Gene Expression and Regulation Program, The Wistar Institute, Philadelphia, PA, USA

Phillip G. Smiraldo, Department of Cell Biology, UT Southwestern Medical Center, Dallas, TX, USA

Jun Tang, Department of Cell Biology, UT Southwestern Medical Center, Dallas, TX, USA

Yehuda Tzfati, Department of Genetics, The Silberman Institute of Life Sciences, The Hebrew University of Jerusalem, Safra Campus, Givat Ram, Jerusalem, Israel

Momchil Vodenicharov, Département de biologie and Département de microbiologie et infectiologie, Université de Sherbrooke, Sherbrooke, Québec, Canada

Raymund Wellinger, Département de biologie and Département de microbiologie et infectiologie, Université de Sherbrooke, Sherbrooke, Québec, Canada

Woodring E. Wright, Department of Cell Biology, UT Southwestern Medical Center, Dallas, TX, USA

1

THE TELOMERASE COMPLEX: AN OVERVIEW

JOHANNA MANCINI AND CHANTAL AUTEXIER

1.1 CONSERVATION OF TELOMERE FUNCTION AND THE DISCOVERY OF TELOMERASE

The concept of a healing factor for chromosome ends or "telomeres" was evoked 80 years ago owing to the recognition by Barbara McClintock and Hermann Muller that the natural end of a linear intact chromosome differs from that of a broken chromosome. Using fruit flies and corn as model organisms, they observed that natural chromosome ends, unlike broken ones, never fuse (McClintock, 1931; Muller, 1938). McClintock reported that during cell division in the embryo a broken chromosome can permanently heal to acquire the functions of a natural chromosome end (McClintock, 1939). One of the healing factors or mechanisms was identified 50 years later, in 1985, by Carol Greider and Elizabeth Blackburn, in the ciliated protozoan, *Tetrahymena thermophila*, and named telomere terminal transferase or telomerase (Greider and Blackburn, 1985).

While the function and essential nature of telomeres is conserved among eukaryotes, the DNA sequences, associated proteins and structures at telomeres, and modes of telomere maintenance vary. Recombination-based mechanisms of telomere maintenance have been reported in telomerase-negative immortalized alternative lengthening of telomere (ALT) human cancer cells and upon telomerase gene deletion in yeast, known as Type I, Type II, and heterochromatin amplification-mediated and telomerase-independent (HAATI) (see Chapters 7, 10, 11, and subsequent sections of this chapter) (Cesare and Reddel, 2010; Jain et al., 2010). Recombination can occur

Telomerases: Chemistry, Biology, and Clinical Applications, First Edition.
Edited by Neal F. Lue and Chantal Autexier.

between telomeric and telomeric, subtelomeric or heterochromatin sequences, and may or may not lead to telomere elongation. In *Drosophila melanogaster*, one of the two organisms in which the special function of chromosome ends first became evident, retrotransposons and specialized "terminin" proteins, which are structurally distinct from the typical telomere nucleoprotein complex, are nevertheless capable of supplying the capping function at chromosome ends (see Chapters 7, 10, 11, and subsequent sections of this chapter) (Mason et al., 2008; Raffa et al., 2009, 2010).

However, the most common mechanism for telomere maintenance is the enzyme telomerase, which is almost universally conserved and active in eukaryotes including ciliated protozoa, yeasts, mammals, and plants (see Chapters 2 and 3) (Autexier and Lue, 2006). Prior to the discovery of telomerase, the first telomere sequences had been identified in *T. thermophila,* by Elizabeth Blackburn and Joseph Gall, to consist of repeats of the hexanucleotide TTGGGG (Blackburn and Gall, 1978). Most eukaryotes which maintain telomeres by telomerase possess G-rich sequences at their chromosome ends (see Chapter 7). The search for an enzyme that can maintain telomeres was spurred by the recognition of the "end replication problem" by James Watson and Alexey Olovnikov in the 1970s (see Chapters 7 and 10) (Olovnikov, 1973; Watson, 1972). Based on the properties of the conventional DNA replication machinery, they postulated that DNA at chromosome ends could not be completely replicated and that terminal sequences would be lost at each cell division. The identification of an enzymatic activity that adds G-rich DNA sequences to synthetic telomeric oligonucleotides *in vitro* led to the discovery of the first cellular reverse transcriptase, a ribonucleoprotein (RNP) composed of both RNA and protein (Greider and Blackburn, 1985, 1987, 1989). Two factors were critical to the development of the activity assay: the use of synthetic oligonucleotides with G-rich telomere-like sequences as substrates and the preparation of extracts from *Tetrahymena* as the source of enzyme. The single-stranded G-rich oligonucleotides mimic the natural substrates for telomerase and can be supplied at high concentrations to drive the reaction (Henderson and Blackburn, 1989; McElligott and Wellinger, 1997). In addition, the enzyme is abundant in *T. thermophila* due to the large number of chromosome ends that are generated and which must be stabilized following the chromosome fragmentation and amplification that occurs during the development of the transcriptionally active somatic macronucleus in this organism (Turkewitz et al., 2002).

The importance of telomere synthesis by telomerase is highlighted by the discovery that this mode of replication at DNA ends is evolutionary conserved. Linear DNA exogenously introduced into yeast cells is typically degraded or rearranged. However, Elizabeth Blackburn and Jack Szostak performed what they later described as an outlandish experiment. They attached *T. thermophila* telomeric sequences to the ends of a linear DNA prior to its introduction into yeast and discovered that the DNA was maintained in a stable linear form due to the addition of yeast telomeric sequences to the *T. thermophila* sequences by a yeast cellular machinery (Blackburn et al., 2006; Szostak and Blackburn, 1982). Moreover, when telomerase activity was identified, Carol Greider and Elizabeth Blackburn also discovered that *T. thermophila* can add *T. thermophila* telomeric sequences to a yeast telomeric substrate *in vitro*, emphasizing the evolutionarily conserved nature of telomere synthesis by telomerase (Blackburn et al., 2006; Greider and Blackburn, 1985). For these pioneering and fundamental discoveries, Blackburn, Greider, and Szostak were awarded the Nobel Prize in Physiology and Medicine in 2009.

1.2 THE DISCOVERY OF THE TWO MINIMAL TELOMERASE COMPONENTS

The RNA component of telomerase (referred to as TR or TER in general) contains a short template region, which is repeatedly reverse transcribed into its complementary telomeric DNA sequence (Table 1.1). Initial proof for this function was elucidated using *in vitro* experiments in which an oligonucleotide complementary to the template region of the *T. thermophila* telomerase RNA was found to inhibit telomerase activity, as did the cleavage of the DNA–RNA hybrid at the RNA template region by RNase H (Greider and Blackburn, 1989). In *T. thermophila* cells, expression of mutant telomerase RNAs leads to the synthesis of the correspondingly mutated telomeric sequences at chromosome ends, confirming the function of telomerase in telomere synthesis (Yu et al., 1990). Phenotypes elicited by the synthesis of mutated telomere sequences include altered telomere length homeostasis, impaired cell division, severe delay or block in completing mitotic anaphase, and senescence (Kirk et al., 1997; Yu et al., 1990). These phenotypes underscore the critical nature of the sequence at the telomeres and the essential nature of telomere maintenance for cell survival. Telomerase RNAs from other eukaryotes were identified using biochemical and genetic approaches, however, some RNAs, for example, those from *Schizosaccharomyces pombe* and *Arabidopsis thaliana*, have only been recently discovered largely due to size divergence and weak primary sequence conservation (see Chapter 2) (Cifuentes-Rojas et al., 2011; Leonardi et al., 2008). Despite the large size variation of the telomerase RNAs (ranging from ~150 nucleotides (nt) in ciliates to over 1300 nt in yeasts), the secondary structures of telomerase RNAs are remarkably well conserved (see Chapter 2).

TABLE 1.1 Nomenclature for the Telomerase Catalytic and RNA Subunits in Various Organisms

Species	Catalytic Subunit	RNA Template
Human	hTRT, hTERT, hEST2, TP2, hTCS1	hTR, hTER, hTERC, TRC3
Budding yeast (*Saccharomyces cerevisiae*)	Est2	TLC1
Fission yeast (*Schizosaccharomyces pombe*)	Trt1	TER1
Holotrichous ciliate (*Tetrahymena thermophila*)	tTERT	TR
Plants (*Arabidopsis thaliana*)	AtTERT	TER1

The search for the protein component of telomerase (TERT) proved as daunting as that of the RNA component. Eventually in 1997, sustained efforts by several laboratories culminated in the identification of TERTs from multiple organisms, including *Saccharomyces cerevisiae*, *Euplotes aediculatus*, and human (originally named hTRT, hEST2, TP2, and hTCS1 in human) (Counter et al., 1997; Harrington et al., 1997; Kilian et al., 1997; Lingner et al., 1997b; Meyerson et al., 1997; Nakamura et al., 1997) (Table 1.1). The *S. cerevisiae* TERT gene had, in fact, been identified in 1996 as EST2 (Ever Shorter Telomeres) in a genetic screen for mutants causing senescence and shortening of telomere length (Lendvay et al., 1996). Genetic and biochemical analyses revealed that conserved amino acids within the reverse transcriptase motifs present in TERT are essential for telomerase activity and telomere synthesis both *in vitro* and *in vivo* (Beattie et al., 1998; Counter et al., 1997; Harrington et al., 1997; Nakamura et al., 1997; Weinrich et al., 1997). More recently, several crystal structures of TERT or TERT domains from various organisms have provided a framework for interpreting existing biochemical and genetic data while allowing further targeted experimentation on this protein (Gillis et al., 2008; Jacobs et al., 2006; Mitchell et al., 2010; Rouda and Skordalakes, 2007) (see Chapter 2). Expression of human TERT (hTERT) mRNA correlated with telomerase activity in cell lines (the telomerase RNA component is constitutively expressed), and was found to be upregulated in tumor cells and during immortalization. Hence, hTERT is believed to be the limiting factor for telomerase activity and to be regulated largely through transcription (see Chapter 5) (Feng et al., 1995; Meyerson et al., 1997). The extent of regulation via posttranslational modification of telomerase by phosphorylation and ubiquitination is currently unclear (see Chapter 6). Nonetheless, inactivation of the c-Abl kinase leads to increased telomerase activity and telomere lengths, while overexpression or downregulation of the ubiquitin ligases Hdm2 and MKN1 alters telomerase activity, telomere lengths, and/or cellular resistance to apoptosis (Kharbanda et al., 2000; Kim et al., 2005; Oh et al., 2010).

Another relatively unexplored and poorly characterized aspect of telomerase regulation is the potential contribution of alternatively spliced TERT variants (see Chapter 5). Analysis of the hTERT gene revealed the potential for complex splicing patterns that may reflect a specific aspect of telomerase regulation in proliferation, differentiation, and apoptosis (Kilian et al., 1997; Sykorova and Fajkus, 2009). A number of alternatively-spliced TERT mRNAs have been identified in vertebrates and plants, yet their role in telomere maintenance and cell survival is poorly characterized. In human development, the specific expression of hTERT splice variants that are predicted to encode catalytically-defective telomerases correlates with telomere shortening, suggesting that these transcripts may have important physiological roles (Ulaner et al., 2001).

1.3 TELOMERASE BEYOND THE MINIMAL COMPONENTS: ASSOCIATED PROTEINS

TERT and TR are sufficient to form an active telomerase enzyme when expressed in a rabbit reticulocyte lysate-based transcription and translation system *in vitro*

(Collins, 2006; Collins and Gandhi, 1998; Weinrich et al., 1997). However, a large number of telomerase-associated proteins have been identified in ciliates, yeast, and vertebrates (Autexier and Lue, 2006) (see Chapter 4). The proteins vary greatly between the species and very few are common to all telomerases. While many have been identified as components of a telomerase holoenzyme, some may be associated only transiently with the complex to regulate telomerase assembly and stability, trafficking, localization, posttranslational modification, and recruitment to and activity at the telomere. Consequently, it is difficult to determine whether the holoenzyme has been described in its entirety. A molecular mass of ~270 or ~500 kDa was determined by chromatography of endogenously assembled ciliate telomerases using glycerol gradient sedimentation or gel filtration, respectively (Collins and Greider, 1993; Wang and Blackburn, 1997; Witkin and Collins, 2004). Human and yeast telomerase complexes appear larger (0.6 MDa for yeast, 0.65–2 MDa for human) possibly due to the larger size of RNAs in these organisms and their ability to act as scaffolds to build complex RNPs (Fu and Collins, 2007; Lingner et al., 1997a; Lustig, 2004; Venteicher et al., 2009).

Adding to the challenges of deciphering the components of the holoenzyme are the difficulties encountered in the purification of telomerase protein complexes, typically in very low abundance in nonciliate organisms. Initial purification strategies based on the use of template-complementary oligonucleotide hybridization in ciliates and human led to disruption of ribonucleoprotein assembly (Lingner and Cech, 1996; Schnapp et al., 1998). Recently, more gentle tandem affinity purification strategies, as first described by the group of Kathleen Collins, have yielded a more complete picture of telomerase RNP organization (Fu and Collins, 2007; Venteicher et al., 2008, 2009; Witkin and Collins, 2004).

Telomerase-associated proteins have been best characterized in a single-celled eukaryotes (Fu and Collins, 2007). The ciliate *T. thermophila* is a good model system owing to its cellular structural and functional complexity, arguably comparable to that of metazoans (Turkewitz et al., 2002). Although many of the fundamental discoveries about telomerase and telomere biology were made using *T. thermophila*, this organism's telomerase appears to have a unique RNP biogenesis pathway that involves the telomerase-specific proteins p65, p45, p75, and p20 (O'Connor and Collins, 2006; Witkin and Collins, 2004; Witkin et al., 2007). More recently, three additional holoenzyme proteins were identified, p19, p50, and p82 (Min and Collins, 2009). The p75, p45, and p19 form a telomere adaptor subcomplex, TASC, whose recruitment to the core enzyme (p65, TERT, and TER) is regulated by the p50 subunit. The p82 subunit is a Replication Protein A (RPA)-related sequence-specific DNA-binding protein, which confers high repeat addition processivity to the telomerase holoenzyme. The RNP biogenesis pathways of yeast and human telomerase employ a set of proteins shared with more abundant RNPs (Collins, 2006). Proteins involved in yeast telomerase RNA processing, stability, trafficking, and biogenesis include importin B, which is involved in nuclear import of mRNA binding proteins, as well as proteins involved in spliceosomal small nuclear (sn) RNP processing (Chapon et al., 1997; Ferrezuelo et al., 2002; Seto et al., 1999). Proteins involved in human telomerase RNA processing and stability, and in RNP trafficking and biogenesis include proteins of H/ACA small nucleolar (sno) and small Cajal body (sca) CAB box-containing RNPs, such as dyskerin, NHP2, NOP10, GAR1, and TCAB1 (telomerase Cajal body protein 1), the chaperone proteins p23 and hsp90, the AAA + ATPases pontin and reptin, the nucleolar acetyltransferase NAT10, and the nucleolar GTPase GNL3L (Cohen et al., 2007; Collins, 2008; Fu and Collins, 2007; Mitchell et al., 1999; Venteicher et al., 2008, 2009). Some of these proteins have been identified using tandem affinity purifications, and it has been proposed that telomerase-associated proteins present at substoichiometric levels might be regulatory as opposed to H/ACA proteins and hTERT, which are required for biological stability and catalytic activity, respectively (Fu and Collins, 2007).

In addition to telomeric proteins, which aid in the recruitment of telomerase to the telomere (see below), a number of other proteins have been identified which have been implicated in the localization and recruitment of telomerase to the nucleus and to the telomere. In yeast, these include Est1 and the Ku70/80 heterodimer, while in human the 14-3-3 regulator of intracellular protein localization, the telomerase inhibitor PinX1, and the heterogenous nuclear RNP family of proteins may regulate localization of telomerase to the nucleus or recruitment to the telomere (Banik and Counter, 2004; Collins, 2006, 2008; Fisher et al., 2004; Ford et al., 2000; Fu and Collins, 2007; Hughes et al., 2000; LaBranche et al., 1998; Seimiya et al., 2000; Zappulla and Cech, 2004; Zhou and Lu, 2001).

1.4 REGULATION OF TELOMERASE BY TELOMERIC PROTEINS AND RNAS

Interestingly, the relationship between telomeres and telomerase extends beyond the role of telomeres as telomerase substrates (see Chapter 7). While disruption of numerous proteins leads to alterations of telomere homeostasis in mammalian cells, including many proteins involved in the maintenance of genomic integrity (e.g., proteins affecting DNA replication, repair, recombination, and the DNA damage response), a six-protein complex known as the "shelterin" complex (TRF1, TRF2, hRAP1, TPP1, POT1, and TIN2), are directly responsible for the protection of mammalian telomeres (d'Adda de Fagagna, 2008; Palm and de Lange, 2008; Slijepcevic, 2008). The shelterin proteins mediate the formation of a t-loop structure at telomeres, which prevents the recognition of the end of the chromosome as a DNA double-strand break and precludes engagement of a DNA damage response. Regulation of telomerase by telomere binding proteins or proteins that associate with telomeres can either be indirect or direct. Proteins that affect access of telomerase to telomeres, including proteins implicated in the generation of the single-stranded G-rich telomere overhang, can be viewed as indirect regulators, while those that recruit telomerase to the telomere and/or modulate telomerase activity are direct regulators. A number of proteins, for example, budding yeast Rif1/2 and mammalian TRF1 and TRF2, regulate telomerase by altering telomere structure and/or length and by increasing telomerase accessibility (see Chapter 7). TPP1 regulates telomerase recruitment to the telomeres and, in concert with Pot1, also regulates activity of telomerase at the telomere (Abreu et al., 2010; Latrick and Cech, 2010; Wang et al., 2007; Xin et al., 2007; Zaug et al., 2010). Similarly, Cdc13, one of the telomeric proteins in budding yeast, participates in the recruitment of telomerase to telomeres, and evidently activates the enzyme as well (Pennock et al., 2001). In fission yeast, Tpz1 (orthologue of the mammalian TPP1) and the associated factors Poz1, Pot1, and Ccq1, are also implicated in telomerase recruitment (Miyoshi et al., 2008; Tomita and Cooper, 2008). Interestingly, TPP1 is a homologue of ciliate TEBP-β, one of the first telomere binding proteins to be identified (Price and Cech, 1989; Xin et al., 2007). The interaction between TPP1/TEBPβ and telomerase appears to be one of the very few conserved interactions between telomeric proteins and telomerase.

Another potentially significant regulator of telomerase at telomeres is the recently discovered telomeric repeat containing RNA (TERRA). These noncoding RNAs are detected at yeast, mammalian, and plant telomeres, and are transcribed from the subtelomeric regions to the chromosome ends (Azzalin et al., 2007; Feuerhahn et al., 2010; Schoeftner and Blasco, 2008; Vrbsky et al., 2010) (see Chapter 6). Interestingly in *A. thaliana*, antisense telomeric transcripts (ARRET) are also reported (Vrbsky et al., 2010). One of the postulated roles for TERRA for which evidence is accumulating, is in the regulation of telomerase. TERRA can bind to telomerase and act as a potent competitive inhibitor for telomeric DNA (Redon et al., 2010; Schoeftner and Blasco, 2008). Increased levels of TERRA are also correlated with shorter telomeres (Luke et al., 2008).

1.5 TELOMERASE, TELOMERE MAINTENANCE, CANCER, AND AGING

In 1989, shortly following the identification of telomerase activity in the human cell line—HeLa, numerous studies were performed to assess the status of telomerase activity and of telomere length in various types of human cells (Morin, 1989). The telomere hypothesis of cellular aging and immortalization emerged as a consequence of the correlation found in these studies between telomere length and telomerase activity in human cells (Harley, 1991) (see Chapter 10). Briefly, because telomerase was active in immortal, transformed human cells and in tumor cell lines, but not in normal somatic cells, and because telomere lengths were maintained with increasing numbers of cell division in the former cells, but not in the latter cells, it was postulated that telomere length serves as a mitotic clock in normal human somatic cells. Telomere shortening in normal human somatic cells occurs in a cell division-dependent fashion, eventually triggering replicative senescence and exit from the cell cycle. The presence of telomerase and the maintenance of telomere length in immortal, transformed human cells and in tumor cell lines support the concept that telomere maintenance is a key requirement for unlimited replication of tumor cells (Hanahan and Weinberg, 2000; Harley, 1991). In 1997, a survey of more than 3500 tumor and control samples showed that telomerase is detected in approximately 85% of cancers, but is absent or weakly expressed in primary cells (Shay and Bacchetti, 1997). This and other studies, as well as the telomere hypothesis for cellular aging and immortalization led to testable predictions, and to the identification of telomerase as an attractive target for anticancer therapy.

Addressing if telomere shortening is a cell division clock that limits cellular lifespan became possible following the identification of hTERT. Elegant experiments by the groups of Woodring Wright and Jerry Shay demonstrated that expression of hTERT in normal human fibroblast cells with limited lifespan led to the induction of telomerase activity, telomere maintenance, and extension of lifespan (Bodnar et al., 1998; Counter et al., 1998; Vaziri and Benchimol, 1998). Importantly, the cells did not adopt characteristics of cancer cells (Jiang et al., 1999; Morales et al., 1999). It was noted however, that telomerase activation was not sufficient to immortalize some normal human cell types, suggesting that other factors besides telomere length, for example, the levels of the tumor suppressor p16, contributed to replicative senescence in human cells (Kiyono et al., 1998). Several pioneering studies addressed the role of telomerase in tumorigenesis, and demonstrated that telomerase activation is essential but not sufficient for transformation of human cells (Hahn et al., 1999, 2002). In these experiments, normal human fibroblasts were converted to tumorigenic cells capable of forming tumors in immunodeficient mice. This conversion required the expression of hTERT and alterations in key cellular genes including the tumor suppressors pRB, p53, the protooncogene Ras, and protein phosphatase 2A.

While the disruption of the telomerase RNA in ciliate and yeast model organisms provided early evidence for an important role of telomerase in cell survival (Singer and Gottschling, 1994; Yu et al., 1990), the potential of telomerase inhibition as a therapeutic approach for treating human cancer was first demonstrated by the expression of antisense hTR in immortal HeLa cells (Feng et al., 1995). Transfection of HeLa cells with an antisense hTR led to loss of telomerase activity, telomere shortening, and cell death after 20–26 population doublings. Since then, several approaches for targeting telomerase and also telomeres have been developed and tested, with several ongoing clinical trials (see Chapter 10) (Harley, 2008).

The first evidence for a role of telomerase and telomere length in organismal aging came from studies in telomerase knockout mouse models (see Chapter 9). Loss of telomere function in aging late generation $mTR^{-/-}$ mice did not elicit a full spectrum of classical pathophysiological symptoms of aging. However, age-dependent telomere shortening and accompanying genetic instability were associated with shortened life span, hair loss and graying, as well as a reduced capacity to respond to stresses such as wound healing and hematopoietic ablation (Rudolph et al., 1999). Premature aging is also characteristic of patients with a rare multisystem disorder, dyskeratosis congenita (DC), who present with three distinctive clinical characteristics: abnormal skin pigmentation, nail dystrophy, and mucosal leukoplakia (Kirwan and Dokal, 2008, 2009). The underlying molecular defect in many DC patients turns out to be abnormally short telomeres due to mutations in the telomerase holoenzyme components dyskerin, TERC, TERT, NOP10, and NHP2. Mutations in the shelterin component, TIN2, have also been identified. Three different subtypes have been described: X-linked recessive, autosomal dominant, autosomal recessive, with the most common fatal complications related to bone marrow failure, pulmonary fibrosis, and cancer.

The link between telomerase and DC was first made in X-linked DC, which is caused by mutations in the gene encoding dyskerin (Mitchell et al., 1999). Due to the role of dyskerin in H/ACA snoRNP biogenesis, DC was initially believed to be due to defects in ribosomal RNA processing. However, dyskerin was found to bind to a previously unidentified H/ACA RNA motif within hTR, and DC patients with mutant dyskerin have decreased hTR levels, decreased telomerase activity, and shorter telomeres.

Mutations in hTERT and hTERC have also been described in other diseases, including other bone marrow failure syndromes such as aplastic anemia (AA), pancytopenia, and myelodysplastic syndrome (MDS), as well as in diseases not typically associated with blood disorders, such as idiopathic pulmonary fibrosis (IPF) and liver disorders (Armanios, 2009; Armanios et al., 2007; Kirwan and Dokal, 2009; Savage and Alter, 2009).

1.6 TELOMERASE BEYOND TELOMERE SYNTHESIS

The initially defined biological function of a protein may limit the identification or assessment of less well characterized roles for the protein (Blackburn, 2005). First identified as having an essential role in the maintenance of telomere length and protection of genetic information, it was not until the late 1990s that evidence of additional telomere synthesis-independent roles for telomerase began to emerge (Blackburn, 2000, 2005; Bollmann, 2008; Martinez and Blasco, 2011) (see Chapter 8). TERT overexpression studies suggested a possible role for TERT in the promotion of tumorigenesis and tumor dissemination (Artandi et al., 2002; Canela et al., 2004; Gonzalez-Suarez et al., 2001, 2002), and in the resistance to cell inhibition and death, in certain instances, of postmitotic, nondividing cells (Lee et al., 2008; Rahman et al., 2005). TERT overexpression leads to rapid induction of growth-promoting genes (Smith et al., 2003), stimulation of hair follicle stem cell proliferation which in some studies was independent of the telomerase RNA component (Choi et al., 2008; Flores et al., 2005; Martinez and Blasco, 2011; Sarin et al., 2005), and activation of the Myc and Wnt pathways (Choi et al., 2008; Park et al., 2009). Park et al. showed that TERT modulates Wnt/β-catenin signaling by serving as a cofactor in a β-catenin transcriptional complex, revealing yet another unanticipated role for the catalytic subunit of telomerase. Alteration of histone modification and sensitization of human cells to DNA damage were observed in TERT small interfering (si) RNA knock-down studies (Masutomi et al., 2005). Contrary to the evidence that TERT affects Wnt signaling, Vidal-Cardenas and Greider (2010) reported no change in gene expression or DNA damage response in both $mTR^{-/-}$ G1 and $mTERT^{-/-}$ G1 mice with long telomeres when compared to wild-type mice. More recently, Strong et al. (2011) failed to find evidence of altered Wnt signaling in various adult and embryonic tissues of mTERT-deficient mice. Additional studies which aim to clarify the role of TERT in Wnt signaling will be required. Other potential alternative roles of telomerase, for example, in the mitochondria, continue to be investigated (see Chapter 8) (Martinez and Blasco, 2011).

Most recently, a novel RNA partner for hTERT was discovered, highlighting a new role for telomerase. Maida et al. (2009) showed that hTERT interacts with the RNA component of mitochondrial RNA processing endoribonuclease (RMRP). Together, they form a ribonucleoprotein complex that exhibits RNA-dependent RNA polymerase (RdRP) activity, generating double-stranded RNAs that are processed in a Dicer-dependent manner into siRNA. Mutations in RMRP are found in cartilage-hair hypoplasia (CHH) (Ridanpaa et al., 2001), suggesting a link between the integrity of the hTERT–RMRP complex and disease development and progression (Maida et al., 2009).

1.7 TELOMERE MAINTENANCE WITHOUT TELOMERASE

Most cancers, which are characterized by high rates of proliferation and high rates of genomic instability, have adapted to the high rate of division by upregulating telomerase activity (Shay and Bacchetti, 1997). However, 10–15% of cancers are able to maintain their telomere lengths in the absence of telomerase, using one or more recombination-based mechanisms referred to as ALT (Cesare and Reddel, 2010; Shay and Bacchetti, 1997). An additional alternate mode of telomerase-independent telomere maintenance occurs in *D. melanogaster* via retrotransposon-type mechanisms (Mason et al., 2008) (see Chapter 11).

While a recombination-mediated method to replicate telomeres was suggested by Walmsley et al. (1984), the first evidence of a recombination-dependent telomere length maintenance mechanism was described in survivors of an *est1*-null mutant of *S. cerevisiae* (Bhattacharyya et al., 2010; Lundblad and Blackburn, 1993; Lundblad and Szostak, 1989). Yeasts that survive in the absence of telomerase holoenzyme components present different methods of survival (Lendvay et al., 1996; Lundblad and Blackburn, 1993; Singer and Gottschling, 1994; Teng and Zakian, 1999). Two classes of survivors were initially identified (Teng and Zakian, 1999). Those classified as Type I show drastically amplified Y' DNA elements that are found in the subtelomeric region of most chromosomes and retain very short terminal repeats, while Type II survivors have long heterogeneous telomere tracts, reminiscent of ALT in human cancer cells.

Fission yeast, on the other hand, survive in the absence of telomerase mainly via circularization of their chromosomes. However, "linear survivors," formed via recombination between persisting telomere sequences, are also observed (Nakamura et al., 1998). Most recently, an additional mode of telomerase-null "linear survivors" was characterized in *S. pombe* (Jain et al., 2010). These cells survive the loss of telomeric sequences by continually amplifying and rearranging heterochromatic sequences using the heterochromatin assembly machinery, and are thus referred to as "HAATI". The linearity of HAATI chromosomes is preserved by Pot1 and its interacting partner Ccq1 (Jain et al., 2010; Miyoshi et al., 2008). Pot1 is able to confer its essential end-protection function in the absence of its specific DNA binding sequence, demonstrating that, as in *D. melanogaster*, telomere sequence is dispensable for chromosome linearity in fission yeast (Jain et al., 2010).

Recombination at human telomeres was first proposed based on the observation of rapid telomere lengthening and shortening in telomerase-negative cells (Cesare and

Reddel, 2010; Murnane et al., 1994). The telomeres of ALT cells retain features common to those of telomerase-positive cells, including double- and single-stranded telomeric repeats, the association of shelterin and other proteins, and the t-loops structures (Cesare and Reddel, 2010). However, ALT cells are characterized by the heterogeneous nature of their telomere lengths, ranging from <2 to >50 kb (Bryan et al., 1995; Cesare and Reddel, 2008, 2010). Hallmarks of ALT include the generation of extrachromosomal telomeric DNA and ALT-associated promyelocytic leukemia bodies (APBs, sites of DNA synthesis and possibly recombination), although these features are also detectable in telomerase-positive cells that have undergone trimming of over-lengthened telomeres (Cesare and Reddel, 2010; Draskovic et al., 2009; Nabetani et al., 2004; Yeager et al., 1999). There have also been reports of telomerase-negative cancer cells that do not have all the characteristics typically associated with ALT cells (Cerone et al., 2005; Fasching et al., 2005; Marciniak et al., 2005), highlighting the potential for complex and varied mechanisms of telomere maintenance. Recent studies by Henson et al. (2009) have shown extrachromosomal C-circles, consisting of a complete C-rich strand and an incomplete G-rich strand, to be the best indicator of whether ALT activity is present. Three suggested mechanisms of telomere elongation in ALT cells, which are not mutually exclusive, include telomere sister chromatid exchanges (T-SCEs), homologous recombination-dependent telomere copying, and t-loop junction resolution (Cesare and Reddel, 2008, 2010).

Unlike most organisms, the telomere elongation and capping functions are naturally uncoupled in *D. melanogaster* (Rong, 2008). A distinctive feature of the fruit fly is that it has no telomerase. Instead, its telomere structure is comprised of head-to-tail arrays of three different telomere-specific non-long-terminal-repeat (non-LTR) retrotransposons, HeT-A, TART, and TAHRE found only at the chromosome ends (Mason et al., 2008; Rong, 2008) (see Chapter 11). All organisms possess an end-capping complex to protect the chromosome end from being recognized as a double-stranded break by the DNA repair machinery. *D. melanogaster* uses a sequence-independent mechanism, contrary to the short repeats employed by most organisms. While a number of telomere-capping proteins prevent chromosome end-to-end fusions in *D. melanogaster*, only three proteins have been found to localize exclusively at telomeres and function solely in telomere maintenance. These are the HP1/ORC2-associated protein (HOAP), modigliani (moi), and Verrocchio (Ver) (Cenci et al., 2003; Perrini et al., 2004; Raffa et al., 2009, 2010). These proteins are functional equivalents of the shelterin complex and have been collectively given the name "terminin" (Raffa et al., 2009, 2010). Modigliani encodes a novel protein that binds both HOAP and the heterochromatin protein HP1, which efficiently binds and stabilizes ssDNA much like POT1.

Although the telomerase-based telomere elongation system enhances telomere stability and length control efficacy, the survival of organisms utilizing various forms of ALT and recombination mechanisms suggests that adaptation is possible. Alternative telomere maintenance mechanisms have been observed after telomerase inhibition (Bechter et al., 2004) or genetic deletion of telomerase (Chang et al., 2003; Hande et al., 1999; Morrish and Greider, 2009; Niida et al., 2000).

These observations potentially complicate the development of treatments that target telomerase or telomere function. Studies in model organisms, including yeast and mice reveal increased telomeric recombination after induction of telomere dysfunction through mutation or deletion of telomere-capping proteins (Bechard et al., 2009; Celli et al., 2006; Grandin et al., 2001; He et al., 2006; Iyer et al., 2005; Teng et al., 2000; Underwood et al., 2004; Wu et al., 2006). Recently telomeric recombination was also observed following the induction of telomere dysfunction in telomerase-positive cells, suggesting that telomeric recombination may be a potential adaptation mechanism in response to telomere dysfunction in mammalian cells (Brault and Autexier, 2010).

1.8 CONCLUSION

The discovery of telomerase was the result of a quest to understand a basic biological question: how are the ends of a linear chromosome replicated? The success of this quest led to a range of experimental questions touching on fundamental aspects of cell function and regulation. Even though quite unanticipated at the outset, the study of telomerase also provided critical insights on aging and cancer. The full significance and implication of the discovery of telomerase are only now becoming clear, as the contributions of Elizabeth Blackburn, Carol Greider, and Jack Szostak to the advancement of our knowledge in this field were recognized by the Nobel Foundation in 2009. Our understanding of telomerase regulation and function remains far from complete. The next few years will surely witness new and exciting developments in the field with regard to fundamental mechanisms of telomerase regulation and function. These developments should in turn provide the foundation for designing specific and effective therapeutic strategies to modulate telomerase in disease.

ACKNOWLEDGMENT

Chantal Autexier acknowledges support from the Canadian Institutes of Health Research, the Canadian Cancer Society and Le Fonds en Recherche en Santé du Québec.

REFERENCES

Abreu E, Aritonovska E, Reichenbach P, Cristofari G, Culp B, Terns RM, Lingner J, Terns MP. (2010) TIN2-tethered TPP1 recruits human telomerase to telomeres *in vivo*. *Mol. Cell Biol.* **30**(12): 2971–2982.

Armanios M. (2009) Syndromes of telomere shortening. *Annu. Rev. Genomics Hum. Genet.* **10**: 45–61.

Armanios MY, Chen JJ, Cogan JD, Alder JK, Ingersoll RG, Markin C, Lawson WE, Xie M, Vulto I, Phillips JA, 3rd, Lansdorp PM, Greider CW, Loyd JE. (2007) Telomerase mutations in families with idiopathic pulmonary fibrosis. *N. Engl. J. Med.* **356**(13): 1317–1326.

Artandi SE, Alson S, Tietze MK, Sharpless NE, Ye S, Greenberg RA, Castrillon DH, Horner JW, Weiler SR, Carrasco RD, DePinho RA. (2002) Constitutive telomerase expression promotes mammary carcinogenesis in aging mice. *Proc. Natl. Acad. Sci. USA* **99**: 8191–8196.

Autexier C, Lue N. (2006) The structure and function of telomerase reverse transcriptase. *Ann. Rev. Biochem.* **75**: 493–517.

Azzalin CM, Reichenbach P, Khoriauli L, Giulotto E, Lingner J. (2007) Telomeric repeat containing RNA and RNA surveillance factors at mammalian chromosome ends. *Science* **318**(5851): 798–801.

Banik SSR, Counter CM. (2004) Characterization of interactions between PinX1 and human telomerase subunits hTERT and hTR. *J. Biol. Chem.* **279**: 51745–51748.

Beattie TL, Zhou W, Robinson MO, Harrington L. (1998) Reconstitution of human telomerase activity *in vitro*. *Curr. Biol.* **8**: 177–180.

Bechard LH, Butuner BD, Peterson GJ, McRae W, Topcu Z, McEachern MJ. (2009) Mutant telomeric repeats in yeast can disrupt the negative regulation of recombination-mediated telomere maintenance and create an alternative lengthening of telomeres-like phenotype. *Mol. Cell Biol.* **29**(3): 626–639.

Bechter OE, Zou Y, Walker W, Wright WE, Shay JW. (2004) Telomeric recombination in mismatch repair deficient human colon cancer cells after telomerase inhibition. *Cancer Res.* **64**: 3444–3451.

Bhattacharyya S, Sandy A, Groden J. (2010) Unwinding protein complexes in ALTernative telomere maintenance. *J. Cell Biochem.* **109**(1): 7–15.

Blackburn EH. (2000) Telomere states and cell fates. *Nature* **408**(6808): 53–56.

Blackburn EH. (2005) Cell biology: Shaggy mouse tales. *Nature* **436**(7053): 922–923.

Blackburn EH, Gall JG. (1978) A tandemly repeated sequence at the termini of the extra-chromosomal ribosomal RNA genes in *Tetrahymena*. *J. Mol. Biol.* **120**: 33–53.

Blackburn EH, Greider CW, Szostak JW. (2006) Telomeres and telomerase: the path from maize, *Tetrahymena* and yeast to human cancer and aging. *Nat. Med.* **12**(10): 1133–1138.

Bodnar AG, Ouellette M, Frolkis M, Holt SE, Chiu C-P, Morin GB, Harley CB, Shay JW, Lichtsteiner S, Wright WE. (1998) Extension of life-span by introduction of telomerase into normal human cells. *Science* **279**: 349–352.

Bollmann FM. (2008) The many faces of telomerase: emerging extratelomeric effects. *BioEssays* **30**: 728–732.

Brault ME, Autexier C. (2010) Telomeric recombination induced by dysfunctional telomeres. *Mol. Biol. Cell* **22**: 179–188.

Bryan TM, Englezou A, Gupta J, Bacchetti S, Reddel RR. (1995) Telomere elongation in immortal human cells without detectable telomerase activity. *EMBO J.* **14**: 4240–4248.

Canela A, Martin-Caballero J, Flores JM, Blasco MA. (2004) Constitutive expression of tert in thymocytes leads to increased incidence and dissemination of T-cell lymphoma in Lck-Tert mice. *Mol. Cell Biol.* **24**: 4275–4293.

Celli GB, Denchi EL, de Lange T. (2006) Ku70 stimulates fusion of dysfunctional telomeres yet protects chromosome ends from homologous recombination. *Nat. Cell Biol.* **8**(8): 885–890.

Cenci G, Siriaco G, Raffa GD, Kellum R, Gatti M. (2003) The *Drosophila* HOAP protein is required for telomere capping. *Nat. Cell Biol.* **5**(1): 82–84.

Cerone MA, Autexier C, Londono-Vallejo J-A, Bacchetti S. (2005) A human cell line that maintains telomeres in the absence of telomerase and of key markers of ALT. *Oncogene* **24**: 7893–7901.

Cesare AJ, Reddel RR (2008) Telomere uncapping and alternative lengthening of telomeres. *Mech. Ageing Dev.* **129**(1, 2): 99–108.

Cesare AJ, Reddel RR. (2010) Alternative lengthening of telomeres: models, mechanisms and implications. *Nat. Rev. Genet.* **11**(5): 319–330.

Chang S, Khoo CM, Naylor ML, Maser RS, DePinho RA (2003) Telomere-based crisis: functional differences between telomerase activation and ALT in tumor progression. *Genes Dev.* **17**: 88–100.

Chapon C, Cech TR, Zaug AJ (1997) Polyadenylation of telomerase RNA in budding yeast. *RNA* **3**: 1337–1351.

Choi J, Southworth LK, Sarin KY, Venteicher AS, Ma W, Chang W, Cheung P, Jun S, Artandi MK, Shah N, Kim SK, Artandi SE. (2008) TERT promotes epithelial proliferation through transcriptional control of a Myc- and Wnt-related developmental program. *PLoS Genet.* **4**(1): e10.

Cifuentes-Rojas C, Kannan K, Tseng L, Shippen DE (2011) Two TER subunits for *Arabidopsis* telomerase and the TER binding activity of POT1a. *Proc. Natl. Acad. Sci. U.S.A.* **108**: 73–78.

Cohen SB, Graham ME, Lovrecz GO, Bache N, Robinson PJ, Reddel RR. (2007) Protein composition of catalytically active human telomerase from immortal cells. *Science* **315**: 1850–1853.

Collins K. (2006) The biogenesis and regulation of telomerase holoenzymes. *Nat. Rev. Mol. Cell Biol.* **7**: 484–494.

Collins K. (2008) Physiological assembly and activity of human telomerase complexes. *Mech. Ageing Dev.* **129**(1, 2): 91–98.

Collins K, Gandhi L. (1998) The reverse transcriptase component of the *Tetrahymena* telomerase ribonucleoprotein complex. *Proc. Natl. Acad. Sci. U.S.A.* **95**: 8485–8490.

Collins K, Greider CW. (1993) *Tetrahymena* telomerase catalyzes nucleolytic cleavage and non-processive elongation. *Genes Dev.* **7**: 1364–1376.

Counter CM, Meyerson M, Eaton EN, Ellisen LW, Dickinson Caddle S, Haber DA, Weinberg RA. (1998) Telomerase activity is restored in human cells by ectopic expression of hTERT (hEST2), the catalytic subunit of telomerase. *Oncogene* **16**: 1217–1222.

Counter CM, Meyerson M, Eaton EN, Weinberg RA. (1997) The catalytic subunit of yeast telomerase. *Proc. Natl. Acad. Sci. U.S.A.* **94**: 9202–9207.

d'Adda de Fagagna F. (2008) Living on a break: cellular senescence as a DNA-damage response. *Nat. Rev. Cancer* **8**: 512–522.

Draskovic I, Arnoult N, Steiner V, Bacchetti S, Lomonte P, Londono-Vallejo A. (2009) Probing PML body function in ALT cells reveals spatiotemporal requirements for telomere recombination. *Proc. Natl. Acad. Sci. U.S.A.* **106**(37): 15726–15731.

Fasching CL, Bower K, Reddel RR. (2005) Telomerase-independent telomere length maintenance in the absence of alternative lengthening of telomeres-associated promyelocytic leukemia bodies. *Cancer Res.* **65**: 2722–2729.

Feng J, Funk WD, Wang S-S, Weinrich SL, Avilion AA, Chiu C-P, Adams RR, Chang E, Allsopp RC, Yu J, Le S, West MD, Harley CB, Andrews WH, Greider CW, Villeponteau B. (1995) The RNA component of human telomerase. *Science* **269**: 1236–1241.

Ferrezuelo F, Steiner B, Aldea M, Futcher B. (2002) Biogenesis of yeast telomerase depends on the importin mtr10. *Mol. Cell Biol.* **22**(17): 6046–6055.

Feuerhahn S, Iglesias N, Panza A, Porro A, Lingner J. (2010) TERRA biogenesis, turnover and implications for function. *FEBS Lett.* **584**(17): 3812–3818.

Fisher TS, Taggart AKP, Zakian V. (2004) Cell cycle-dependent regulation of yeast telomerase by Ku. *Nat. Struct. Mol. Biol.* **11**: 1198–1205.

Flores I, Cayuela ML, Blasco MA. (2005) Effects of telomerase and telomere length on epidermal stem cell behavior. *Science* **309**(5738): 1253–1256.

Ford LP, Suh JM, Wright WE, Shay JW. (2000) Heterogeneous nuclear ribonucleoproteins C1 and C2 associate with the RNA component of human telomerase. *Mol. Cell Biol.* **20**: 9084–9091.

Fu D, Collins K. (2007) Purification of human telomerase complexes identifies factors involved in telomerase biogenesis and telomere length regulation. *Mol. Cell* **28**: 773–785.

Gillis AJ, Schuller AP, Skordalakes E. (2008) Structure of the *Tribolium castaneum* telomerase catalytic subunit TERT. *Nature* **455**(7213): 633–637.

Gonzalez-Suarez E, Flores JM, Blasco MA. (2002) Cooperation between p53 mutation and high telomerase transgenic expression in spontaneous cancer development. *Mol. Cell Biol.* **22**: 7291–7301.

Gonzalez-Suarez E, Samper E, Ramirez A, Flores JM, Martin-Caballero J, Jorcano JL, Blasco MA. (2001) Increased epidermal tumors and increased skin wound healing in transgenic mice overexpressing the catalytic subunit of telomerase, mTERT, in basal keratinocytes. *EMBO J.* **20**: 2619–2630.

Grandin N, Damon C, Charbonneau M. (2001) Cdc13 prevents telomere uncapping and Rad50-dependent homologous recombination. *EMBO J.* **20**(21): 6127–6139.

Greider CW, Blackburn EH. (1985) Identification of a specific telomere terminal transferase activity in *Tetrahymena* extracts. *Cell* **43**: 405–413.

Greider CW, Blackburn EH. (1987) The telomere terminal transferase of *Tetrahymena* is a ribonucleoprotein enzyme with two kinds of primer specificity. *Cell* **51**: 887–898.

Greider CW, Blackburn EH. (1989) A telomeric sequence in the RNA of *Tetrahymena* telomerase required for telomere repeat synthesis. *Nature* **337**: 331–337.

Hahn WC, Counter CM, Lundberg AS, Beijersbergen RL, Brooks MW, Weinberg RA. (1999) Creation of human tumour cells with defined genetics elements. *Nature* **400**: 464–468.

Hahn WC, Dessain SK, Brooks MW, King JE, Elenbaas B, Sabatini DM, DeCaprio JA, Weinberg RA. (2002) Enumeration of the simian virus early region elements necessary for human cell transformation. *Mol. Cell Biol.* **22**: 2111–2123.

Hanahan D, Weinberg RA. (2000) The hallmarks of cancer. *Cell* **100**(1): 57–70.

Hande MP, Samper E, Lansdorp P, Blasco MA. (1999) Telomere length dynamics and chromosomal instability in cells derived from telomerase null mice. *J. Cell Biol.* **144**: 589–601.

Harley CB. (1991) Telomere loss: mitotic clock or genetic time bomb? *Mutat. Res.* **256**: 271–282.

Harley CB. (2008) Telomerase and cancer therapeutics. *Nat. Rev. Cancer* **8**(3): 167–179.

Harrington L, Zhou W, McPhail T, Oulton R, Yeung DSK, Mar V, Bass MB, Robinson MO. (1997) Human telomerase contains evolutionarily conserved catalytic and structural subunits. *Genes Dev.* **11**: 3109–3115.

He H, Multani AS, Cosme-Blanco W, Tahara H, Ma J, Pathak S, Deng Y, Chang S. (2006) POT1b protects telomeres from end-to-end chromosomal fusions and aberrant homologous recombination. *EMBO J.* **25**(21): 5180–5190.

Henderson E, Blackburn EH. (1989) An overhang 3′ terminus is a conserved feature of telomeres. *Mol. Cell Biol.* **9**: 345–348.

Henson JD, Cao Y, Huschtscha LI, Chang AC, Au AY, Pickett HA, Reddel RR. (2009) DNA C-circles are specific and quantifiable markers of alternative-lengthening-of-telomeres activity. *Nat. Biotechnol.* **27**(12): 1181–1185.

Hughes TR, Evans SK, Weilbaecher RG, Lundblad V. (2000) The Est3 protein is a subunit of yeast telomerase. *Curr. Biol.* **10**: 809–812.

Iyer S, Chadha AD, McEachern MJ. (2005) A mutation in the STN1 gene triggers an alternative lengthening of telomere-like runaway recombinational telomere elongation and rapid deletion in yeast. *Mol. Cell Biol.* **25**(18): 8064–8073.

Jacobs SA, Podell ER, Cech TR. (2006) Crystal structure of the essential N-terminal domain of telomerase reverse transcriptase. *Nat. Struct. Mol. Biol.* **13**: 218–225.

Jain D, Hebden AK, Nakamura TM, Miller KM, Cooper JP. (2010) HAATI survivors replace canonical telomeres with blocks of generic heterochromatin. *Nature* **467**: 223–227.

Jiang X-R, Jimenez G, Chang E, Frolkis M, Kusler B, Sage M, Beeche M, Bodnar AG, Wahl GM, Tisty TD, Chiu C-P. (1999) Telomerase expression in human somatic cells does not induce changes associated with a transformed phenotype. *Nat. Genet.* **21**: 111–114.

Kharbanda S, Kumar V, Dhar S, Pandey P, Chen C, Majumder P, Yuan Z-M, Whang Y, Strauss W, Pandita TK, Weaver D, Kufe D. (2000) Regulation of the hTERT telomerase catalytic subunit by the c-Abl tyrosine kinase. *Curr. Biol.* **10**: 568–575.

Kilian A, Bowtell DDL, Abud HE, Hime GR, Venter DJ, Keese PK, Duncan EL, Reddel RR, Jefferson RA. (1997) Isolation of a candidate human telomerase catalytic subunit gene, which reveals complex splicing patterns in different cell types. *Hum. Mol. Genet.* **6**: 2011–2019.

Kim JH, Park SM, Kang MR, Oh SY, Lee TH, Muller MT, Chung IK. (2005) Ubiquitin ligase MKRN1 modulates telomere length homeostasis through a proteolysis of hTERT. *Genes Dev.* **19**(7): 776–781.

Kirk KE, Harmon BP, Reichardt IK, Sedat JW, Blackburn EH. (1997) Block in anaphase chromosome separation caused by a telomerase template mutation. *Science* **275**: 1478–1481.

Kirwan M, Dokal I. (2008) Dyskeratosis congenita: a genetic disorder of many faces. *Clin. Genet.* **73**: 103–112.

Kirwan M, Dokal I. (2009) Dyskeratosis congenita, stem cells and telomeres. *Biochim. Biophys. Acta* **1792**(4): 371–379.

Kiyono T, Foster SA, Koop JI, McDougall JK, Galloway DA, Klingelhutz AJ. (1998) Both Rb/$p16^{INK4a}$ inactivation and telomerase activity are required to immortalize human epithelial cells. *Nature* **396**: 84–88.

LaBranche H, Dupuis S, Ben-David Y, Bani M-R, Wellinger RJ, Chabot B. (1998) Telomere elongation by hnRNPA1 and a derivative that interacts with telomeric repeats and telomerase. *Nat. Genet.* **19**: 199–202.

Latrick CM, Cech TR. (2010) POT1-TPP1 enhances telomerase processivity by slowing primer dissociation and aiding translocation. *EMBO J.* **29**(5): 924–933.

Lee J, Sung YH, Cheong C, Choi YS, Jeon HK, Sun W, Hahn WC, Ishikawa F, Lee H-W. (2008) TERT promotes cellular and organismal survival independently of telomerase activity. *Oncogene* **27**: 3754–3760.

Lendvay TS, Morris DK, Sah J, Balasubramanian B, Lundblad V. (1996) Senescence mutants of *Saccharomyces cerevisiae* with a defect in telomere replication identify three additional *EST* genes. *Genetics* **144**: 1399–1412.

Leonardi J, Box JA, Bunch JT, Baumann P. (2008) TER1, the RNA subunit of fission yeast telomerase. *Nat. Struct. Mol. Biol.* **15**(1): 26–33.

Lingner J, Cech TR. (1996) Purification of telomerase from *Euplotes aediculatus*: requirement of a 3′ overhang. *Proc. Natl. Acad. Sci. U.S.A.* **93**: 10712–10717.

Lingner J, Cech TR, Hughes TR, Lundblad V. (1997a) Three ever shorter telomere (*EST*) genes are dispensable for *in vitro* yeast telomerase activity. *Proc. Natl. Acad. Sci. U.S.A.* **94**: 11190–11195.

Lingner J, Hughes TR, Shevchenko A, Mann M, Lundblad V, Cech TR. (1997b) Reverse transcriptase motifs in the catalytic subunit of telomerase. *Science* **276**: 561–567.

Luke B, Panza A, Redon S, Iglesias N, Li Z, Lingner J. (2008) The Rat1p 5′ to 3′ exonuclease degrades telomeric repeat-containing RNA and promotes telomere elongation in *Saccharomyces cerevisiae*. *Mol. Cell* **32**(4): 465–477.

Lundblad V, Blackburn EH. (1993) An alternative pathway for yeast telomere maintenance rescues *est1*-senescence. *Cell* **73**: 347–360.

Lundblad V, Szostak JW. (1989) A mutant with a defect in telomere elongation leads to senescence in yeast. *Cell* **57**: 633–643.

Lustig AJ. (2004) Telomerase RNA: a flexible RNA scaffold for telomerase biosynthesis. *Curr. Biol.* **14**(14): R565–R567.

Maida Y, Yasukawa M, Furuuchi M, Lassmann T, Possemato R, Okamoto N, Kasim V, Hayashizaki Y, Hahn WC, Masutomi K. (2009) An RNA-dependent RNA polymerase formed by TERT and the RMRP RNA. *Nature* **461**(7261): 230–235.

Marciniak RA, Cavazos D, Montellano R, Chen Q, Guarente L, Johnson FB. (2005) A novel telomere structure in a human alternative lengthening of telomeres cell line. *Cancer Res.* **65**: 2730–2737.

Martinez P, Blasco MA. (2011) Telomeric and extra-telomeric roles for telomerase and the telomere-binding proteins. *Nat. Rev. Cancer* **11**(3): 161–176.

Mason JM, Frydrychova RC, Biessmann H. (2008) *Drosophila* telomeres: an exception providing new insights. *Bioessays* **30**(1): 25–37.

Masutomi K, Possemato R, Wong JM, Currier JL, Tothova Z, Manola JB, Ganesan S, Lansdorp PM, Collins K, Hahn WC. (2005) The telomerase reverse transcriptase regulates chromatin state and DNA damage responses. *Proc. Natl. Acad. Sci. U.S.A.* **102**(23): 8222–8227.

McClintock B. (1931) Cytological observations of deficiencies involving known genes, translocations and an inversion in *Zea mays*. *Mo. Agric. Exp. Res. Stn. Res. Bull.* **163**: 4–30.

McClintock B. (1939) The behavior in successive nuclear divisions of a chromosome broken at meiosis. *Proc. Natl. Acad. Sci. U.S.A.* **25**: 405–416.

McElligott R, Wellinger RJ. (1997) The terminal DNA structure of mammalian chromosomes. *EMBO J.* **16**: 3705–3714.

Meyerson M, Counter CM, Eaton EN, Ellisen LW, Steiner P, Dickinson SC, Ziaugra L, Beijersbergen RL, Davidoff MJ, Liu Q, Bacchetti S, Haber DA, Weinberg RA. (1997) *hEST2*, the putative human telomerase catalytic subunit gene, is up-regulated in tumor cells and during immortalization. *Cell* **90**: 785–795.

Min B, Collins K. (2009) An RPA-related sequence-specific DNA-binding subunit of telomerase holoenzyme is required for elongation processivity and telomere maintenance. *Mol. Cell* **36**(4): 609–619.

Mitchell JR, Wood E, Collins K. (1999) A telomerase component is defective in the human disease dyskeratosis congenita. *Nature* **402**: 551–555.

Mitchell M, Gillis A, Futahashi M, Fujiwara H, Skordalakes E. (2010) Structural basis for telomerase catalytic subunit TERT binding to RNA template and telomeric DNA. *Nat. Struct. Mol. Biol.* **17**(4): 513–518.

Miyoshi T, Kanoh J, Saito M, Ishikawa F. (2008) Fission yeast Pot1-Tpp1 protects telomeres and regulates telomere length. *Science* **320**(5881): 1341–1344.

Morales CP, Holt SE, Ouellette M, Kaur KJ, Yan Y, Wilson KS, White MA, Wright WE, Shay JW. (1999) Absence of cancer-associated changes in human fibroblasts immortalized with telomerase. *Nat. Genet.* **21**: 115–118.

Morin GB. (1989) The human telomere terminal transferase enzyme is a ribonucleoprotein that synthesizes TTAGGG repeats. *Cell* **59**: 521–529.

Morrish TA, Greider CW. (2009) Short telomeres initiate telomere recombination in primary and tumor cells. *PLoS Genet.* **5**(1): e1000357.

Muller HJ. (1938) The remaking of chromosomes. *Collecting Net* **13**: 181–198.

Murnane JP, Sabatier L, Marder BA, Morgan WF. (1994) Telomere dynamics in an immortal human cell line. *EMBO J.* **13**: 4953–4962.

Nabetani A, Yokoyama O, Ishikawa F. (2004) Localization of hRad9, hHus1, hRad1, and hRad17 and caffeine-sensitive DNA replication at the alternative lengthening of telomeres-associated promyelocytic leukemia body. *J. Biol. Chem.* **279**(24): 25849–25857.

Nakamura TM, Cooper JP, Cech TR. (1998) Two modes of survival of fission yeast without telomerase. *Science* **282**: 493–496.

Nakamura TM, Morin GB, Chapman KB, Weinrich SL, Andrews WH, Lingner J, Harley CB, Cech TR. (1997) Telomerase catalytic subunit homologs from fission yeast and human. *Science* **277**: 955–959.

Niida H, Shinkai Y, Hande MP, Matsumoto T, Takehara S, Tachibana M, Oshimura M, Lansdorp PM, Furuichi Y. (2000) Telomere maintenance in telomerase-deficient mouse embryonic stem cells: characterization of an amplified telomeric DNA. *Mol. Cell Biol.* **20**(11): 4115–4127.

O'Connor CM, Collins K. (2006) A novel RNA binding domain in tetrahymena telomerase p65 initiates hierarchical assembly of telomerase holoenzyme. *Mol. Cell Biol.* **26**(6): 2029–2036.

Oh W, Lee EW, Lee D, Yang MR, Ko A, Yoon CH, Lee HW, Bae YS, Choi CY, Song J. (2010) Hdm2 negatively regulates telomerase activity by functioning as an E3 ligase of hTERT. *Oncogene* **29**(28): 4101–4112.

Olovnikov AM. (1973) A theory of marginotomy. *J. Theor. Biol.* **41**: 181–190.

Palm W, de Lange T. (2008) How shelterin protects mammalian telomeres. *Annu. Rev. Genet.* **42**: 301–334.

Park JI, Venteicher AS, Hong JY, Choi J, Jun S, Shkreli M, Chang W, Meng Z, Cheung P, Ji H, McLaughlin M, Veenstra TD, Nusse R, McCrea PD, Artandi SE. (2009) Telomerase modulates Wnt signalling by association with target gene chromatin. *Nature* **460**(7251): 66–72.

Pennock E, Buckley K, Lundblad V. (2001) Cdc13 delivers separate complexes to the telomere for end protection and replication. *Cell* **104**: 387–396.

Perrini B, Piacentini L, Fanti L, Altieri F, Chichiarelli S, Berloco M, Turano C, Ferraro A, Pimpinelli S. (2004) HP1 controls telomere capping, telomere elongation, and telomere silencing by two different mechanisms in *Drosophila*. *Mol. Cell* **15**(3): 467–476.

Price CM, Cech TR. (1989) Properties of the telomeric DNA binding protein from *Oxytricha nova*. *Biochemistry* **28**: 769–774.

Raffa GD, Raimondo D, Sorino C, Cugusi S, Cenci G, Cacchione S, Gatti M, Ciapponi L. (2010) Verrocchio, a *Drosophila* OB fold-containing protein, is a component of the terminin telomere-capping complex. *Genes Dev.* **24**(15): 1596–1601.

Raffa GD, Siriaco G, Cugusi S, Ciapponi L, Cenci G, Wojcik E, Gatti M. (2009) The *Drosophila modigliani* (moi) gene encodes a HOAP-interacting protein required for telomere protection. *Proc. Natl. Acad. Sci. U. S. A.* **106**(7): 2271–2276.

Rahman R, Latonen L, Wiman KG. (2005) hTERT antagonizes p53-induced apoptosis independently of telomerase activity. *Oncogene* **24**: 1320–1327.

Redon S, Reichenbach P, Lingner J. (2010) The non-coding RNA TERRA is a natural ligand and direct inhibitor of human telomerase. *Nucleic Acids Res.* **38**(17): 5797–5806.

Ridanpaa M, van Eenennaam H, Pelin K, Chadwick R, Johnson C, Yuan B, vanVenrooij W, Pruijn G, Salmela R, Rockas S, Makitie O, Kaitila I, de la Chapelle A. (2001) Mutations in the RNA component of RNase MRP cause a pleiotropic human disease, cartilage-hair hypoplasia. *Cell* **104**(2): 195–203.

Rong YS. (2008) Telomere capping in *Drosophila*: dealing with chromosome ends that most resemble DNA breaks. *Chromosoma* **117**(3): 235–242.

Rouda S, Skordalakes E. (2007) Structure of the RNA-binding domain of telomerase: implications for RNA recognition and binding. *Structure* **15**(11): 1403–1412.

Rudolph KL, Chang S, Lee H-W, Blasco M, Gottlieb GJ, Greider C, DePinho RA. (1999) Longevity, stress response and cancer in aging telomerase-deficient mice. *Cell* **96**: 701–712.

Sarin KY, Cheung P, Gilison D, Lee E, Tennen RI, Wang E, Artandi MK, Oro AE, Artandi SE. (2005) Conditional telomerase induction causes proliferation of hair follicle stem cells. *Nature* **436**: 1048–1052.

Savage SA, Alter BP. (2009) Dyskeratosis congenita. *Hematol. Oncol. Clin. North Am.* **23**(2): 215–231.

Schnapp G, Rodi H-P, Rettig WJ, Schnapp A, Damm K. (1998) One-step affinity purification protocol for human telomerase. *Nucl. Acids Res.* **26**: 3311–3313.

Schoeftner S, Blasco MA. (2008) Developmentally regulated transcription of mammalian telomeres by DNA-dependent RNA polymerase II. *Nat. Cell Biol.* **10**(2): 228–236.

Seimiya H, Sawada H, Muramatsu Y, Shimizu M, Ohko K, Yamane K, Tsuruo T. (2000) Involvement of 14-3-3 proteins in nuclear localization of telomerase. *EMBO J.* **19**: 2652–2661.

Seto AG, Zaug AJ, Sobel SG, Wolin SL, Cech TR. (1999) *Saccharomyces cerevisiae* telomerase is an Sm small nuclear ribonucleoprotein particle. *Nature* **401**: 177–180.

Shay JW, Bacchetti S. (1997) A survey of telomerase activity in human cancer. *Eur. J. Cancer* **33**(5): 787–791.

Singer MS, Gottschling DE. (1994) *TLC1*: template RNA component of *Saccharomyces cerevisiae* telomerase. *Science* **266**: 404–409.

Slijepcevic P. (2008) DNA damage response, telomere maintenance and ageing in light of the integrative model. *Mech. Ageing Dev.* **129**(1, 2): 11–16.

Smith LL, Coller HA, Roberts JM. (2003) Telomerase modulates expression of growth-controlling genes and enhances cell proliferation. *Nat. Cell Biol.* **5**: 474–479.

Strong MA, Vidal-Cardenas SL, Karim B, Yu H, Guo N, Greider CW. (2011) Phenotypes in mTERT$^{+/-}$ and mTERT$^{-/-}$ mice are due to short telomeres, not telomere-independent functions of TERT. *Mol. Cell Biol.* **31**(12): 2369–2379.

Sykorova E, Fajkus J. (2009) Structure–function relationships in telomerase genes. *Biol. Cell* **101**(7): 375–392.

Szostak JW, Blackburn EH. (1982) Cloning yeast telomeres on linear plasmid vectors. *Cell* **29**: 245–255.

Teng S, Zakian V. (1999) Telomere-telomere recombination is an efficient bypass pathway for telomere maintenance in *Saccharomyces cerevisiae. Mol. Cell Biol.* **19**: 8083–8093.

Teng SC, Chang J, McCowan B, Zakian VA. (2000) Telomerase-independent lengthening of yeast telomeres occurs by an abrupt Rad50p-dependent, Rif-inhibited recombinational process. *Mol. Cell* **6**(4): 947–952.

Tomita K, Cooper JP. (2008) Fission yeast Ccq1 is telomerase recruiter and local checkpoint controller. *Genes Dev.* **22**(24): 3461–3474.

Turkewitz AP, Orias E, Kapler G. (2002) Functional genomics: the coming of age for *Tetrahymena thermophila. Trends Genet.* **18**(1): 35–40.

Ulaner GA, Hu J-F, Vu TH, Ciudice LC, Hoffman AR. (2001) Tissue-specific alternate splicing of human telomerase reverse transcriptase (hTERT) influences telomere lengths during human development. *Int. J. Cancer* **91**: 644–649.

Underwood DH, Carroll C, McEachern MJ. (2004) Genetic dissection of the Kluyveromyces lactis telomere and evidence for telomere capping defects in TER1 mutants with long telomeres. *Eukaryot. Cell* **3**(2): 369–384.

Vaziri H, Benchimol S. (1998) Reconstitution of telomerase activity in normal human cells leads to elongation of telomeres and extended replicative life span. *Curr. Biol.* **8**: 279–282.

Venteicher AS, Abreu EB, Meng Z, McCann KE, Terns RM, Veenstra TD, Terns MP, Artandi SE. (2009) A human telomerase holoenzyme protein required for Cajal body localization and telomere synthesis. *Science* **323**(5914): 644–648.

Venteicher AS, Meng Z, Mason PJ, Veenstra TD, Artandi SE. (2008) Identification of ATPases pontin and reptin as telomerase components essential for holoenzyme assembly. *Cell* **132**: 945–957.

Vidal-Cardenas SL, Greider CW. (2010) Comparing effects of mTR and mTERT deletion on gene expression and DNA damage response: a critical examination of telomere length maintenance-independent roles of telomerase. *Nucleic Acids Res.* **38**(1): 60–71.

Vrbsky J, Akimcheva S, Watson JM, Turner TL, Daxinger L, Vyskot B, Aufsatz W, Riha K. (2010) siRNA-mediated methylation of *Arabidopsis* telomeres. *PLoS Genet.* **6**(6): e1000986.

Walmsley RW, Chan CSM, Tye B-K, Petes TD. (1984) Unusual DNA sequences associated with the ends of yeast chromosomes. *Nature* **310**: 157–160.

Wang F, Podell ER, Zaug AJ, Yang Y, Baciu P, Cech TR, Lei M. (2007) The Pot1-TPP1 telomere complex is a telomerase processivity factor. *Nature* **445**: 506–510.

Wang H, Blackburn EH. (1997) *De novo* telomere addition by *Tetrahymena* telomerase *in vitro*. *EMBO J.* **16**: 866–879.

Watson JD. (1972) Origin of concatameric T4 DNA. *Nat. New Biol.* **239**: 197–201.

Weinrich SL, Pruzan R, Ma L, Ouellette M, Tesmer VM, Holt SE, Bodnar AG, Lichtsteiner S, Kim NW, Trager JB, Taylor RD, Carlos R, Andrews WH, Wright WE, Shay JW, Harley CB, Morin GB. (1997) Reconstitution of human telomerase with the template RNA component hTR and the catalytic protein subunit hTRT. *Nat. Genet.* **17**: 498–502.

Witkin KL, Collins K. (2004) Holoenzyme proteins required for physiological assembly and activity of telomerase. *Genes Dev.* **18**: 1107–1118.

Witkin KL, Prathapam R, Collins K. (2007) Positive and negative regulation of *Tetrahymena* telomerase holoenzyme. *Mol. Cell Biol.* **27**(6): 2074–2083.

Wu L, Multani AS, He H, Cosme-Blanco W, Deng Y, Deng JM, Bachilo O, Pathak S, Tahara H, Bailey SM, Behringer RR, Chang S. (2006) Pot1 deficiency initiates DNA damage checkpoint activation and aberrant homologous recombination at telomeres. *Cell* **126**(1): 49–62.

Xin H, Liu D, Wan M, Safari A, Kim H, Sun W, O'Connor MS, Songyang Z. (2007) TPP1 is a homologue of ciliate TEBP-b and interacts with Pot1 to recruit telomerase. **445**: 559–562.

Yeager TR, Neumann AA, Englezou A, Huschtscha LI, Noble JR, Reddel RR. (1999) Telomerase-negative immortalized human cells contain a novel type of promyelocytic leukemia (PML) body. *Cancer Res.* **59**(17): 4175–4179.

Yu G-L, Bradley JD, Attardi LD, Blackburn EH. (1990) *In vivo* alteration of telomere sequences and senescence caused by mutated *Tetrahymena* telomerase RNAs. *Nature* **344**: 126–132.

Zappulla DC, Cech TR. (2004) Yeast telomerase RNA: a flexible scaffold for protein subunits. *Proc. Natl. Acad. Sci. U. S. A.* **101**(27): 10024–10029.

Zaug AJ, Podell ER, Nandakumar J, Cech TR. (2010) Functional interaction between telomere protein TPP1 and telomerase. *Genes Dev.* **24**(6): 613–622.

Zhou XZ, Lu KP. (2001) The Pin2/TRF1-interacting protein PinX1 is a potent telomerase inhibitor. *Cell* **107**: 347–359.

2

TELOMERASE RNA: STRUCTURE, FUNCTION, AND MOLECULAR MECHANISMS

Yehuda Tzfati and Julian J.-L. Chen

2.1 INTRODUCTION—TELOMERASE RNA: AN ESSENTIAL COMPONENT OF TELOMERASE

Telomerase has evolved to synthesize DNA repeats of short, species-specific sequences onto eukaryotic chromosome ends. This highly specialized task distinguishes telomerase from other DNA- or RNA-dependent DNA polymerases, which copy long extrinsic DNA or RNA templates without significant sequence preference. Another unique feature of telomerase is its integral RNA moiety, termed telomerase RNA. Accumulating data support the notion that this RNA is responsible for many unique features of the telomerase enzyme. Studying the structure and function of telomerase RNA is thus a key to understanding this specialized ribonucleoprotein (RNP) complex.

When telomerase activity was first identified in the ciliated protozoa *Tetrahymena thermophila* by Carol Greider and Elizabeth Blackburn (1985), the authors did not identify any template that could direct telomeric repeat synthesis, and therefore suggested that it was template-independent. Indeed telomerase activity is independent of an extrinsic template, but, as became evident later, it uses an intrinsic RNA template. Greider and Blackburn (1987) reported that telomerase is a RNP enzyme containing an essential RNA component. Even before cloning and sequencing of this RNA component, they suggested several possible roles for this RNA, which are as follows:

Telomerases: Chemistry, Biology, and Clinical Applications, First Edition.
Edited by Neal F. Lue and Chantal Autexier.

1. Providing a scaffold for the assembly of proteins in the active enzyme complex.
2. Providing a template directing telomere synthesis.
3. Recognizing the telomerase substrate—the 3′-end of the telomere.
4. Participating in the polymerization reaction itself.

Later, in 1989, they reported on the cloning of the 159 nt long telomerase RNA (TER) from *T. thermophila* (Greider and Blackburn, 1989). This RNA contained the sequence CAACCCCAA, which is complementary to one and a half copies of the telomeric repeat (TTGGGG). The finding of this sequence immediately suggested that it serves as a template for telomeric repeat synthesis. Furthermore, this putative template contained one and a half telomeric repeats, i.e., the sequence in the beginning of the template is repeated in the end, suggesting a mechanistic explanation for the ability of telomerase to synthesize multiple repeats. According to this model, once the end of the template is copied onto the end of the telomere, this telomere end can unwind, translocate, realign to the beginning of the template, and be extended again by copying the rest of the template. This hypothesis was later confirmed *in vitro* by using various primer substrates in a telomerase reaction performed with cell extracts from another ciliate, *Euplotes crassus* (Shippen-Lentz and Blackburn, 1990), and *in vivo* by introducing template mutations into the *T. thermophila* TER gene and detecting the incorporation of the corresponding mutant telomeric repeats onto the endogenous telomeres (Yu et al., 1990). Finally, telomerase was shown to synthesize multiple repeats processively, indicating a translocation step as predicted by the model (Greider, 1991).

While the role of TER in providing a template for telomere synthesis was already well established by 1990, the roles of its nontemplating domains remained elusive. Nevertheless, data accumulated over the ensuing years support all additional roles of TER originally suggested by Greider and Blackburn: providing a scaffold for the assembly of telomerase proteins, recognizing the telomerase substrate, and participating in telomerase action. Since telomerase acts as an RNP enzyme, it is crucial to understand how TER contributes to the unique features of telomerase.

During the 1990s, more TERs were identified in ciliates, yeast, and vertebrates. It was soon realized that TER length and sequence are extremely variable compared to other noncoding RNAs (such as ribosomal RNAs), accumulating changes almost as fast as intergenic regions. Such high divergence hindered the identification and functional study of conserved elements in TER. In this chapter, we will examine how this obstacle has been tackled and what is presently known about TER structure, function, and mechanism of action.

2.2 THE UNUSUAL DIVERSITY OF TELOMERASE RNA

2.2.1 Size, Sequence, and Secondary Structure

TER is evolutionarily divergent in size, sequence, and even secondary structure (Fig. 2.1). Homologues of TER identified from a variety of organisms exhibit a

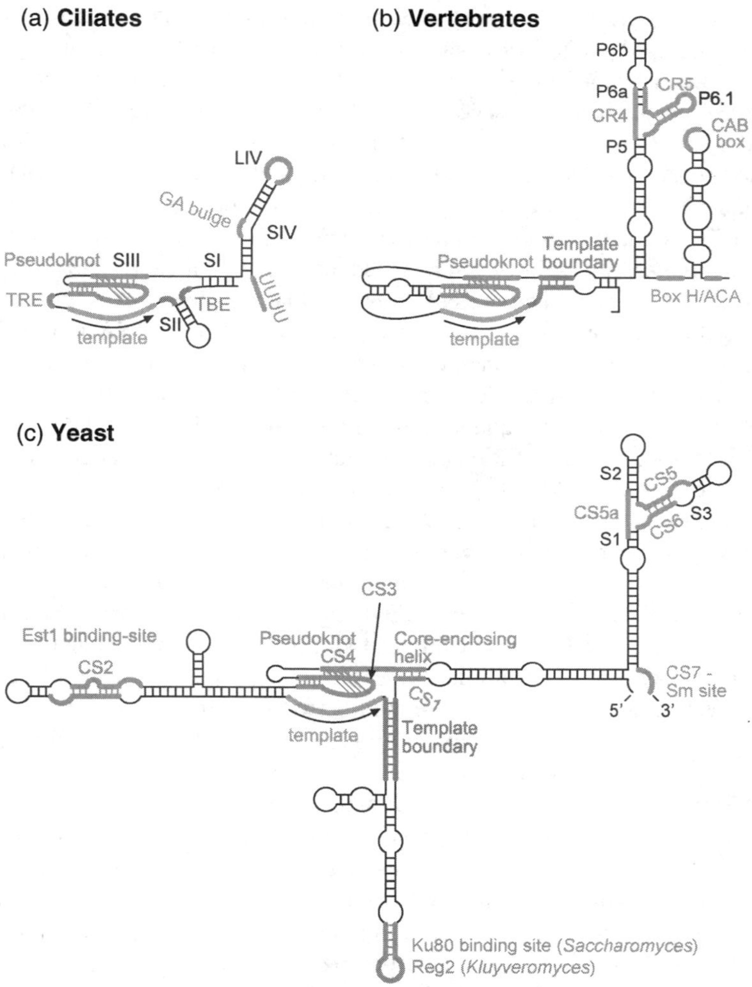

FIGURE 2.1 Common secondary structure models for ciliates, vertebrates, and budding yeast TERs. Indicated are the conserved regions/sequences (CR or CS), pairings/stems (P or S), loops (L), template recognition element (TRE), and template boundary element (TBE). (See the color version of this figure in Color Plates section.)

remarkable size variation ranging from 147–205 nt in ciliates (Lingner et al., 1994; McCormick-Graham and Romero, 1995; Romero and Blackburn, 1991; ten Dam et al., 1991), to 312–559 nt in vertebrates (Chen et al., 2000; Feng et al., 1995; Xie et al., 2008), and 779–2030 nt in yeasts (Chappell and Lundblad, 2004; Dandjinou et al., 2004; Gunisova et al., 2009; Hsu et al., 2007; Kachouri-Lafond et al., 2009; Lin et al., 2004; Tzfati et al., 2003; Zappulla and Cech, 2004). In addition to the nearly 15-fold difference in size, the primary sequence of TER is also extremely divergent, lacking overall sequence similarity even between closely related groups of species. For example, TER sequences from tetrahymenine ciliates (e.g., *Tetrahymena* and *Glaucoma*) and hypotrichous ciliates (e.g., *Euplotes* and *Oxytricha*), or from the *Saccharomyces sensu stricto* species (e.g., *S. cerevisiae*) and *Saccharomyces sensu latu* species (e.g., *S. castellii* or *S. kluyveri*) cannot be aligned reliably based solely on sequence conservation (Chappell and Lundblad, 2004; Lingner et al., 1994). Such high sequence diversity seemed inconsistent with the critical roles of nontemplate regions of TER and the notion that the mechanism of telomerase action is conserved across eukaryotes. It was, however, believed that while TER sequences are variable, conservation should be found in RNA structures that are necessary for the conserved functions. Furthermore, it was suggested that not the entire RNA is functionally important but rather small conserved "beads," which are connected by variable, flexible, and largely dispensable "strings" (Zappulla and Cech, 2004).

The unusual size and sequence diversity became less puzzling when the secondary structures of TER started to take shapes. For inferring secondary structures of large RNAs, phylogenetic comparison has proven to be the most reliable approach (Pace et al., 1989). This comparative analysis searches in the aligned sequences of homologous RNAs for phylogenetic covariation, that is, changes of bases in concert while maintaining base pairing potential in conserved helical regions. This phylogenetic method requires a large number of homologous sequences that manifest sufficient similarity for reliable alignment and provide two or more independent covariations for each helix proposed. Although TER sequences were too divergent to be aligned between groups of species, secondary structure models of TERs have been inferred independently using the phylogenetic comparison for closely related species of ciliates and vertebrates (Chen et al., 2000; Lingner et al., 1994; McCormick-Graham and Romero, 1995; Romero and Blackburn, 1991). In yeast, however, the TER sequences identified from *Kluyveromyces* and *Saccharomyces* species were too few, too long, and too divergent for reliable structure prediction that is based solely on the manual phylogenetic approach. Instead, a computational approach was used, which combines free energy calculation and phylogenetic covariation and requires less TER homologues for structure prediction. Using programs such as *ClustalX* for multiple sequence alignment (Chenna et al., 2003) and *RNAalifold* and *X2s* for secondary structure prediction (Hofacker et al., 2002; Juan and Wilson, 1999), secondary structure models were predicted for these yeast TERs (Brown et al., 2007; Chappell and Lundblad, 2004; Dandjinou et al., 2004; Lin et al., 2004; Tzfati et al., 2003; Zappulla and Cech, 2004). Based on the predicted secondary structure models, it is now understood that the dramatic size variation of TER resulted from acquisition and deletion of a variety of species-specific structural domains in different groups of species. Some of these species-specific domains serve as binding sites for species-specific telomerase-associated proteins, which have distinct functions in telomerase RNP biogenesis and regulation *in vivo* but are dispensable for the enzymatic activity *in vitro,* and other domains seem to be dispensable *in vivo* as well as *in vitro* (Roy et al., 1998; Zappulla and Cech, 2004; Zappulla et al., 2005).

2.2.2 Transcription and Biogenesis

The transcription and biogenesis pathways of TER are also divergent. The ciliate TER is transcribed by RNA polymerase III, which terminates after copying a stretch of adenines and leaving a short poly-U tail at the 3′-end of the transcript. In contrast, the yeast and vertebrate TERs are transcribed by RNA polymerase II, which allows synthesis of larger RNAs containing internal uridine-rich regions. While it is not clear how the transition of transcription machinery took place during evolution, it presumably coincided with a transition of the biogenesis pathways for the processing and assembly of yeast and vertebrate telomerases. All budding yeast TERs examined contain an Sm site close to the 3′-end of the RNA and share the biogenesis pathway with the small nuclear RNAs (snRNAs) (Seto et al., 1999). The Sm site is bound by the seven Sm protein subunits that form a ring around the RNA and facilitate TER maturation and stability. The fission yeast *Schizosaccharomyces pombe* TER also contains an Sm site (Leonardi et al., 2008; Webb and Zakian, 2008). Remarkably, vertebrate TERs share the biogenesis and trafficking pathways with a different group of noncoding RNAs—the small nucleolar RNAs (snoRNAs), or more precisely a subgroup of snoRNAs termed small Cajal body RNA (scaRNAs). Vertebrate TERs share the common vertebrate-specific snoRNA structural domain called box H/ACA domain, located in the 3′ half of the RNA (Jady et al., 2004; Mitchell et al., 1999a) (see details in Section 2.5.1). This box H/ACA domain provides binding sites for a number of accessory proteins that mediate proper processing and nuclear trafficking of TER (Cristofari et al., 2007; Fu and Collins, 2003, 2007; Theimer et al., 2007).

Interestingly, the *S. pombe* TER contains a 5′ splice site and an intron branch site downstream of the Sm site, which facilitates 3′-end processing by the spliceosome in a partial splicing reaction (Box et al., 2008a). These consensus sequences are conserved in the same position also in seven *Candida* TERs, indicating that this novel 3′ processing pathway is not limited to *S. pombe* but is shared by some budding yeast species as well (Gunisova et al., 2009).

TER has rapidly evolved to utilize different biogenesis pathways in different species through acquisition of species-specific structural domains, which recruit different proteins. Identification and study of TER homologues from other groups of species beyond ciliates, yeast, and vertebrates, would provide a comprehensive understanding of TER evolution. While acquiring new RNA domains for new functions, TER preserves a core structure for functional assembly with the catalytic reverse transcriptase protein. We will discuss below the structure and function of both the common core and the species-specific structural elements of TERs from ciliate, yeast, and vertebrate.

2.3 THE COMMON CORE OF TER

Despite the remarkable diversity in size and sequence, all TERs appear to share a common core, also named the pseudoknot-template domain, which is essential for the telomerase catalytic function (Chen and Greider, 2004, 2005; Forstemann and

Lingner, 2005). In addition to the template and a pseudoknot structure downstream, this core domain contains also a template boundary element and, in most species, a long-range helical region that positions the pseudoknot in close proximity to the template. Finally, the core domain has structural elements that are required for the binding of the catalytic telomerase reverse transcriptase protein (TERT; Est2 in budding yeast) (Miller et al., 2000; Mitchell and Collins, 2000; Moriarty et al., 2005). Each of these elements is discussed in further detail below.

2.3.1 Template Boundary Definition

One of the specialized features of telomerase is its ability to reverse transcribe accurately a short, well-defined template. How does telomerase define the template? In *T. thermophila*, the sequences at the base of stem II and the single-stranded 5′ and 3′ flanking sequences are important for defining the 5′ boundary of the template where reverse transcription terminates (Fig. 2.2) (Autexier and Greider, 1995;

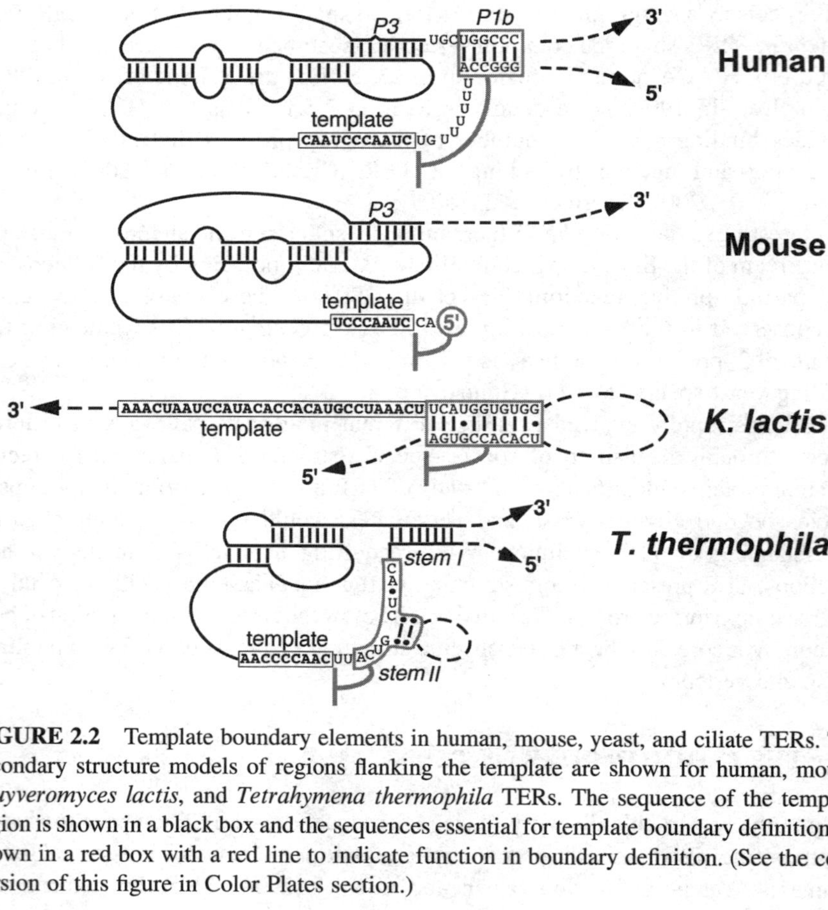

FIGURE 2.2 Template boundary elements in human, mouse, yeast, and ciliate TERs. The secondary structure models of regions flanking the template are shown for human, mouse, *Kluyveromyces lactis*, and *Tetrahymena thermophila* TERs. The sequence of the template region is shown in a black box and the sequences essential for template boundary definition are shown in a red box with a red line to indicate function in boundary definition. (See the color version of this figure in Color Plates section.)

Lai et al., 2002). The ciliate template boundary element (TBE) also serves as a high-affinity TERT binding site, suggesting that the mere binding of TERT forms a block to DNA synthesis and specifies the 5′ boundary. Although stem-loop II is absent in some ciliates, the sequence of the TBE is highly conserved, presumably reflecting the sequence-dependent interaction with TERT.

The budding yeast TBE is a stable helix located 0–2 nucleotides 5′ to the template (Seto et al., 2003; Tzfati et al., 2000). The location of this helix suggests that it limits polymerization simply by providing a physical barrier (Fig. 2.2). Indeed, disruption of this helix in both *K. lactis* and *S. cerevisiae* causes DNA synthesis to proceed beyond the normal template boundary *in vivo* and *in vitro*. Restoring the TBE helix by compensatory mutations reestablishes the normal boundary. Thus, unlike the ciliate TBE, which requires a specific sequence, the budding yeast TBE requires the base-paired structure to specify the template boundary. Nevertheless, the effects of some mutations in the *S. cerevisiae* TBE cannot be fully explained by the simple model of a barrier to DNA synthesis but rather suggest that this RNA element has additional functions in determining enzyme processivity and template usage (Seto et al., 2003). Given the later discovery of the core-enclosing helix, which connects the TBE to the pseudoknot (see below), it is plausible that disruption of the TBE structure not only impairs the boundary definition but may also disrupt the precise positioning of the template at the catalytic site and thus affects processivity and template usage.

The *S. pombe* TBE resembles that of budding yeasts (Wang et al., 2009; Webb and Zakian, 2008). However, unlike other characterized TBEs, this helical region partially overlaps with the template, suggesting that pairing and unwinding of base pairs at the base of the boundary element generate the telomeric repeat heterogeneity found in *S. pombe* (Box et al., 2008b).

In vertebrates, two distinct mechanisms were suggested for template boundary definition. For human and most vertebrate TERs, the core-enclosing helix P1b (Fig. 2.2), located 6–8 bases upstream of the template, is essential for preventing telomerase from copying beyond the template 5′-end (Chen and Greider, 2003). The distance of the vertebrate TBE from the template suggests that it serves as an anchor site that limits the movement of the template in the active site during elongation, rather than providing a physical barrier as suggested for the yeast TBEs (Fig. 2.2). It has also been suggested that TERT binding to helix P1b plays a role in defining the 5′-boundary of the template (Moriarty et al., 2005). Remarkably, the mouse, rat, and hamster TERs lack helix P1b and their 5′-end is located only two residues upstream of the template (Hinkley et al., 1998). Extension or shortening of the 5′-end of the mouse TER alters the 5′ boundary of the template, indicating that mouse telomerase terminates two residues downstream of the 5′-end of the RNA (Fig. 2.2) (Chen and Greider, 2003).

In addition to the TBE, which specifies the 5′ boundary of the template, another sequence identified in *T. thermophila* and termed "template recognition element" (TRE) specifies the position where the 3′-end of the telomeric DNA substrate anneals and the initiation of nucleotide addition takes place. In budding yeasts, a 7–10 nt sequence immediately 3′ of the template can form additional imperfect pairing with the telomeric substrate (Wang et al., 2009). This pairing is important for telomerase function *in vivo*, perhaps by stabilizing the interaction of the telomerase substrate in the correct positioning for reverse transcription initiation.

2.3.2 Pseudoknot

2.3.2.1 Ciliate Within the common core, all TERs examined contain an H-type (H for hairpin loop) pseudoknot structure formed by base-pairing interaction between the loop sequence of a hairpin and a complementary sequence flanking this hairpin. A pseudoknot structure was first identified in ciliate TERs (ten Dam et al., 1991). The ciliate pseudoknot is the smallest and least stable of the telomerase pseudoknot elements (Fig. 2.1). While the ciliate pseudoknot is important for ciliate telomerase RNP assembly *in vivo* (Gilley and Blackburn, 1999), it is not essential for *in vitro* telomerase activity or binding to TERT (Berman et al., ; Bhattacharyya and Blackburn, 1994; Sperger and Cech, 2001). Structure probing of TER showed that the ciliate pseudoknot structure does not readily form in the naked RNA (Bhattacharyya and Blackburn, 1994), but is greatly stabilized upon the binding of the TERT protein and the assembly of the telomerase RNP (Sperger and Cech, 2001). A recent study also shows that the accessory protein p65 can fully rescue the activity defects of pseudoknot mutants (Berman et al., 2010), suggesting that the P65-assisted RNP assembly stabilizes the pseudoknot structure. Since a p65 ortholog has not been found in yeast or vertebrates, such a role for stabilizing the pseudoknot may be specific for ciliates.

2.3.2.2 Vertebrate The identification of a similar but larger pseudoknot structure in vertebrate TERs suggests that the pseudoknot structure is an evolutionarily conserved feature of TER and likely plays a crucial role in telomerase function (Chen et al., 2000). The vertebrate pseudoknot also adopts an H-type structure, consisting of two stems, P2 and P3 (Fig. 2.2). In mammalian TERs, the P2 contains three consecutive helices, P2a, P2b, and P2a.1, separated by two internal loops. The vertebrate pseudoknot is formed essentially by the P3 long-range base-pairing interaction. Similar to the ciliate pseudoknot, the vertebrate pseudoknot is also unstable, as revealed by the modification of the sequence in the P3 region by chemicals specific to single-stranded residues (Antal et al., 2002). It was also proposed that the pseudoknot might be dynamic with a conformational balance between the pseudoknot (i.e., P3 pairing) and an extended stem P2b with intraloop noncanonical U–U base-pairing (Comolli et al., 2002; Theimer et al., 2003). However, mutations that disrupt the intraloop interactions without affecting the pseudoknot conformation do not reduce telomerase activity, suggesting that a static pseudoknot structure is sufficient for telomerase function (Chen and Greider, 2005). Nuclear magnetic resonance (NMR) studies of an isolated RNA pseudoknot containing only P2b and P3 revealed a unique triple helix in which the J2b/3 single-stranded region invades the major groove of the P3 helix and forms U:A–U Watson–Crick–Hoogsteen base triples (Kim et al., 2008; Theimer et al., 2005). This triple helix structure is critical for telomerase activity, as mutations that destabilize the Hoogsteen interactions decrease telomerase activity (Theimer et al., 2005). More recent NMR studies of helices P2a, P2b, and P2a.1 of human TER, together with the previously determined P2b/P3 structure, have lead to a high-resolution structure model for the entire pseudoknot core domain (Zhang et al., 2010). In this structure, the conserved 5-nucleotide asymmetric internal loop J2a/2b in mammalian TERs forms a defined S-shape structure, generating a sharp 90° bend between helices P2a and P2b. Moreover, the structure of internal loop J2a/2b was suggested to be dynamic, permitting conformational changes within the core domain and potentially facilitating template translocation during processive repeat addition (Zhang et al., 2010).

2.3.2.3 Yeast The yeast telomerase pseudoknot was first predicted by comparative analysis of the *Kluyveromyces* TERs, and confirmed by disruptive and compensatory mutations introduced into the *K. lactis* TER gene (*TER1*) *in vivo* (Tzfati et al., 2003). Identification of the TER genes (termed *TLC1* in *Saccharomyces*) in several *Saccharomyces sensu stricto* species permitted covariation-supported secondary-structure prediction and several models for the catalytic core structure with alternative pseudoknot configurations were proposed (Chappell and Lundblad, 2004; Dandjinou et al., 2004; Lin et al., 2004; Zappulla and Cech, 2004). Domain-swapping experiments showed that the *Tetrahymena* and the human pseudoknots can substitute for the *S. cerevisiae* element, supporting telomerase function *in vivo,* albeit partially (Chappell and Lundblad, 2004). These experiments suggest that the pseudoknot elements are functionally homologous. Accordingly, they are predicted to exhibit conserved sequence and/or structural features that facilitate their common function. While the *Tetrahymena* pseudoknot can replace the yeast pseudoknot, a comparable replacement of the human pseudoknot by the *Tetrahymena* element enabled telomerase activity only in an *in vitro* reconstitution assay but not *in vivo* (Marie-Egyptienne et al., 2005). Furthermore, the reconstituted *in vitro* activity was compromised both in total activity and in repeat-addition processivity. Thus, while the pseudoknot elements share a common function, they also have some species-specific features and requirements.

Interestingly, the predicted yeast pseudoknots are much larger than those found in ciliates and vertebrates (Lin et al., 2004; Tzfati et al., 2003) (Fig. 2.3b). In *K. lactis,* it was not obvious how a pseudoknot structure could accommodate a short six-nt loop (loop 1), connecting both ends of a longer 10 base-pair stem (stem 2) that presumably spans a full helical turn. Mutational analysis and computer modeling revealed that the short loop 1 must invade the major groove of stem 2 of the pseudoknot (Shefer et al., 2007). This invasion requires the formation of U:A–U base triples with Hoogsteen-type hydrogen bonds between the A residue of the duplex and the U residue of the third strand, in order to counterbalance the repulsing forces between the negatively charged nucleotides. Thus, the formation of a triplex enables the short loop to follow the helical path of the major groove of stem 2 and be accommodated within the large and tight structure of the yeast pseudoknot. In *K. lactis*, the model was supported by mutational analysis *in vivo*; introducing small mutations into each of the three strands of the triplex severely impaired or abolished telomerase function *in vivo*, while triple compensatory mutations replacing three

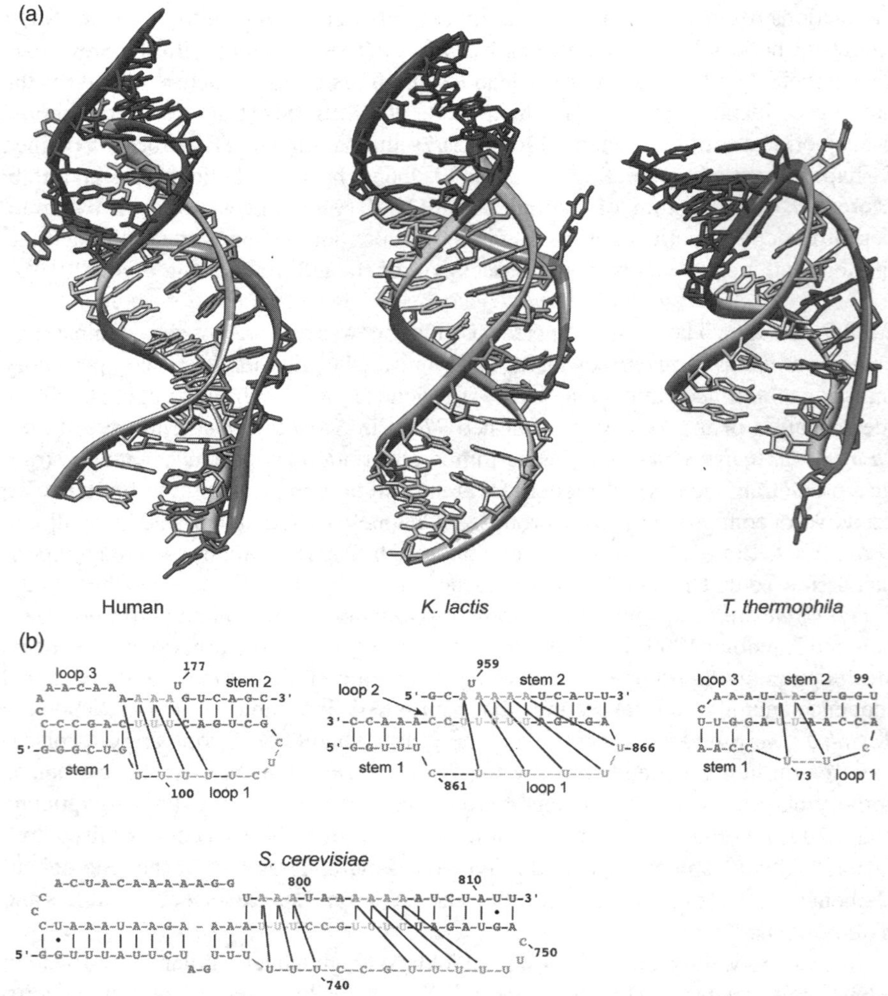

FIGURE 2.3 The pseudoknot structures of human, *K. lactis*, and *Tetrahymena thermophila* telomerase RNAs. (a) Ribbon representations of the three-dimensional solution structure of the human (Kim et al., 2008) pseudoknot, and the computer models of the *K. lactis* (Shefer et al., 2007) and *T. thermophila* (Ulyanov et al., 2007) pseudoknot, illustrated using the computer program *Chimera* (Couch et al., 2006). Stem 1 is shown in gray, residues of stem 2 not participating in base triples are shown in blue. Residues of stem 2 that are part of the triplex, are shown in orange (purines) and yellow (pyrimidines). Bulged-out U residues are shown in red. Residues of loop 1 that are part of the triplex, are shown in cyan. The rest of loop 1, as well as loop 2 if present, are shown in green. Loop 3 is shown in magenta. (b) A schematic representation of base pairing in the pseudoknot, including also the predicted scheme for the *S. cerevisiae* pseudoknot (Gunisova et al., 2009; Qiao and Cech, 2008). Vertical lines represent Watson–Crick interactions; tilted lines, Hoogsteen hydrogen bonds; and "•," a G:U wobble pair. Note that in the *K. lactis* pseudoknot, the region of the junction between stem 1 and 2 is illustrated as unpaired, since the interactions among these nucleotides are unknown. (See the color version of this figure in Color Plates section.)

U–A:U base triples with C–G:C$^+$ (the protonation of the C residue is required for the formation of the Hoogsteen hydrogen bonds) restored telomerase function (Shefer et al., 2007). Further confirmation of the triple helix model came from mutational analyses and biochemical analysis of *in vitro*-reconstituted *S. cerevisiae* telomerase activity (Qiao and Cech, 2008). Finally, one to three base triples were also predicted to form in the small pseudoknot elements of ciliates (Ulyanov et al., 2007), suggesting that base triples within a pseudoknot is a conserved feature of all telomerases.

2.3.2.4 The Function of the Triple Helix The remarkable conservation of major groove base triples in stem 2 of the vertebrates, yeast, and ciliates TER pseudoknots, despite the widely different stem lengths and stabilities of these elements, argues against a simple role in stabilizing the structure and rather suggest a more specific function conserved across all telomerases. What is this function? Several observations suggest that the pseudoknot is intimately involved in catalysis: (1) In *K. lactis*, even a relatively minor alteration of the triplex structure affects telomerase activity, and in particular its fidelity and processivity *in vivo*, resulting in short telomeres containing nucleotide misincorporations and aberrantly truncated telomeric repeats (Shefer et al., 2007; Tzfati et al., 2003). (2) Crosslinking experiments revealed that the triple helix is positioned in close proximity to the 3′-end of the telomeric substrate where catalysis takes place (Qiao and Cech, 2008). (3) 2′-OH groups protruding from both the triple and double helix portions of stem 2 are important for telomerase action; modifying these groups to 2′-H impairs the *in vitro*-reconstituted telomerase activity (Qiao and Cech, 2008). Based on these observations, Qiao and Cech suggested that these hydroxyl groups form hydrogen bonds that are important for orienting the 3′-end of the telomerase substrate in the active site. To play such a role, the pseudoknot must be anchored precisely to the catalytic protein. Indeed, it was shown that the duplex part of stem 2 (which is adjacent to the triplex part) and the core-enclosing helix are important for Est2 binding (Chappell and Lundblad, 2004; Lin et al., 2004; Qiao and Cech, 2008). However, some other results are puzzling and the precise role of the triple helix is not yet understood. While the 2′-H modification of three residues participating in the predicted triplex (A804, A805, and A806 altogether) impaired *S. cerevisiae* telomerase activity, in human, when each of the three A nucleotides of the triplex was modified separately to either 2′-H or 2′-O-methyl, only the 3′-most nucleotide had a significant effect (Qiao and Cech, 2008). In addition, 2′-H modification of residues in the duplex part of stem 2 (U809, A810, and U811) also severely reduced *S. cerevisiae* telomerase activity. An elegant approach developed to target specific nucleotides for 2′-O-methylation *in vivo* demonstrated that the modification of U809 reduced telomerase activity (Huang and Yu, 2010), consistent with the results obtained *in vitro*. Surprisingly, however, 2′-O-methylation of A804 and A805 enhanced telomerase activity and caused telomere elongation. These results suggest that (1) not all the base-triples have the same role, and (2) 2′-OH groups in the duplex part of stem 2 may also be involved in the enzymatic reaction. In addition, loop 1 of the pseudoknot showed sequence conservation across budding yeasts, suggesting that it may also play a crucial, yet unknown role in telomerase action (Gunisova et al., 2009; Kachouri-Lafond et al., 2009).

Given the length of the yeast template (30 nt in the case of *K. lactis*), the active site must travel a considerable distance along the template. It is tempting to speculate that the triple helix slides together with the protein active site along the template and maintains the association of the 3′-end of the telomeric substrate with the catalytic site during elongation. Alternatively, the triple helix may stay fixed in its position with respect to the template and the protein moves during elongation. In the latter case, the role of the triple helix may be required only at a specific stage in the reaction cycle, for example, at the initiation or translocation step. The finding that pseudoknot mutations cause nucleotide misincorporation at different positions along the template and truncated telomeric repeats (Shefer et al., 2007; Tzfati et al., 2003) is more consistent with the former possibility. The correlation between the large size of the pseudoknot and the large template may indicate that the yeast pseudoknot is required to be more stable and to withstand stronger strains induced during polymerization along the long telomerase template.

2.3.3 Core-Enclosing Helix

The core-enclosing helix, present in all examined species except for some rodents, brings the template and the crucial pseudoknot structure into close proximity (Fig. 2.1). In human, the core-enclosing helix P1b potentially interacts with TERT and serves as a TBE (Chen and Greider, 2003; Moriarty et al., 2005). However, the lack of helix P1b in mouse TER suggests the presence of alternative mechanisms for the positioning of pseudoknot and template within the catalytic core. In ciliates, the binding of TERT to the template-adjacent TBE is presumably sufficient for positioning the template at the catalytic site without the core-enclosing helix (stem I). Indeed, a recent study suggested that in the presence of the accessory protein p65, stem I is not essential for telomerase activity (Berman et al., 2010). The core-enclosing helix appears to be more important for the large yeast TER. Mutational analysis of TER in both *S. cerevisiae* and *K. lactis* confirmed the importance of the core-enclosing helix for telomerase function *in vivo* ((Lin et al., 2004) and Gorkovoi and Tzfati, unpublished results).

2.4 THE ASSEMBLY/ACTIVATION STEM-LOOP ELEMENT

In each of the three groups of species, ciliates, budding yeast, and vertebrates, another structural element distant from the template-pseudoknot core domain (at least as appears in the secondary structure models) was found to be conserved and critical for telomerase activity both *in vivo* and *in vitro*. This element is predicted to form a stem with conserved apical and/or asymmetrical internal loops, termed stem-loop IV in ciliates, stem 2 of a three-way junction in yeast, and stem-loop P6.1 within the vertebrates CR4–CR5 domain. Several lines of evidence described below suggest that stem-loop IV, stem 2, and p6.1 are homologous elements that provide a common role in the assembly of an active telomerase RNP complex. Therefore, we suggest naming this element "the assembly/activation stem-loop."

2.4.1 The Ciliate Stem-Loop IV

2.4.1.1 Identification of a Conserved Structural Element The first TER secondary structure model was reported for ciliates by Romero and Blackburn (1991). Based on a phylogenetic analysis of six *Tetrahymena* and one *Glaucoma* TER sequences, the authors identified four conserved helices (I–IV; see Fig. 2.1). They noted that helix IV and its apical loop have higher sequence conservation than that observed for helices II and III. This conservation of stem-loop IV is maintained even among more distant ciliates (Lingner et al., 1994; McCormick-Graham and Romero, 1995). Within helix IV, they identified a conserved GA bulge and hypothesized that it may function to increase the number of conformations possible for this helix. Bhattacharyya and Blackburn (1994) confirmed the secondary structure model by chemical and enzymatic probing and suggested that the GA bulge introduces a kink in stem IV. They also suggested that stem IV, together with stem I, form a helical scaffold important for binding of proteins and the assembly of the telomerase complex. These three hypotheses were confirmed over a decade later. The three-dimensional structure of stem loop IV, solved by NMR, revealed a highly structured apical loop with a conserved noncanonical C–U pair at its base (Chen et al., 2006; Richards et al., 2006). The structure was kinked at the conserved GA bulge and displayed a conformational flexibility, which, presumably allows precise positioning of the conserved and highly structured apical loop in the catalytic core.

2.4.1.2 The Role of Stem-Loop IV in the Binding of TERT to TER Mutational analyses of *Tetrahymena* telomerase reconstituted *in vitro* revealed a role for stem-loop IV, and in particular the conserved GA bulge and the apical loop, in the association of TERT with TER (Lai et al., 2003; Sperger and Cech, 2001). While it was suggested that TERT interacts directly with loop IV, the affinity of this interaction is much lower than the affinity of TERT to its high-affinity site, TBE (Lai et al., 2003). Interestingly, stem-loop IV can execute its function even when added in trans (Lai et al., 2003). These observations suggest that stem-loop IV indirectly promotes the interaction of TERT with TBE, though the precise mechanism is still not understood.

2.4.1.3 The Assembly Role of Stem-Loop IV is Facilitated by the Protein p65 Purification of the telomerase ribonucleoprotein complexes from the ciliate *Euplotes aediculatus* revealed another telomerase protein of 43 kDa, in addition to TERT (Lingner and Cech, 1996). This protein, termed p43, contains an RNA recognition motif (RRM) and a La motif, known to facilitate the binding of proteins of the La family to a stretch of uridines typically present at the 3′-end of RNA polymerase III transcripts. p43 binds stem I and stem IV of TER and enhances the activity and repeat addition processivity of telomerase assembled *in vitro* (Aigner and Cech, 2004;

Aigner et al., 2003). A *Tetrahymena* La protein ortholog was identified and termed p65 (Prathapam et al., 2005; Witkin and Collins, 2004). p65 forms extensive interactions with the RNA: the p65 N-terminal domain with stem I, the La motif with the 3′-poly(U) tract, the RRM with the template-proximal part of stem IV, and the C-terminal domain with the central stem IV including the GA bulge (O'Connor and Collins, 2006). Interestingly, the binding of p65 enhances the subsequent binding of TERT (O'Connor and Collins, 2006). Elegant single-molecule fluorescence resonance energy transfer (smFRET) experiments revealed that the binding of p65 induces a conformational change, which is dependent on the GA bulge, and in turn this conformational change enables the binding of TERT (Stone et al., 2007). Furthermore, nuclease protection assays and mutational analysis in an *in vitro* reconstitution system revealed that p65 interacts with and stimulates conformational changes in regions of TER beyond stem IV (Berman et al., 2010). These experiments revealed the importance of p65 in presenting TERT with the active conformation of TER and mutually stabilizing TER and TERT in catalytically active conformations (Berman et al., 2010).

2.4.1.4 The Role of Stem-Loop IV in Telomerase Action Mutational analyses of *Tetrahymena* telomerase reconstituted *in vitro* from recombinant TER and TERT, revealed the importance of stem-loop IV, and in particular the conserved nucleotides in the apical loop, for the activity of telomerase (Autexier and Greider, 1998; Cunningham and Collins, 2005; Lai et al., 2003; Licht and Collins, 1999; Mason et al., 2003; Sperger and Cech, 2001). However, loop IV did not appear to be an essential determinant in TER–TERT association; a complete deletion of stem-loop IV, as well as smaller substitutions of its conserved nucleotides, had a partial effect on TERT binding. The moderate reduction in TERT–TER association could not fully explain the more sever effect of these mutations on the activity (Sperger and Cech, 2001). Based on these results, Sperger and Cech suggested that stem-loop IV has two distinct roles in addition to its role in TERT binding to TER. One is to enable a conformational change at the pseudoknot domain, which occurs upon the binding of TERT, and another distinct role is more directly involved with the catalytic reaction itself (Sperger and Cech, 2001). Indeed, mutations in stem-loop IV affect nucleotide addition and repeat addition processivity (Lai et al., 2003), perhaps reflecting such a role in the polymerization reaction. Consistently, the La protein ortholog p43 (in *Euplotes*), enhances telomerase activity and processivity *in vitro* (Aigner and Cech, 2004).

The addition of p65 to the Tetrahymena telomerase *in vitro* reconstitution assay greatly improves complex assembly and partially compensates for the effect of loop IV mutations (O'Connor and Collins, 2006). In contrast to the results obtained with the minimal TER–TERT RNP, in the TER–TERT–p65 RNP the effects of single nucleotide substitutions in loop IV on the TERT–TER association correlated well with the reduction in telomerase activity (Robart et al., 2010). These results suggest that one role of p65 is to place loop IV in the correct position for TERT binding and stabilize their otherwise weak interaction. Moreover, according to the proposed model, loop IV induces a conformational change in TERT, which stabilizes the active conformation of the enzyme (Robart et al., 2010).

In conclusion, stem-loop IV is an essential conserved domain of ciliate TERs, which can act in *trans* to promote the assembly of an active telomerase RNP. The accumulating data suggest that the conserved U-rich loop IV is precisely positioned to interact with TERT by the binding of p65 and the consequent conformational change in stem IV that is enabled by the conserved GA bulge. In turn, the functional loop interacts with TERT to stabilize the correct conformation of the RNP complex, which is required for catalysis.

2.4.2 The Vertebrate CR4–CR5 Domain

The CR4–CR5 domain is a structural element conserved in all vertebrate TERs, which contains two conserved regions (CRs), CR4 and CR5, defined originally by sequence alignment of 35 vertebrate TER sequences (Chen et al., 2000). The secondary structure of the CR4–CR5 domain was later refined by mutational analysis, revealing a highly conserved stem-loop termed P6.1 (Chen et al., 2002). Together with the pseudoknot-template domain and the TERT protein, the CR4–CR5 domain is required for reconstituting active telomerase both *in vitro* and *in vivo*. More remarkably, the pseudoknot-template and CR4–CR5 domains bind to TERT independently and, like the stem-loop IV in ciliates, can function in *trans* as two separate RNA molecules (Mitchell and Collins, 2000; Tesmer et al., 1999). In a reconstitution system using separate CR4–CR5 and pseudoknot-template TER fragments, a lower concentration of CR4–CR5 RNA is required for the assembly of an active telomerase complex, suggesting a higher binding affinity of TERT to the CR4–CR5 domain than the pseudoknot domain (Xie et al., 2008).

Stem-loop P6.1 is the most conserved element in the CR4–CR5 domain, and it is essential for telomerase action (Chen et al., 2002; Mitchell and Collins, 2000). The loop sequence (5′-YUVGN-3′; Y=C or U; V=A, C, or G; and N= any base) contains invariant uridine and guanine residues, that when mutated abolish the activity of telomerase reconstituted *in vitro*, but not the physical binding of the CR4–CR5 domain to TERT. Similarly, extension of the P6.1 stem that changes the spatial position of the loop affects only telomerase activity, but not TERT binding (Chen et al., 2002). In contrast, mutations that disrupt the P6.1 stem affect the TERT–TER interaction. These results suggest that the loop of P6.1 needs to be positioned precisely within the RNP complex to generate an active telomerase RNP. Interestingly, two highly conserved uridines in the loop of P6.1 are potentially modified into pseudouridines (Kim et al., 2010). Pseudouridinylation of these uridines alters the loop structure and increases its stability. It also attenuates telomerase activity *in vitro* but slightly increases its processivity (Kim et al., 2010).

2.4.3 The Yeast Three-Way Junction

The first indication for a conserved element close to the 3′-end of a yeast TER came from phylogenetic analysis of six *Kluyveromyces* TERs, revealing two short conserved sequences termed CS5 and CS6 (Tzfati et al., 2003). Comparing secondary structure models predicted for *Saccharomyces sensu stricto* and *Kluyveromyces*

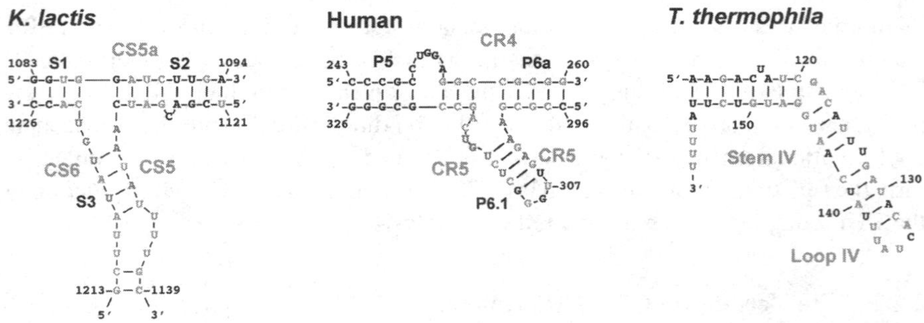

FIGURE 2.4 Conservation of the assembly/activation stem-loop elements in budding yeast, vertebrates, and *Tetrahymena* species. Shown are secondary structure models for the *K. lactis* TWJ, the human CR4–CR5 domain, and the *T. thermophila* stem-loop IV. Indicated are the conserved regions/sequences (CR or CS) and pairings/stems (P or S). Indicated in blue are nucleotides conserved in at least five of six *Kluyveromyces* spp., 33 of 35 vertebrate TERs, or in all six *Tetrahymena* and one *Glaucoma* spp. examined. Indicated in red are 7 nt conserved across yeast and vertebrates in at least 51 of the 55 TER sequences examined (>93%). Indicated in orange are U residues that are important for telomerase activity. (See the color version of this figure in Color Plates section.)

TERs revealed that CS5 and CS6 fall within similar three-way junction (TWJ) structures in both groups of species, as well as in other *Saccharomyces* and *Candida* species (Brown et al., 2007; Gunisova et al., 2009). Interestingly, the yeast TWJ shows sequence and structure similarity to the critical CR4–CR5 domain of vertebrate TERs. In particular, seven nucleotides are conserved in the same positions of stem 3 of the yeast TWJ and stem p6.1 of the vertebrate CR4–CR5 domain, with respect to the adjacent junction (Fig. 2.4). In addition, a stretch of U residues is found in the context of an asymmetrical internal loop (*Kluyveromyces* and *Saccharomyces*) or an apical loop (*Candida*), resembling the critical U residues in the ciliate loop IV and in the apical loop of the vertebrate p6.1 (Brown et al., 2007; Gunisova et al., 2009).

Mutational analysis confirmed that the *K. lactis* TWJ is critical for telomerase activity both *in vivo* and *in vitro* using telomerase fractions partially purified from cell extracts (Brown et al., 2007). In particular, small (1–3 nt) mutations in the residues conserved across yeast and vertebrates and in the U-rich internal loop almost completely abolished telomerase activity, indicating that the formation of the three stems, the precise angles between the stems (dictated by the linker nucleotides), and the U-rich internal loop, are all essential for the function of TWJ. Coimmunoprecipitation of TER and Myc-tagged Est2 from whole cell extracts revealed that small mutations in TWJ reduced the association of Est2 with TER by 50–80%, suggesting that also in *K. lactis* one role of this element is to facilitate the binding of Est2 to TER (Brown and Tzfati, unpublished results). However, as in *Tetrahymena*, the reduction in TER–Est2 association cannot fully explain the severe effect of at least some of the mutations. Consistently, overexpression of Est2 partially (but not fully) suppressed the short telomere phenotype of TWJ mutants; presumably, the increased protein concentration compensated for the reduced binding affinity (Brown et al., 2007).

Interestingly, Est3 overexpression also partially suppressed the short telomere phenotype of TWJ mutants without improving the TER–Est2 association, suggesting that Est3 is functionally associated with a role of TWJ in telomerase activation, which is distinct from its potential role in Est2 binding ((Brown et al., 2007) and unpublished results). In conclusion, while the function of the *K. lactis* TWJ is still unclear, the observations are consistent with two distinct roles: one, in facilitating the binding of Est2 to TER and another in Est3-related activation of telomerase. While TWJ is conserved also in *Saccharomyces* species, it appears to be less critical; a large TLC1 deletion that includes this element caused only a moderate effect on telomere length in *S. cerevisiae* (Zappulla et al., 2005).

2.4.3.1 Do the Ciliate Stem-Loop IV, the Vertebrate p6.1 and the Yeast Stem 3 Represent Homologous Elements? Several lines of evidence are consistent with the proposal that these three elements are functional homologues that can be termed assembly/activation stem-loop. First, specific nucleotides are conserved in the same structural context of a three-way junction across yeast and vertebrates. While a three-way junction was not found in ciliates, a highly structured apical loop with conserved and critical U residues is found in both vertebrates and ciliates, and a possibly analogous asymmetrical internal loop with critical U residues is present in yeast. The precise positioning of this loop appears to be critical for telomerase action in all three groups. In both human and *Tetrahymena* the function of this element can be provided in *trans*. In all species this element is important for the association of TER with Est2/TERT, and in all species it appears that this element is also important for telomerase activation or the stabilization of an active telomerase conformation.

2.5 BINDING SITES FOR TELOMERASE ACCESSORY/REGULATORY PROTEINS

2.5.1 Vertebrate

2.5.1.1 Dyskerin Protein Complex In addition to the pseudoknot-template and the CR4–CR5 domain, vertebrate TERs contain the box H/ACA domain and thus belong to the box H/ACA subfamily of snoRNAs, and more specifically to a subgroup of snoRNAs termed scaRNAs (Chen et al., 2000; Jady et al., 2004; Mitchell et al., 1999a). This domain consists of two stem-loops separated by a single-stranded hinge region, and three highly conserved motifs, box H, ACA, and the scaRNA-specific motif CAB box. The box H/ACA domain is a binding site for the dyskerin protein complex, consisting of four proteins: dyskerin, Nop10, Nhp2, and Gar1. Binding of the dyskerin protein complex to the box H/ACA domain is crucial for telomerase RNP biogenesis *in vivo* (Meier, 2006; Mitchell et al., 1999a, 1999b). A recent biochemical study shows that the box H/ACA domain is occupied by two full sets of dyskerin complexes (Egan and Collins, 2010).

While dyskerin is a component of the catalytically active telomerase complex (Cohen et al., 2007), its assembly with TERT and TER *in vivo* requires other accessory proteins. Through affinity purification of human telomerase complex followed by mass spectrometry, two ATPases, pontin and reptin, were identified as telomerase accessory proteins and shown to form a complex with dyskerin, assisting the telomerase RNP assembly *in vivo* (Venteicher et al., 2008). In addition to pontin and reptin, a number of proteins such as SmB, SmD3, hnRNP proteins (hnRNP C and hnRNP U), NTPase proteins (NAT10 and GNL3L), and chaperon proteins (HSP90 and p23) have also been found to associate with the vertebrate telomerase complex (Fu and Collins, 2006, 2007) (Holt et al., 1999). While these telomerase-associated proteins have been suggested to be involved in telomerase biogenesis or telomere length regulation, their exact functions await further characterization.

2.5.1.2 TCAB1/WDR79 The CAB box, located in the distal loop (L8) of the 3′ stem loop (CR7), is crucial for the localization of TER to the Cajal bodies *in vivo* (Jady et al., 2004). While Cajal body localization of TER is not essential for assembly of active telomerase, it is crucial for telomerase recruitment to the telomeres *in vivo* (Cristofari et al., 2007). The CAB box has been long suspected to provide a binding site for a specific protein that regulates subnuclear localization and/or biogenesis of TER. Indeed, an RNA-binding protein called TCAB1 (telomerase Cajal body protein 1) or WDR79 was recently found to bind the CAB box (Tycowski et al., 2009; Venteicher et al., 2009). While not essential for hTR accumulation or telomerase catalytic activity, TCAB1/WDR79 is important for hTR localization to the Cajal bodies (Venteicher et al., 2009). Interestingly, SmB and SmD3, which are associated with TER, as well as with other scaRNAs, require the CAB box for their association (Fu and Collins, 2006). In addition to the CAB box, a separate element in the CR7 hairpin loop was suggested to be critical for the 3′-end processing of TER (Theimer et al., 2007). Since this processing signal is partially overlapping with the CAB box and might interact with TCAB1/WDR79, it would be interesting to test if TCAB1/WDR79 also plays a role in the 3′-end processing of TER.

2.5.2 Budding Yeast

2.5.2.1 Est1 Binding Domain Est1 was first identified in yeast as a gene whose deletion resulted in the loss of telomerase activity *in vivo*, a phenotype termed ever shorter telomeres (EST) (Lundblad and Szostak, 1989). Est1 is essential for *S. cerevisiae* telomerase activity *in vivo* but not *in vitro* (Lingner et al., 1997), and it was also identified as essential for telomerase function in *Candida albicans* (Singh et al., 2002) and *S. pombe* (Beernink et al., 2003). Est1 was shown to mediate the recruitment of telomerase to the telomere (Evans and Lundblad, 1999; Pennock et al., 2001), but it may perform additional yet unclear roles in telomerase activation (Evans and Lundblad, 2002; Taggart et al., 2002). The region of TLC1 where Est1 binds was first defined by screening of a library of small TLC1 deletions to a fragment of 172 nt in the central region of TLC1 (Livengood et al., 2002). Later, through comparative analysis of six *Kluyveromyces* TERs and TLC1, the Est1-binding site was identified as a conserved bulged stem within this central region (Seto et al., 2002). The importance of this bulged stem for the binding of Est1 was confirmed by mutational analysis *in vivo*, binding assays *in vitro*, and by demonstrating that overexpression of Est1 suppresses mutations in the bulged stem, presumably by compensating for the reduced affinity of the protein for the RNA (Seto et al., 2002).

Surprisingly, while Est1 is essential for telomerase activity in *C. albicans* (Singh et al., 2002), the bulged-stem does not appear to be conserved in *Candida* TERs (Gunisova et al., 2009). Phylogenetic analysis of this domain revealed another sequence (termed CS2a) that is conserved in all budding yeast TERs examined, and is predicted to form one side of a large internal loop upstream of the bulged stem (Gunisova et al., 2009). Substitutions introduced into CS2a abolished telomerase activity *in vivo*, in both *S. cerevisiae* and *K. lactis* (Gunisova et al., 2009). It is still unknown whether CS2a is important for Est1 binding or if it has a different essential role in telomerase action.

2.5.2.2 The Ku80-Binding Stem-Loop in S. cerevisiae and Reg2 in K. lactis The Ku heterodimer (Ku70–Ku80) was found to play multiple roles at telomeres (Bertuch and Lundblad, 2003; Fisher and Zakian, 2005). In *S. cerevisiae*, Ku was proposed to contribute to the recruitment of telomerase to the telomeres, in particular in the G1 phase of the cell cycle (Fisher et al., 2004; Grandin et al., 2000). Ku80 interacts with an RNA domain close to the 5′-end of TLC1, termed the 48 nt stem-loop (Peterson et al., 2001; Stellwagen et al., 2003). This stem-loop is highly conserved in *Saccharomyces sensu stricto* species and in *Candida glabrata*, all of which share the ancestral whole genome duplication, but not in other budding or fission yeasts (Gunisova et al., 2009; Kabaha et al., 2008; Kachouri-Lafond et al., 2009; Webb and Zakian, 2008) (Fig. 2.5). Consistently, no experimental evidence was found for the association of *K. lactis* or *S. pombe* TERs with Ku (Calado and Young, 2008; Kabaha et al., 2008; Webb and Zakian, 2008). Ku was suggested to interact also with the human TER (Ting et al., 2005), though the functional relevance of this interaction is not clear. Why does Ku bind TER in some species and not others? Is its function dispensable in some species? Or is it performed by another factor? In *K. lactis*, a conserved stem loop (termed Reg2) was identified in a location within TER corresponding to the 48 nt stem loop in TLC1 (Kabaha et al., 2008). The apical loop of Reg2 contains the sequence 5′-CGGA, which is conserved in all six *Kluyveromyces* species (Fig. 2.5). Strikingly, even a single nucleotide substitution within this conserved loop causes severe telomere shortening (Kabaha and Tzfati, unpublished results). Whether Reg2 mediates a recruitment function analogous to the Ku-binding site in *S. cerevisiae* is an interesting question for future investigation.

2.5.3 Two TERs with Different Interacting Proteins in *Arabidopsis thaliana*

Interestingly, two different TER moieties, TER1 and TER2, encoded by two different genes, were identified recently in the plant *A. thaliana* (Cifuentes-Rojas et al., 2011). Both TER1 and TER2 copurify with telomerase activity and serve as templates for telomerase in a reconstitution assay *in vitro*. TER1 is essential for telomerase function

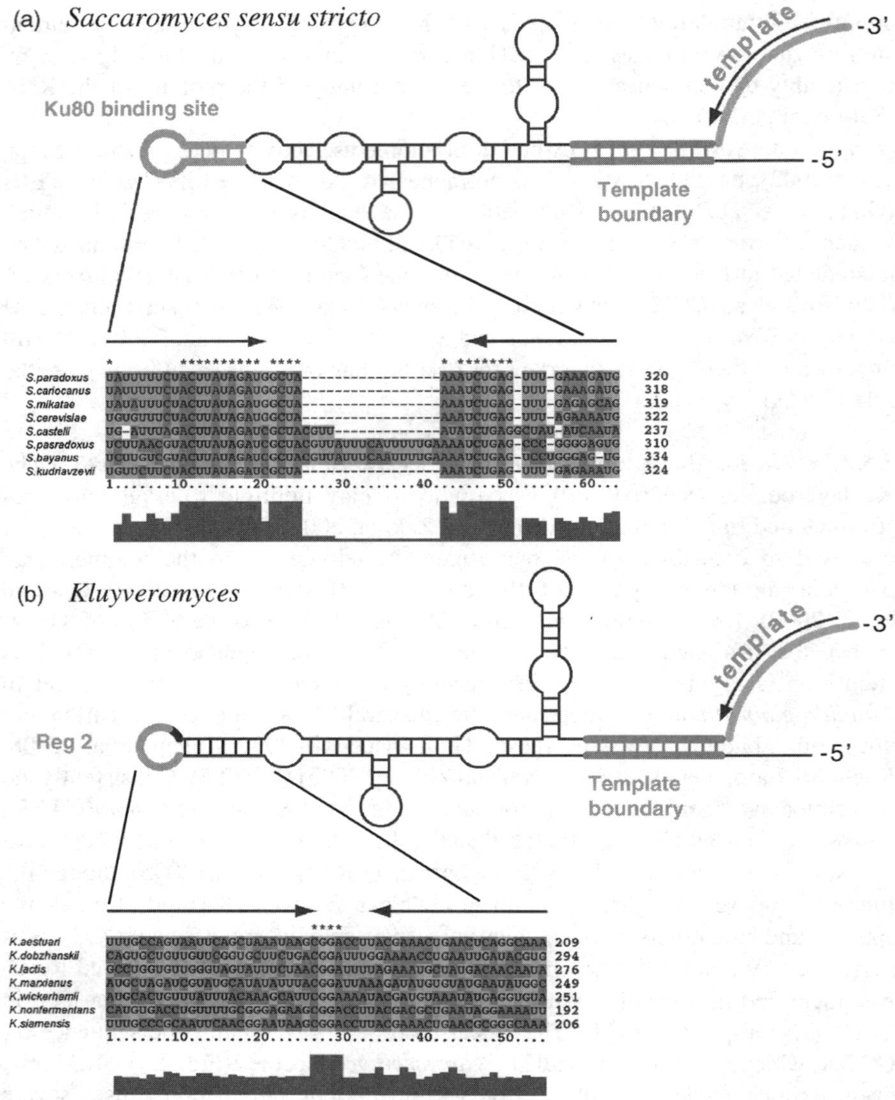

FIGURE 2.5 The 5′ arm of *Saccharomyces* and *Kluyveromyces* telomerase RNAs. Schemes illustrate the 5′ arm, template, and template boundary of *Saccharomyces sensu stricto* (a) and *Kluyveromyces* (b) TERs. Blue lines indicate sequence conservation. Sequence alignments showing the conservation of the Ku80-binding site (48 nt stem-loop) in *Saccharomyces* and a CGGA sequence motif in the *Kluyveromyces* Reg 2 element were made by the computer program *ClustalX* (Chenna et al., 2003). (See the color version of this figure in Color Plates section.)

in vivo, but the function of TER2 is unknown. While the two TERs are highly divergent in sequence and in the position of the template within the RNA, they have a conserved core sequence of about 150 nt. An RNA fragment containing this conserved sequence is sufficient to reconstitute with TERT active telomerase, demonstrating the ability of TER to incorporate changes in its sequence and evolve while maintaining the essential catalytic core functions.

The diverging sequences of the two TER molecules suggest that they may interact with different auxiliary proteins (Cifuentes-Rojas et al., 2011). In addition to TERT and dyskerin, also POT1a (one of three POT1 paralogs in *A. thaliana*) associates with TER1, but not with TER2. In other organisms, the OB fold containing POT1 homologues bind the telomeric single-stranded overhang and function in chromosome end protection and telomerase regulation. Cifuentes-Rojas et al. suggested that coduplication of TER and POT1 genes in *Arabidopsis* enabled novel protein–RNA interactions to form between TER1 and POT1a. Consequently, POT1a migrated from the telomere to the telomerase RNP and acquired an essential function in telomerase action, demonstrating the versatility of the telomere–telomerase system (Cifuentes-Rojas et al., 2011).

2.6 TELOMERASE RNA MUTATIONS IN HUMAN DISEASES

A number of human diseases have been linked to specific mutations in telomerase genes, validating the importance of telomerase in telomere length maintenance and cellular functions (Armanios, 2009). The mutations identified in human TER span all essential domains, the pseudoknot-template, CR4–CR5, and the box H/ACA, and result in various degrees of reduction in telomerase level or activity (Calado and Young, 2008). While the mutations in the pseudoknot-template and the CR4–CR5 domain affect mainly the telomerase enzymatic activity, the mutations in the box H/ACA domain lead to defects in the assembly of TER with the telomerase accessory proteins such as the box H/ACA protein complex and TCAB1, which mediate the biogenesis and subnuclear localization of TER (Fu and Collins, 2003; Theimer et al., 2007; Venteicher et al., 2009). Consistently, mutations in the proteins that bind these domains have also been linked to the telomere-mediated diseases (Mitchell et al., 1999b; Vulliamy et al., 2008; Walne et al., 2007, 2008; Zhong et al., 2011). Telomerase insufficiency leads to progressive telomere shortening and genomic instability in proliferating cells, and eventual cell death. The implications of telomerase mutations in diseases are discussed in more detail in Chapter 9.

2.7 CONCLUDING REMARKS

In the vast majority of eukaryotes, telomerase serves to elongate the 3′-end of the telomeres, compensating for telomere shortening caused by incomplete DNA replication and nuclease action. The fundamental arrangement of a reverse transcriptase protein copying a short template within an intrinsic RNA moiety is common to all known telomerases, from single cellular organisms such as ciliates and yeasts to mammals. Furthermore, specific domains and functional elements are conserved in both TERT and TER from yeast and ciliates to mammals. This conservation suggests that telomerase was present as an RNP enzyme, in which both RNA and protein moieties play essential roles, already in the ancestral eukaryote when linear chromosomes appeared and their ends required a special replication machinery. The conservation of reverse transcriptase motifs in TERT indicates its closest evolutionary relationship with retrotransposon RTs, which are encoded by their template RNA (see Chapter 11 for a more detailed discussion of telomerase evolution). The observation that retrotransposons maintain telomeres in some insects further supports this evolutionary relationship and suggests that the telomerase genes evolved from a single gene whose RNA was reverse transcribed by its encoded protein onto the telomeres to maintain the ends of linear chromosomes (reviewed in Curcio and Belfort (2007)). According to this hypothesis, the ancestral RT gene evolved into two genes, encoding an RNA and a protein subunit of the telomerase RNP. Once telomerase was split into two genes, its RNA moiety was free of protein-related evolutionary constrains and adapted rapidly to provide various functions in the telomerase RNP.

The telomerase core components TER and TERT coevolved together with telomerase accessory proteins to perform the universal task of elongating telomeres. However, telomerase in different organisms acquired different specifications. The variations in telomere sequences and length among different species also required the coevolution of telomere binding proteins, which function in telomere protection and length regulation. Such coevolution of the telomere system enabled telomerase to adapt to different constraints, requirements, and regulatory pathways associated with telomeres in different eukaryotes and in tissues with different cell proliferation potential. Telomerase RNA remained an essential component of telomerase, indicating that it has central roles in the assembly of the telomerase RNP complex and its mechanism of action. At the heart of this fascinating versatility is the flexible conformation of TER and its ability to accommodate changes in sequence and length, and provide alternative RNA structures and binding sites for different proteins.

The mechanism of telomerase action is still unclear. To fully understand it, more efficient methods should be developed for purification and reconstitution of active telomerase complexes at higher concentration, homogeneity and specific activity. It is also essential to understand the changes in conformation and interactions within the telomerase RNP, which presumably take place during the telomerase reaction cycle. To this end, structural methods that can follow such changes, such as single molecule FRET, should be employed and further developed. Such studies are not only important for understanding telomerase but would also advance our understanding of the structure, function, and evolution a rapidly growing number of other noncoding RNAs and RNPs.

REFERENCES

Aigner S, Cech TR. (2004) The Euplotes telomerase subunit p43 stimulates enzymatic activity and processivity *in vitro*. *RNA (New York, NY)* **10**: 1108–1118.

Aigner S, Postberg J, Lipps HJ, Cech TR. (2003) The Euplotes La motif protein p43 has properties of a telomerase-specific subunit. *Biochemistry* **42**: 5736–5747.

Antal M, Boros E, Solymosy F, Kiss T. (2002) Analysis of the structure of human telomerase RNA *in vivo*. *Nucleic Acids Res.* **30**: 912–920.

Armanios M. (2009) Syndromes of telomere shortening. *Annu. Rev. Genomics Hum. Genet.* **10**: 45–61.

Autexier C, Greider CW. (1995) Boundary elements of the *Tetrahymena* telomerase RNA template and alignment domains. *Genes Dev.* **9**: 2227–2239.

Autexier C, Greider CW. (1998) Mutational analysis of the *Tetrahymena* telomerase RNA: identification of residues affecting telomerase activity *in vitro*. *Nucleic Acids Res.* **26**: 787–795.

Beernink HT, Miller K, Deshpande A, Bucher P, Cooper JP. (2003) Telomere maintenance in fission yeast requires an Est1 ortholog. *Curr. Biol.* **13**: 575–580.

Berman AJ, Gooding AR, Cech TR. (2010) *Tetrahymena* telomerase protein p65 induces conformational changes throughout telomerase RNA (TER) and rescues telomerase reverse transcriptase and TER assembly mutants. *Mol. Cell Biol.* **30**: 4965–4976.

Bertuch AA, Lundblad V. (2003) The Ku heterodimer performs separable activities at double-strand breaks and chromosome termini. *Mol. Cell Biol.* **23**: 8202–8215.

Bhattacharyya A, Blackburn EH. (1994) Architecture of telomerase RNA. *EMBO J.* **13**: 5721–5731.

Box JA, Bunch JT, Tang W, Baumann P. (2008a) Spliceosomal cleavage generates the 3′ end of telomerase RNA. *Nature* **456**: 910–914.

Box JA, Bunch JT, Zappulla DC, Glynn EF, Baumann P. (2008b) A flexible template boundary element in the RNA subunit of fission yeast telomerase. *J. Biol. Chem.* **283**: 24224–24233.

Brown Y, Abraham M, Pearl S, Kabaha MM, Elboher E, Tzfati Y. (2007) A critical three-way junction is conserved in budding yeast and vertebrate telomerase RNAs. *Nucleic Acids Res.* **35**: 6280–6289.

Calado RT, Young NS. (2008) Telomere maintenance and human bone marrow failure. *Blood* **111**: 4446–4455.

Chappell AS, Lundblad V. (2004) Structural elements required for association of the *Saccharomyces cerevisiae* telomerase RNA with the Est2 reverse transcriptase. *Mol. Cell Biol.* **24**: 7720–7736.

Chen JL, Blasco MA, Greider CW. (2000) Secondary structure of vertebrate telomerase RNA. *Cell* **100**: 503–514.

Chen JL, Greider CW. (2003) Template boundary definition in mammalian telomerase. *Genes Dev.* **17**: 2747–2752.

Chen JL, Greider CW. (2004) An emerging consensus for telomerase RNA structure. *Proc. Natl. Acad. Sci. U. S. A.* **101**: 14683–14684.

Chen JL, Greider CW. (2005) Functional analysis of the pseudoknot structure in human telomerase RNA. *Proc. Natl. Acad. Sci. U. S. A.* **102**: 8080–8085; discussion 8077–8089.

Chen JL, Opperman KK, Greider CW. (2002) A critical stem-loop structure in the CR4–CR5 domain of mammalian telomerase RNA. *Nucleic Acids Res.* **30**: 592–597.

Chen Y, Fender J, Legassie JD, Jarstfer MB, Bryan TM, Varani G. (2006) Structure of stem-loop IV of *Tetrahymena* telomerase RNA. *EMBO J.* **25**: 3156–3166.

Chenna R, Sugawara H, Koike T, Lopez R, Gibson TJ, Higgins DG, Thompson JD. (2003) Multiple sequence alignment with the Clustal series of programs. *Nucleic Acids Res.* **31**: 3497–3500.

Cifuentes-Rojas C, Kannan K, Tseng L, Shippen DE. (2011) Two RNA subunits and POT1a are components of *Arabidopsis* telomerase. *Proc. Natl. Acad. Sci. U. S. A.* **108**(1): 73–78.

Cohen SB, Graham ME, Lovrecz GO, Bache N, Robinson PJ, Reddel RR. (2007) Protein composition of catalytically active human telomerase from immortal cells. *Science (New York, NY)* **315**: 1850–1853.

Comolli LR, Smirnov I, Xu L, Blackburn EH, James TL. (2002) A molecular switch underlies a human telomerase disease. *Proc. Natl. Acad. Sci. U. S. A.* **99**: 16998–17003.

Couch GS, Hendrix DK, Ferrin TE. (2006) Nucleic acid visualization with UCSF Chimera. *Nucleic Acids Res.* **34**: e29.

Cristofari G, Adolf E, Reichenbach P, Sikora K, Terns RM, Terns MP, Lingner J. (2007) Human telomerase RNA accumulation in Cajal bodies facilitates telomerase recruitment to telomeres and telomere elongation. *Mol. Cell* **27**: 882–889.

Cunningham DD, Collins K. (2005) Biological and biochemical functions of RNA in the tetrahymena telomerase holoenzyme. *Mol. Cell Biol.* **25**: 4442–4454.

Curcio MJ, Belfort M. (2007) The beginning of the end: links between ancient retroelements and modern telomerases. *Proc. Natl. Acad. Sci. U. S. A.* **104**: 9107–9108.

Dandjinou AT, Levesque N, Larose S, Lucier JF, Abou Elela S, Wellinger RJ. (2004) A phylogenetically based secondary structure for the yeast telomerase RNA. *Curr. Biol.* **14**: 1148–1158.

Egan ED, Collins K. (2010) Specificity and stoichiometry of subunit interactions in the human telomerase holoenzyme assembled *in vivo*. *Mol. Cell Biol.* **30**: 2775–2786.

Evans SK, Lundblad V. (1999) Est1 and Cdc13 as comediators of telomerase access. *Science* **286**: 117–120.

Evans SK, Lundblad V. (2002) The Est1 subunit of *Saccharomyces cerevisiae* telomerase makes multiple contributions to telomere length maintenance. *Genetics* **162**: 1101–1115.

Feng J, Funk WD, Wang SS, Weinrich SL, Avilion AA, Chiu CP, Adams RR, Chang E, Allsopp RC, Yu J, Le S, West MD, Harley CB, Andrews WH, Greider CW, Villeponteau B. (1995) The RNA component of human telomerase. *Science* **269**: 1236–1241.

Fisher TS, Taggart AK, Zakian VA. (2004) Cell cycle-dependent regulation of yeast telomerase by Ku. *Nat. Struct. Mol. Biol.* **11**: 1198–1205.

Fisher TS, Zakian VA. (2005) Ku: A multifunctional protein involved in telomere maintenance. *DNA Repair (Amst)* **4**(11): 1215–1226.

Forstemann K, Lingner J. (2005) Telomerase limits the extent of base pairing between template RNA and telomeric DNA. *EMBO Rep.* **6**: 361–366.

Fu D, Collins K. (2003) Distinct biogenesis pathways for human telomerase RNA and H/ACA small nucleolar RNAs. *Mol. Cell* **11**: 1361–1372.

Fu D, Collins K. (2006) Human telomerase and Cajal body ribonucleoproteins share a unique specificity of Sm protein association. *Genes Dev.* **20**: 531–536.

Fu D, Collins K. (2007) Purification of human telomerase complexes identifies factors involved in telomerase biogenesis and telomere length regulation. *Mol. Cell* **28**: 773–785.

Gilley D, Blackburn EH. (1999) The telomerase RNA pseudoknot is critical for the stable assembly of a catalytically active ribonucleoprotein. *Proc. Natl. Acad. Sci. U. S. A.* **96**: 6621–6625.

Grandin N, Damon C, Charbonneau M. (2000) Cdc13 cooperates with the yeast Ku proteins and Stn1 to regulate telomerase recruitment. *Mol. Cell Biol.* **20**: 8397–8408.

Greider CW. (1991) Telomerase is processive. *Mol. Cell Biol.* **11**: 4572–4580.

Greider CW, Blackburn EH. (1985) Identification of a specific telomere terminal transferase activity in *Tetrahymena* extracts. *Cell* **43**: 405–413.

Greider CW, Blackburn EH. (1987) The telomere terminal transferase of *Tetrahymena* is a ribonucleoprotein enzyme with two kinds of primer specificity. *Cell* **51**: 887–898.

Greider CW, Blackburn EH. (1989) A telomeric sequence in the RNA of *Tetrahymena* telomerase required for telomere repeat synthesis. *Nature* **337**: 331–337.

Gunisova S, Elboher E, Nosek J, Gorkovoy V, Brown Y, Lucier JF, Laterreur N, Wellinger RJ, Tzfati Y, Tomaska L. (2009) Identification and comparative analysis of telomerase RNAs from *Candida* species reveal conservation of functional elements. *RNA* **15**: 546–559.

Hinkley CS, Blasco MA, Funk WD, Feng J, Villeponteau B, Greider CW, Herr W. (1998) The mouse telomerase RNA 5″-end lies just upstream of the telomerase template sequence. *Nucleic Acids Res.* **26**: 532–536.

Hofacker IL, Fekete M, Stadler PF. (2002) Secondary structure prediction for aligned RNA sequences. *J. Mol. Biol.* **319**: 1059–1066.

Holt SE, Aisner DL, Baur J, Tesmer VM, Dy M, Ouellette M, Trager JB, Morin GB, Toft DO, Shay JW, Wright WE, White MA. (1999) Functional requirement of p23 and Hsp90 in telomerase complexes. *Genes Dev.* **13**: 817–826.

Hsu M, McEachern MJ, Dandjinou AT, Tzfati Y, Orr E, Blackburn EH, Lue NF. (2007) Telomerase core components protect *Candida* telomeres from aberrant overhang accumulation. *Proc. Natl. Acad. Sci. U. S. A.* **104**: 11682–11687.

Huang C, Yu YT. (2010) Targeted 2′-O methylation at a nucleotide within the pseudoknot of telomerase RNA reduces telomerase activity *in vivo*. *Mol. Cell Biol.* **30**: 4368–4378.

Jady BE, Bertrand E, Kiss T. (2004) Human telomerase RNA and box H/ACA scaRNAs share a common Cajal body-specific localization signal. *J. Cell Biol.* **164**: 647–652.

Juan V, Wilson C. (1999) RNA secondary structure prediction based on free energy and phylogenetic analysis. *J. Mol. Biol.* **289**: 935–947.

Kabaha MM, Zhitomirsky B, Schwartz I, Tzfati Y. (2008) The 5′ arm of *Kluyveromyces lactis* telomerase RNA is critical for telomerase function. *Mol. Cell Biol.* **28**: 1875–1882.

Kachouri-Lafond R, Dujon B, Gilson E, Westhof E, Fairhead C, Teixeira MT. (2009) Large telomerase RNA, telomere length heterogeneity and escape from senescence in *Candida glabrata*. *FEBS Lett.* **583**(22): 3605–3610.

Kim NK, Theimer CA, Mitchell JR, Collins K, Feigon J. (2010) Effect of pseudouridylation on the structure and activity of the catalytically essential P6.1 hairpin in human telomerase RNA. *Nucleic Acids Res.* **38**(19): 6746–6756.

Kim NK, Zhang Q, Zhou J, Theimer CA, Peterson RD, Feigon J. (2008) Solution structure and dynamics of the wild-type pseudoknot of human telomerase RNA. *J. Mol. Biol.* **384**: 1249–1261.

Lai CK, Miller MC, Collins K. (2002) Template boundary definition in *Tetrahymena* telomerase. *Genes Dev.* **16**: 415–420.

Lai CK, Miller MC, Collins K. (2003) Roles for RNA in telomerase nucleotide and repeat addition processivity. *Mol. Cell* **11**: 1673–1683.

Leonardi J, Box JA, Bunch JT, Baumann P. (2008) TER1, the RNA subunit of fission yeast telomerase. *Nat. Struct. Mol. Biol.* **15**: 26–33.

Licht JD, Collins K. (1999) Telomerase RNA function in recombinant *Tetrahymena* telomerase. *Genes Dev.* **13**: 1116–1125.

Lin J, Ly H, Hussain A, Abraham M, Pearl S, Tzfati Y, Parslow TG, Blackburn EH. (2004) A universal telomerase RNA core structure includes structured motifs required for binding the telomerase reverse transcriptase protein. *Proc. Natl. Acad. Sci. U. S. A.* **101**: 14713–14718.

Lingner J, Cech TR. (1996) Purification of telomerase from *Euplotes aediculatus*: requirement of a primer 3′ overhang. *Proc. Natl. Acad. Sci. U. S. A.* **93**: 10712–10717.

Lingner J, Cech TR, Hughes TR, Lundblad V. (1997) Three ever shorter telomere (EST) genes are dispensable for *in vitro* yeast telomerase activity. *Proc. Natl. Acad. Sci. U. S. A.* **94**: 11190–11195.

Lingner J, Hendrick LL, Cech TR. (1994) Telomerase RNAs of different ciliates have a common secondary structure and a permuted template. *Genes Dev.* **8**: 1984–1998.

Livengood AJ, Zaug AJ, Cech TR. (2002) Essential regions of *Saccharomyces cerevisiae* telomerase RNA: separate elements for Est1p and Est2p interaction. *Mol. Cell Biol.* **22**: 2366–2374.

Lundblad V, Szostak JW. (1989) A mutant with a defect in telomere elongation leads to senescence in yeast. *Cell* **57**: 633–643.

Marie-Egyptienne DT, Cerone MA, Londono-Vallejo JA, Autexier C. (2005) A human-*Tetrahymena* pseudoknot chimeric telomerase RNA reconstitutes a nonprocessive enzyme *in vitro* that is defective in telomere elongation. *Nucleic Acids Res.* **33**: 5446–5457.

Mason DX, Goneska E, Greider CW. (2003) Stem-loop IV of tetrahymena telomerase RNA stimulates processivity in trans. *Mol. Cell Biol.* **23**: 5606–5613.

McCormick-Graham M, Romero DP. (1995) Ciliate telomerase RNA structural features. *Nucleic Acids Res.* **23**: 1091–1097.

Meier UT. (2006) How a single protein complex accommodates many different H/ACA RNAs. *Trends Biochem. Sci.* **31**: 311–315.

Miller MC, Liu JK, Collins K. (2000) Template definition by *Tetrahymena* telomerase reverse transcriptase. *EMBO J.* **19**: 4412–4422.

Mitchell JR, Cheng J, Collins K. (1999a) A box H/ACA small nucleolar RNA-like domain at the human telomerase RNA 3′ end. *Mol. Cell Biol.* **19**: 567–576.

Mitchell JR, Collins K. (2000) Human telomerase activation requires two independent interactions between telomerase RNA and telomerase reverse transcriptase. *Mol. Cell* **6**: 361–371.

Mitchell JR, Wood E, Collins K. (1999b) A telomerase component is defective in the human disease dyskeratosis congenita. *Nature* **402**: 551–555.

Moriarty TJ, Marie-Egyptienne DT, Autexier C. (2005) Regulation of 5′ template usage and incorporation of noncognate nucleotides by human telomerase. *RNA* **11**: 1448–1460.

O'Connor CM, Collins K. (2006) A novel RNA binding domain in tetrahymena telomerase p65 initiates hierarchical assembly of telomerase holoenzyme. *Mol. Cell Biol.* **26**: 2029–2036.

Pace NR, Smith DK, Olsen GJ, James BD. (1989) Phylogenetic comparative analysis and the secondary structure of ribonuclease P RNA—a review. *Gene* **82**: 65–75.

Pennock E, Buckley K, Lundblad V. (2001) Cdc13 delivers separate complexes to the telomere for end protection and replication. *Cell* **104**: 387–396.

Peterson SE, Stellwagen AE, Diede SJ, Singer MS, Haimberger ZW, Johnson CO, Tzoneva M, Gottschling DE. (2001) The function of a stem-loop in telomerase RNA is linked to the DNA repair protein Ku. *Nat. Genet.* **27**: 64–67.

Prathapam R, Witkin KL, O'Connor CM, Collins K. (2005) A telomerase holoenzyme protein enhances telomerase RNA assembly with telomerase reverse transcriptase. *Nat. Struct. Mol. Biol.* **12**: 252–257.

Qiao F, Cech TR. (2008) Triple-helix structure in telomerase RNA contributes to catalysis. *Nat. Struct. Mol. Biol.* **15**: 634–640.

Richards RJ, Wu H, Trantirek L, O'Connor CM, Collins K, Feigon J. (2006) Structural study of elements of *Tetrahymena* telomerase RNA stem-loop IV domain important for function. *RNA (New York, NY)* **12**: 1475–1485.

Robart AR, O'Connor CM, Collins K. (2010) Ciliate telomerase RNA loop IV nucleotides promote hierarchical RNP assembly and holoenzyme stability. *RNA* **16**: 563–571.

Romero DP, Blackburn EH. (1991) A conserved secondary structure for telomerase RNA. *Cell* **67**: 343–353.

Roy J, Fulton TB, Blackburn EH. (1998) Specific telomerase RNA residues distant from the template are essential for telomerase function. *Genes Dev.* **12**: 3286–3300.

Seto AG, Livengood AJ, Tzfati Y, Blackburn EH, Cech TR. (2002) A bulged stem tethers Est1p to telomerase RNA in budding yeast. *Genes Dev.* **16**: 2800–2812.

Seto AG, Umansky K, Tzfati Y, Zaug AJ, Blackburn EH, Cech TR. (2003) A template-proximal RNA paired element contributes to *Saccharomyces cerevisiae* telomerase activity. *RNA* **9**: 1323–1332.

Seto AG, Zaug AJ, Sobel SG, Wolin SL, Cech TR. (1999) *Saccharomyces cerevisiae* telomerase is an Sm small nuclear ribonucleoprotein particle. *Nature* **401**: 177–180.

Shefer K, Brown Y, Gorkovoy V, Nussbaum T, Ulyanov NB, Tzfati Y. (2007) A triple helix within a pseudoknot is a conserved and essential element of telomerase RNA. *Mol. Cell Biol.* **27**: 2130–2143.

Shippen-Lentz D, Blackburn EH. (1990) Functional evidence for an RNA template in telomerase. *Science* **247**: 546–552.

Singh SM, Steinberg-Neifach O, Mian IS, Lue NF. (2002) Analysis of telomerase in *Candida albicans*: potential role in telomere end protection. *Eukaryot. Cell* **1**: 967–977.

Sperger JM, Cech TR. (2001) A stem-loop of *Tetrahymena* telomerase RNA distant from the template potentiates RNA folding and telomerase activity. *Biochemistry* **40**: 7005–7016.

Stellwagen AE, Haimberger ZW, Veatch JR, Gottschling DE. (2003) Ku interacts with telomerase RNA to promote telomere addition at native and broken chromosome ends. *Genes Dev.* **17**: 2384–2395.

Stone MD, Mihalusova M, O'Connor CM, Prathapam R, Collins K, Zhuang X. (2007) Stepwise protein-mediated RNA folding directs assembly of telomerase ribonucleoprotein. *Nature* **446**: 458–461.

Taggart AK, Teng SC, Zakian VA. (2002) Est1p as a cell cycle-regulated activator of telomere-bound telomerase. *Science* **297**: 1023–1026.

ten Dam E, van Belkum A, Pleij K. (1991) A conserved pseudoknot in telomerase RNA. *Nucleic Acids Res.* **19**: 6951.

Tesmer VM, Ford LP, Holt SE, Frank BC, Yi X, Aisner DL, Ouellette M, Shay JW, Wright WE. (1999) Two inactive fragments of the integral RNA cooperate to assemble active telomerase with the human protein catalytic subunit (hTERT) *in vitro*. *Mol. Cell Biol.* **19**: 6207–6216.

Theimer CA, Blois CA, Feigon J. (2005) Structure of the human telomerase RNA pseudoknot reveals conserved tertiary interactions essential for function. *Mol. Cell* **17**: 671–682.

Theimer CA, Finger LD, Trantirek L, Feigon J. (2003) Mutations linked to dyskeratosis congenita cause changes in the structural equilibrium in telomerase RNA. *Proc. Natl. Acad. Sci. U. S. A.* **100**: 449–454.

Theimer CA, Jady BE, Chim N, Richard P, Breece KE, Kiss T, Feigon J. (2007) Structural and functional characterization of human telomerase RNA processing and Cajal body localization signals. *Mol. Cell* **27**: 869–881.

Ting NS, Yu Y, Pohorelic B, Lees-Miller SP, Beattie TL. (2005) Human Ku70/80 interacts directly with hTR, the RNA component of human telomerase. *Nucleic Acids Res.* **33**: 2090–2098.

Tycowski KT, Shu MD, Kukoyi A, Steitz JA. (2009) A conserved WD40 protein binds the Cajal body localization signal of scaRNP particles. *Mol. Cell* **34**: 47–57.

Tzfati Y, Fulton TB, Roy J, Blackburn EH. (2000) Template boundary in a yeast telomerase specified by RNA structure. *Science* **288**: 863–867.

Tzfati Y, Knight Z, Roy J, Blackburn EH. (2003) A novel pseudoknot element is essential for the action of a yeast telomerase. *Genes Dev.* **17**: 1779–1788.

Ulyanov NB, Shefer K, James TL, Tzfati Y. (2007) Pseudoknot structures with conserved base triples in telomerase RNAs of ciliates. *Nucleic Acids Res.* **35**: 6150–6160.

Venteicher AS, Abreu EB, Meng Z, McCann KE, Terns RM, Veenstra TD, Terns MP, Artandi SE. (2009) A human telomerase holoenzyme protein required for Cajal body localization and telomere synthesis. *Science* **323**: 644–648.

Venteicher AS, Meng Z, Mason PJ, Veenstra TD, Artandi SE. (2008) Identification of ATPases pontin and reptin as telomerase components essential for holoenzyme assembly. *Cell* **132**: 945–957.

Vulliamy T, Beswick R, Kirwan M, Marrone A, Digweed M, Walne A, Dokal I. (2008) Mutations in the telomerase component NHP2 cause the premature ageing syndrome dyskeratosis congenita. *Proc. Natl. Acad. Sci. U. S. A.* **105**: 8073–8078.

Walne AJ, Vulliamy T, Marrone A, Beswick R, Kirwan M, Masunari Y, Al-Qurashi FH, Aljurf M, Dokal I. (2007) Genetic heterogeneity in autosomal recessive dyskeratosis congenita with one subtype due to mutations in the telomerase-associated protein NOP10. *Hum. Mol. Genet.* **16**: 1619–1629.

Walne AJ, Vulliamy TJ, Beswick R, Kirwan M, Dokal I. (2008) TINF2 mutations result in very short telomeres: analysis of a large cohort of patients with dyskeratosis congenita and related bone marrow failure syndromes. *Blood* **112**: 3594–3600.

Wang ZR, Guo L, Chen L, McEachern MJ. (2009) Evidence for an additional base-pairing element between the telomeric repeat and the telomerase RNA template in *Kluyveromyces lactis* and other yeasts. *Mol. Cell Biol.* **29**: 5389–5398.

Webb CJ, Zakian VA. (2008) Identification and characterization of the *Schizosaccharomyces pombe* TER1 telomerase RNA. *Nat. Struct. Mol. Biol.* **15**: 34–42.

Witkin KL, Collins K. (2004) Holoenzyme proteins required for the physiological assembly and activity of telomerase. *Genes Dev.* **18**: 1107–1118.

Xie M, Mosig A, Qi X, Li Y, Stadler PF, Chen JJ-L. (2008) Structure and function of the smallest vertebrate telomerase RNA from teleost fish. *J. Biol. Chem.* **283**: 2049–2059.

Yu GL, Bradley JD, Attardi LD, Blackburn EH. (1990) *In vivo* alteration of telomere sequences and senescence caused by mutated *Tetrahymena* telomerase RNAs. *Nature* **344**: 126–132.

Zappulla DC, Cech TR. (2004) Yeast telomerase RNA: a flexible scaffold for protein subunits. *Proc. Natl. Acad. Sci. U. S. A.* **101**: 10024–10029.

Zappulla DC, Goodrich K, Cech TR. (2005) A miniature yeast telomerase RNA functions *in vivo* and reconstitutes activity *in vitro*. *Nat. Struct. Mol. Biol.* **12**: 1072–1077.

Zhang Q, Kim NK, Peterson RD, Wang Z, Feigon J. (2010) Structurally conserved five nucleotide bulge determines the overall topology of the core domain of human telomerase RNA. *Proc. Natl. Acad. Sci. U. S. A.* **107**: 18761–18768.

Zhong F, Savage SA, Shkreli M, Giri N, Jessop L, Myers T, Chen R, Alter BP, Artandi SE. (2011) Disruption of telomerase trafficking by TCAB1 mutation causes dyskeratosis congenita. *Genes Dev.* **25**: 11–16.

3

TERT STRUCTURE, FUNCTION, AND MOLECULAR MECHANISMS

EMMANUEL SKORDALAKES AND NEAL F. LUE

3.1 INTRODUCTION

A major milestone in telomerase research was the discovery of TERT (telomerase reverse transcriptase), the catalytic protein subunit of the complex (Lingner et al., 1997). In keeping with the ever-important contributions of model systems to the field, TERT was initially uncovered through a series of seminal experiments that involve both genetic screens in *Saccharomyces cerevisiae* and biochemical purification of the *Euplotes aediculatus* telomerase complex (Lendvay et al., 1996; Lingner et al., 1997). Subsequent studies led to the identification of more than 100 orthologues in all five supergroups of the eukaryotic kingdom, which encompass fungi, plants, protozoa, mammals, as well as the deeply divergent heterotrophic protist (Podlevsky et al., 2008). Evidently the protein arose early in evolution and has been conserved owing to its critical role in maintaining the linear chromosomes of eukaryotic organisms.

As the name implies, TERT is a reverse transcriptase (RT) and bears unmistakable resemblance to other RTs, such as those of retroviral and retrotransposon origins. A more thorough discussion of the evolutionary relationship of TERTs to other RTs is provided in Chapter 11 of this volume. However, TERTs also exhibit distinguishing features both in and outside of the canonical RT domains, some of which are, not surprisingly, responsible for unique enzymatic properties of telomerase (Autexier and Lue, 2006). As noted elsewhere, the two major enzymatic features that distinguish TERT from other RTs are its tight association with telomerase RNA

Telomerases: Chemistry, Biology, and Clinical Applications, First Edition.
Edited by Neal F. Lue and Chantal Autexier.

and its ability to repetitively reverse transcribe the template segment of the RNA (referred to as repeat addition processivity or RAP). These features can now be attributed at least in part to specific structural elements of the TERT protein.

The mechanisms of TERT have been investigated through analysis of isolated proteins or various RNP complexes or subcomplexes. The most extensive analyses have been performed on proteins derived from ciliated protozoa, yeast, and mammals. A multitude of assays ranging from primer extension, direct nucleic-acid binding, cross-linking, and single-molecule studies have been applied to the wild-type protein as well as to numerous truncation and substitution variants. Regions or residues important for specific functions were identified through these analyses. Moreover, several atomic resolution structures of TERT or TERT domains are now available, either alone or with nucleic acids and nucleotides (Gillis et al., 2008; Jacobs et al., 2006; Mitchell et al., 2010; Rouda and Skordalakes, 2007). These structures provide a much needed framework for interpreting existing biochemical and genetic data as well as motivations for further mechanistic analysis of this remarkable polymerase. In the following sections, we will first provide an overview of the architecture of the TERT protein, and then discuss available evidence for the roles of individual structural elements in mediating specific TERT functions, including RNP assembly, nucleotide addition, template recognition and utilization, and RAP. TERT mutations that result in a variety of human disorders will also be described.

3.2 DOMAIN ORGANIZATION AND STRUCTURES

3.2.1 Overall Organization

In most organisms, TERTs are ~600 to 1300 amino acids long and contain core domains that resemble prototypical RTs, which constitute one of the seven distinct families of DNA polymerases (Johansson and Macneill, 2010). Like other DNA polymerases, the prototypical RT domains found in retroviruses and retrotransposons (~300 amino acids long) are often characterized as having the configuration of a right hand, comprising the "fingers," palm, and thumb domains. Sequence, functional, and structural studies support the existence of all of these domains in the TERT protein (Autexier and Lue, 2006; Gillis et al., 2008). The "fingers" and the palm of TERTs contain the signature RT motifs (1, 2, A, B′, C, D, and E) that mediate critical interactions with metal ions, nucleic acids, and nucleotides (Autexier and Lue, 2006; Bryan et al., 1998; Lee et al., 2003; Peng et al., 2001; Fig. 3.1a). Moreover, two new motifs proposed to be TERT-specific have been noted in the "fingers" and palm region, one between conserved motifs 2 and A, and the other between motifs A and B′. They have been designated motif 3 and IFD, respectively (Lue et al., 2003; Xie et al., 2010). The thumb of TERT, while bearing no sequence similarity to other RTs, nevertheless occupies the equivalent position and performs functions predicted for this structure (Hossain et al., 2002; Huard et al., 2003). In addition to these RT-equivalent domains, TERT in most organisms contains two telomerase-specific domains, that is, the telomerase essential N-terminal (TEN) domain and the telomerase RNA binding

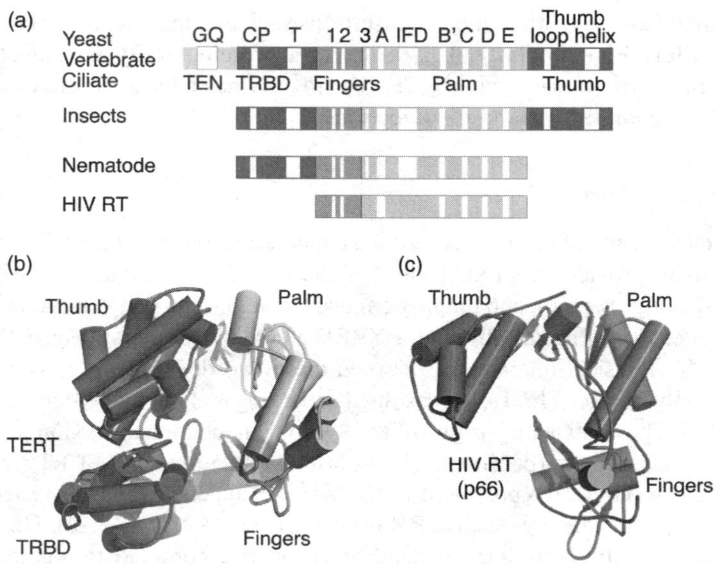

FIGURE 3.1 (a) The domain organizations of TERTs from different species are illustrated. (b) The structure of *T. castaneum* TERT (PDB ID: 3DU6): the TRBD, fingers, palm, and thumb domains are colored in blue, orange, wheat, and red, respectively. (c) The structure of the p66 subunit of HIV-1 RT without the nuclease domain (PDB ID: 1RTD) is shown. (See the color version of this figure in Color Plates section.)

domain (TRBD), both of which are located at the N-terminal side of the RT domains (Fig. 3.1a). The major functions of these two domains are to interact with telomeric DNA and telomerase RNA, respectively (Bosoy et al., 2003; Jacobs et al., 2006; Lai et al., 2001; Lue, 2005; Moriarty et al., 2002, 2005b). The two domains appear to be connected through a flexible linker; this putative linker region is extremely variable in length, bears little conservation at the sequence level, and is known to be sensitive to proteolytic cleavage (Autexier and Lue, 2006).

3.2.2 Evolutionary Conservation and Diversity

Although the five-domain organization of the *TERT* genes is largely conserved, there are notable exceptions. For example, the insect *TERT*s from *Tribolium casteneum* and *Bombyx mori* are both devoid of the TEN domain (Osanai et al., 2006). Likewise, several roundworm members (*Caenorhabditis elegans*, *Caenorhabditis briggsae*, and *Caenorhabditis remanei*) appear to lack much of the TEN and thumb domains (Fig. 3.1a; Meier et al., 2006). Other instances of structural variations, while less dramatic, also provide interesting illustrations of the flexibility and adaptability of TERT family members. For example, a recognizable T-motif, which is believed to be a crucial element of the TRBD domain has yet to be identified in TERT from the deeply branching eukaryote *Giardia lamblia* (Malik et al., 2000). The TERTs from *Plasmodium spp*., while retaining all conserved motifs, are decorated with numerous insertions such that they are twice the size of a typical TERT (Figueiredo et al., 2005). As a protein family with a conserved function in telomere maintenance, TERT evidently has a basic architecture that is amenable to elaborations and modifications.

3.2.3 Architecture

The crystal structure of the *Tc*TERT, which contains all conserved domains except the TEN domain, provides a striking view of the overall architecture of this protein. The TRBD, fingers, palm, and thumb domains are organized into a ring-like structure through extensive contacts between the TRBD and thumb domains (Fig. 3.1b) (Gillis et al., 2008). Physical interactions between these two domains have been detected biochemically in other TERTs, suggesting that the ring architecture is conserved (Arai et al., 2002). The relative disposition of the RT-equivalent domains is similar to that of the polymerase domain (p66 minus the RNase H domain) of HIV-1 RT (Sarafianos et al., 2002), the viral RNA polymerases (Di Marco et al., 2005), and the bacteriophage B-family DNA polymerases such as RB69 (Wang et al., 1997; Fig. 3.1c). The cavity in the interior of the ring is sufficient in depth and width to accommodate double-stranded nucleic acids approximately seven to eight base pairs long, which roughly corresponds to the length of the RNA–DNA duplex formed between telomerase and the DNA substrate (Forstemann and Lingner, 2005; Hammond and Cech, 1998). The interior of the TERT ring is lined with residues from ten conserved motifs (T, 1, 2, A, B′, C, D, E, thumb loop, and thumb helix) that are hallmarks of this family of polymerases and are implicated in nucleic acid association, nucleotide binding, and DNA synthesis (see below). Together, these motifs form a spiral in the interior of the ring which matches the geometry of the backbone of double-stranded nucleic acids.

3.2.4 The TRBD Domain

The RNA-binding domain of telomerase (TRBD), located between the TEN and RT domains, is universally conserved, and is essential for telomerase function both *in vitro* and *in vivo* (Bosoy et al., 2003; Lai et al., 2001). Three conserved motifs, designated CP, QFP, and T, have been recognized within this domain (Bosoy et al., 2003). The TRBD–TER interactions are largely responsible for the stable assembly of the core complex (O'Connor et al., 2005). Moreover, in *Tetrahymena*, the binding of TRBD to TER is thought to be necessary for template boundary definition (Lai et al., 2002). The first structure of TRBD was obtained from *Tetrahymena thermophila* (Rouda and Skordalakes, 2007; Fig. 3.2a). The second structure, derived from the crystal of the full length *Tc*TERT, is quite similar (root mean square deviation of 2.7 Å) (Gillis et al., 2008). These structures revealed a novel nucleic acid binding fold consisting mostly of α-helices linked together by several loops and two short β-strands. The body of TRBD is connected to the fingers domain through an extended helix (connector helix) and the two are related to each other at a ~120° angle, creating a deep indentation on the surface of the protein. Within this cavity is located a β-hairpin formed by a long insertion that connects the main body of TRBD to the

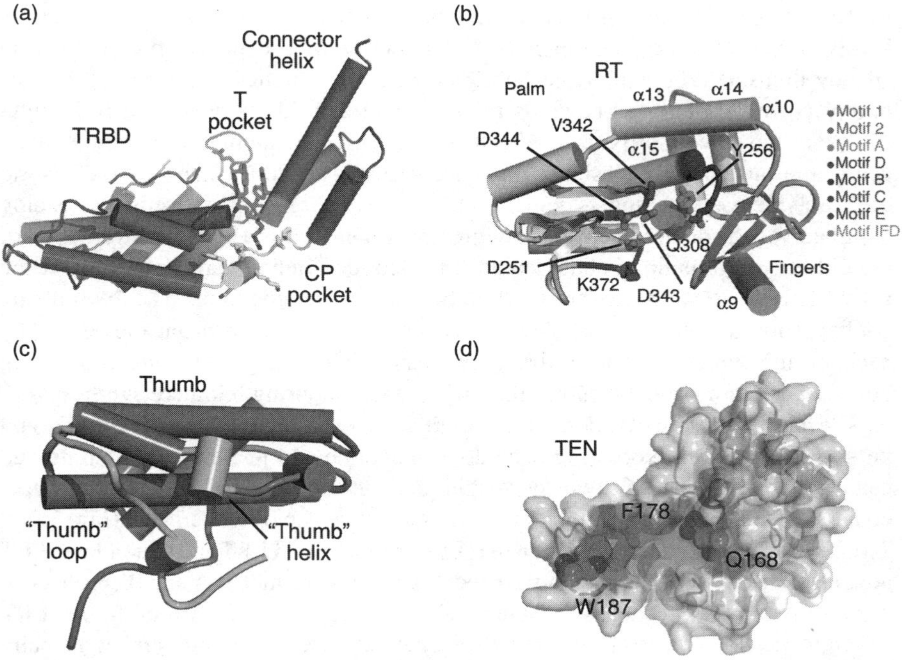

FIGURE 3.2 (a) The RNA-binding domain of *Tetrahymena thermophila* TERT with motif T in cyan and CP in yellow: conserved residues that comprise these motifs are shown in the stick representation. (b) The fingers (orange) and palm (wheat) subdomains of *Tc*TERT: conserved motifs implicated in nucleotide and nucleic acid binding and catalysis are displayed in the designated colors. (c) The thumb domain of *Tc*TERT with the two DNA-binding structural elements (the thumb loop and helix) highlighted in green. (d) The TEN domain of *T. thermophila* TERT is displayed in a surface representation; the putative DNA-binding groove and the residues implicated in DNA binding (Q168, F178, and W187) are accented. (See the color version of this figure in Color Plates section.)

connector helix. The β-hairpin protrudes from the base of this indentation and stands at a 45° angle to the plane of the connector helix (Fig. 3.2a). This hairpin appears to be held in place by a short helix and a succeeding loop positioned at the back of the hairpin. The potential importance of this β-hairpin will be discussed later in connection with TERT mechanisms.

3.2.5 The Reverse Transcriptase Domain: Fingers, Palm, and Thumb

The RT domain of TERT consists of a mix of α-helices and β-strands organized into two subdomains that are structurally analogous to the "fingers" and "palm" subdomains of retroviral RTs, viral RNA polymerases, and the bacteriophage B-family DNA polymerases (Fig. 3.2b; Di Marco et al., 2005; Sarafianos et al., 2002; Wang et al., 1997). The key signature motifs that are diagnostic of RTs are all present in the TERT family members (Lingner et al., 1997).

The fingers domain of telomerase, similar to other comparable domains, is involved in nucleic acid and nucleotide binding as well as processivity regulation (Bosoy and Lue, 2001; Miller et al., 2000). These activities are mediated by two highly conserved motifs known as motifs 1 and 2. In the structure of full-length *Tc*TERT, these two motifs form an extended β-hairpin that contains several solvent accessible and conserved residues (Fig. 3.2b; Gillis et al., 2008). In retroviral RTs and other polymerases, the fingers domain is known to undergo large movements, leading alternately to the open and closed configurations (Ding et al., 1998). The formation of the closed conformation is coupled with nucleotide binding and positioning at the active site of the enzyme to form a tight catalytic complex. The closed-ring configuration adopted by TERT would appear to restrict significant movement by each of the domains (Fig. 3.1b), thus raising interesting questions concerning potential conformational changes that may transpire during telomere synthesis.

The palm domain of telomerase is structurally similar to the HIV-1 RT and possesses conserved motifs implicated in nucleotide and nucleic acid binding and catalysis. Motifs A and C, two short rigid loops located in proximity to each other, contain three invariant aspartates that form the active site of the enzyme (Fig. 3.2b). Two significant differences between the "palm" domain of TERT and that of HIV-1 RT are longer insertions between motifs 2 and A, and between motifs A and B′, referred to as motif 3 and IFD, respectively (Lue et al., 2003; Xie et al., 2010). In the *Tc*TERT structure, motif 3 consists of two α-helices that flank the "upstream" side of the palm domain. By contrast, motif IFD, also consisting of two α-helices is located on the periphery of the TERT ring. These helices make extensive contacts with the rest of the palm domain and probably play an important role in the structural organization of this domain. Both motif 3 and IFD have also been implicated in processivity control. Interestingly, motifs 3 and IFD of most TERTs are considerably longer than those in *Tc*TERT, and possibly have more elaborate structures (Xie et al., 2010).

The thumb domain of retroviral RTs, viral RNA and B-family DNA polymerases are well characterized and known to be crucial for nucleic acid binding during the elongation process. The thumb domain of telomerase is an elongated helical bundle that bears little similarity to the equivalent domain in other families (Fig. 3.2c). Nevertheless, structural comparison of TERT with other polymerases places the "thumb" domain of these enzymes in the same spatial position with respect to the "fingers" and "palm" subdomains, suggesting that the telomerase "thumb" is also involved in nucleic acid binding during the elongation process, a finding that is supported by studies of yeast and human TERT (Hossain et al., 2002; Huard et al., 2003; Xie et al., 2010).

3.2.6 The TEN Domain

With few exceptions, the TERT family members are distinguished from other RTs by the presence of an autonomously folded, ~200 amino acid N-terminal domain referred to as the GQ or TEN domain in general (Autexier and Lue, 2006). This domain is also designated region I in yeast TERT (Est2), and RID1 in human TERT. Definite or possible loss of the TEN domain has transpired in the insect and worm phyla. Notably, the only full-length TERT for which a crystal structure is available, namely the protein from *Tribolium castaneum* (flour beetle), evidently lacks the TEN domain (Gillis et al., 2008). In addition, the N-terminus of *Caenorhabditis* TERTs are quite small and unlikely to possess typical TEN domains (Malik et al., 2000; Meier et al., 2006). On the other hand, the prevalence of the TEN domains in TERTs from protozoa, plants, yeast, and vertebrates argues for an early origin and robust functional significance. Crystal structure of the TEN domain from *T. thermophila* has been solved at 2.2 Å, revealing a single globular domain with a novel fold consisting of four β strands and seven α helices (Jacobs et al., 2006; Fig. 3.2d). Relative insertions or deletions in the TEN domains of other family members occur mostly in surface loops, suggesting that the general architecture of the *T. thermophila* domain is shared by other homologues. This high-resolution structure has provided a crucial platform for interpreting existing biochemical data and motivating additional studies (Jacobs et al., 2006; Lue and Li, 2007; Romi et al., 2007; Sealey et al., 2010; Wyatt et al., 2009).

3.3 TELOMERASE RNP ASSEMBLY

3.3.1 The Role of the TRBD Domain in Telomerase RNP Assembly

Telomerase exists as a stable ribonucleoprotein complex and multiple domains of TERT are believed to interact with various regions of the TER RNA. Although the detailed requirements for stable binding to TERs appear to vary somewhat for different TERTs, the TRBD domain evidently acts as a conserved determinant for high-affinity RNA interaction (Bachand and Autexier, 2001; Beattie et al., 2000; Bosoy et al., 2003; Bryan et al., 2000; Lai et al., 2001). Three universally conserved motifs, designated CP, QFP, and T, have been recognized and confirmed to be functionally important through mutagenesis of the ciliate, yeast, and mammalian proteins (Bosoy et al., 2003; Bryan et al., 2000; Lai et al., 2002). In addition, the region N-terminal to the CP motif, though less well conserved, nevertheless appears to participate in RNA-binding and contains residues shared by TERTs in specific lineages (e.g., the CP2 motif shared by ciliates and the VSR motif shared by mammals) (Lai et al., 2002; Moriarty et al., 2002). The interaction between the TRBD domain and its RNA target is best understood in *T. thermophila* owing to extensive biochemical and mutagenesis studies. These studies revealed the importance of conserved amino acid residues in TRBD and two structural elements in TER (stem I and the template boundary element (TBE)) in mediating protein–RNA interaction (Lai et al., 2001; O'Connor et al., 2005). The recently determined structures of the TRBD domain have provided some insights into the potential mechanisms of TRBD–TER association (Gillis et al., 2008; Rouda and Skordalakes, 2007). The TRBD domain can be described as a "boomerang" with two asymmetric lobes connected to each other through several long loops. On the surface of the domain are two well-defined cavities that contain conserved and solvent-exposed residues (named the CP- and T-pockets in accordance with conserved motifs lining these pockets; Fig. 3.2a). The T-pocket is narrow and deep, and located at the junction of the two halves of the molecule. Part of the pocket is hydrophobic in nature while another part, located in proximity to the CP-pocket, is positively charged. The width and shape of the hydrophobic portion of the T-pocket suggest that it binds ssRNA, perhaps the single-stranded TBE. Mutating the conserved and mostly hydrophobic residues that comprise the T-motif of *Tt*TERT resulted in severe loss of RNA-binding affinity and telomerase activity (Bryan et al., 2000; Lai et al., 2002). Because the side chains of these residues are generally solvent accessible, it is tempting to suggest that they mediate direct contacts with the RNA substrate, possibly through base-stacking interactions.

In contrast to the T-pocket, the CP-pocket is a positively charged, shallow cavity located on the side of the molecule and forms an extension of the T-pocket (Fig. 3.2a). The hydrophilic part of the T- and the CP-pockets are lined with multiple lysines and arginines, the side chains of which are solvent exposed. The width and the chemical nature of this pocket suggest that it binds double-stranded RNA, perhaps stem I or stem II of the TBE in the case of *T. thermophila*. The functional importance of the residues lining the pockets is supported by extensive mutagenesis analyses. Single- and double- as well as stretches of 4–10 amino acid alanine substitutions within these two motifs showed moderate to severe loss (20–100%) of RNA-binding affinity and polymerase activity when compared to the wild-type enzyme (Bryan et al., 2000; Lai et al., 2002).

The protein/nucleic acid interactions mediated by the CP and T pockets are at least partly responsible for the formation of a stable RNP and may guide the binding of TERT to a specific target site in TER. Interestingly, rodent TERs do not contain the TBE found in most organisms, but rather a 2-nt overhang 5′ to the RNA template (Blasco et al., 1995). A third motif within the TRBD domain, known as the QFP motif, comprises mostly hydrophobic residues that are buried within the core of the domain (Bosoy et al., 2003; Rouda and Skordalakes, 2007). These residues are thus likely to be required for the proper folding of the TRBD domain.

3.3.2 TRBD–TER Association and Template Utilization

As with most DNA and RNA polymerases, nucleic acid synthesis by telomerase requires pairing of the RNA template region of TER with the incoming DNA primer (Baran et al., 2002; Lingner and Cech, 1996). RNA–DNA pairing is a prerequisite of telomere synthesis in that it brings the 3′-end of the incoming DNA primer in proximity to the active site of the enzyme for nucleotide addition while the RNA component of the heteroduplex provides the template for the addition of identical DNA repeats at the ends of chromosomes. The close proximity of the TRBD and thumb domains and the location of the TRBD RNA-binding pocket with respect to the central cavity of the ring can readily account for the placement of the RNA template near the active site of the enzyme upon TERT–TER assembly (Gillis et al., 2008). Of particular significance is the arrangement of the β-hairpin that forms part of the T-motif. This motif extends from the RNA-binding pocket and makes extensive contacts with the "thumb" loop and motifs 1 and 2. Contacts between this hairpin and both the "fingers" and the "thumb" domains place the opening of the TRBD pocket and in particular the T-motif that faces the interior of the ring in proximity to the active site of the enzyme. The spatial arrangement of these structural elements is well suited to facilitate the formation of the RNA–DNA duplex required for telomerase activity. Supporting evidence for this theory is provided by the structure of TERT in complex with an RNA–DNA heteroduplex (discussed in detail in the section below) bound in the interior of the TERT ring (Mitchell et al., 2010). TERT–RNA/DNA interactions place the 5′-end of the RNA substrate at the entry of the RNA-binding pocket, and the 3′-end of the incoming DNA primer at the active site of TERT, thus allowing the reaction to take place.

3.4 INTERACTION WITH NUCLEIC ACID AND NUCLEOTIDE NEAR THE ACTIVE SITE

The cocrystal structure of the full length and catalytically active *Tc*TERT with an RNA–DNA hairpin containing the putative RNA template region (5′-rCrUr-GrArCrCrU-3′) and the complementary telomeric DNA (5′-GTCAGGT-3′) was recently determined (Fig. 3.3a; Mitchell et al., 2010). The RNA–DNA hairpin was designed to contain a three-nucleotide overhang at the 5′-end of the RNA template. The crystals were grown in the presence of the slowly hydrolysable nucleotide analogues dNTPαS and Mg^{++} ions. Remarkably, the structure revealed extra density for three nucleotides at the 3′-end of the telomeric DNA, suggesting that TERT had extended the 3′-end of the DNA substrate in the crystallization drop. That the structure likely reflects features of the native telomerase–DNA complex is supported by both its apparent enzymatic activity and its resemblance to the HIV-1 RT–substrate complex. In the nucleic acid bound TERT structure, the four domains are organized into a ring configuration similar to that observed for the substrate-free enzyme (Figs. 3.1b and 3.3a). The interior of the TERT ring is lined with many positively charged residues and is ~22 Å wide and 21 Å deep. Within this cavity binds one molecule of the RNA–DNA duplex that is similar in structure to both the DNA–DNA and RNA–DNA substrate bound to the HIV-1 RT (Huang et al., 1998; Sarafianos et al., 2001). As described below, the detailed features of the complex suggest that it is not in the catalytic state. Nevertheless, the interactions revealed by the cocrystal structure provide interesting insights on the mechanisms of nucleotide and nucleic acid binding by TERT.

3.4.1 Binding to the Template Region of the RNA

Interactions between the protein and the RNA template region are mediated by the fingers, the palm, and thumb domains (Fig. 3.3b). The 5′-end cytosine and uracil residues (rC1 and rU2) are located at the interface of the fingers and palm domains and are involved in a network of interactions with conserved residues of motifs 2 and B′, both of which are located near the active site of the enzyme. In particular, the 2′-OH and the base carbonyl of rC1 is within hydrogen bonding distance of the backbone

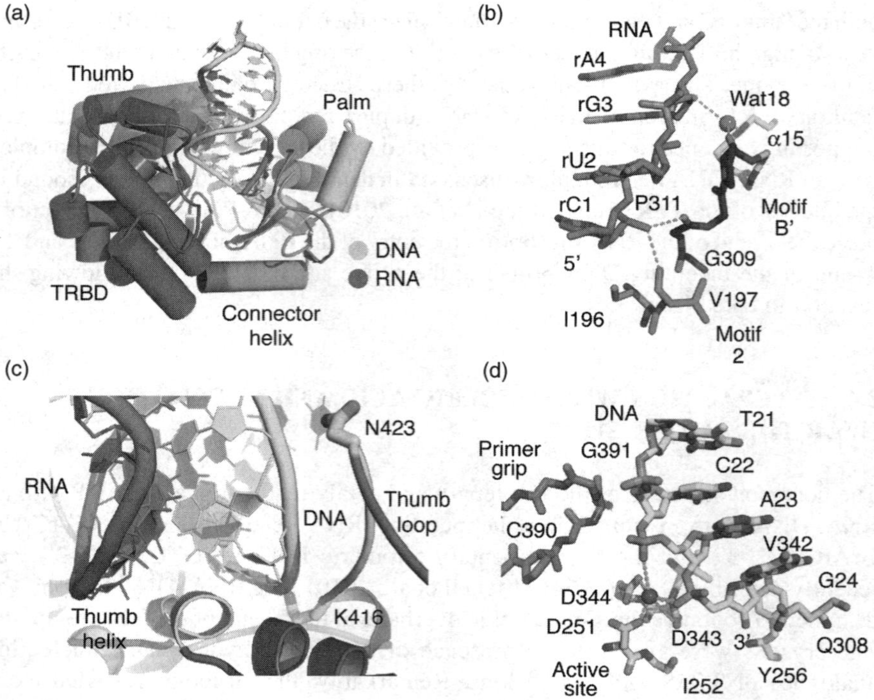

FIGURE 3.3 (a) A complex between *Tc*TERT and an RNA–DNA hairpin (PDB ID: 3KYL): the domain orientation and color scheme are similar to those shown in Figure 3.1b. (b) A close view of the contacts between the RNA template and motifs 2 and B′ of *Tc*TERT. (c) A close view of the contacts between the RNA–DNA hybrid and the thumb helix (light blue) and thumb loop (light blue) in the complex. (d) The primer grip region (motif E) is juxtaposed to the 3′-end of the DNA primer at the active site of the enzyme. (See the color version of this figure in Color Plates section.)

carbonyls of Val197 of motif 2 and Gly309 of motif B′, whereas the pyrimidine base sits over the otherwise solvent exposed, hydrophobic side chain of the conserved Ile196 that also forms part of motif 2. Contacts between rU2 and the protein are mediated by the short aliphatic side chain of Pro311 and the ribose group. Interactions between rC1 and rU2 with motifs 2 and B′ place the cytosine base in proximity of the active site of the enzyme, where it is well positioned for Watson–Crick base-pairing with the incoming nucleotide substrate. Stabilization and placement of the 5′ bases of the template region above the active site of the enzyme is in large part mediated by the interactions of the remaining five ribonucleotides with the DNA primer. Limited contacts between this part of the RNA and the protein are mediated via a water molecule which coordinates the 2′-OH of rG3 with the backbone of helix α15. This helix packs against the two α-helices (α13 and α14) that form part of the IFD motif, which provides one potential explanation for why mutations in this motif lead to loss of telomerase function (Gillis et al., 2008; Lue et al., 2003; Xie et al., 2010).

3.4.2 Binding to Telomeric DNA

Contacts between TERT and the DNA substrate are mediated in large part via backbone interactions with the thumb loop and helix (Fig. 3.3c; Gillis et al., 2008; Mitchell et al., 2010). The thumb helix sits in the minor groove of the RNA–DNA heteroduplex, making extensive contacts with the phosphodiester backbone and the ribose groups of the RNA–DNA hybrid. The mode of action of the thumb helix appears to be similar to that proposed for the equivalent helix (helix H) of retroviral RTs (Jacobo-Molina et al., 1993; Kohlstaedt et al., 1992). The thumb loop closely tracks the curvature of the DNA primer and the two are involved in a network of backbone and solvent-mediated interactions. Interactions between the DNA and the thumb loop include the side chains of Lys416 and Asn423, both of which extend towards the center of the ring and are within hydrogen bonding distance of the DNA backbone. Contacts between the thumb domain and the DNA position the nucleotides located at its 3′-end within coordinating distance of the primer grip region (motif E), a short, rigid loop located at the interface of the palm and thumb domains and in proximity to the active site of the enzyme (Figs. 3.2b and 3.3d). The backbone of the tip of this loop formed by the conserved residues Cys390 and Gly391 abuts the ribose group of C22 and this interaction guides the 3′-end of the DNA towards the active site of the enzyme.

3.4.3 Interactions with the Nucleotide

The telomerase nucleotide binding pocket is located at the interface of the “fingers” and “palm” domains of TERT and surrounded by conserved residues that form part of motifs 1, 2, A, B′, C, and D implicated in template and nucleotide binding (Figs. 3.2a and 3.3d; Bosoy and Lue, 2001; Haering et al., 2000). This notion is supported by mutagenesis experiments and structural comparisons of TERT with HIV-1 RT bound to TTP (Huang et al., 1998). Surrounding the pocket are three invariant aspartic acid residues in motifs A and C (D251, D343, and D344) that play critical roles in catalysis. Alanine substitutions of these residues resulted in complete loss of telomerase activity (Lingner et al., 1997; Mitchell et al., 2010). Adjacent to the aspartates are two highly conserved and surface exposed residues (Y256 of motif A and V342 of motifs C) that probably make contact with the base of the nucleotide substrate. Binding of the nucleotide in this pocket places the triphosphate moiety in proximity of the active site of the enzyme for coordination with one of the Mg^{++}. It also positions the ribose group within coordinating distance of an invariant glutamine (Q308 motif B′) thought to be important for substrate specificity (Smith et al., 2006). Protein contacts with the triphosphate moiety of the nucleotide are also mediated by motif D, a long loop located beneath the active site of the enzyme. In particular, the side chain of the invariant lysine in this motif (K372 in *Tc*TERT) is within coordinating distance of the γ-phosphate of the nucleotide, and probably helps position and stabilize the triphosphate group during catalysis. The side chains of the highly conserved K189 and R192 of motifs 1 and 2, which together form a long β-hairpin within the “fingers” subdomain, are also within coordinating distance of both the sugar and triphosphate

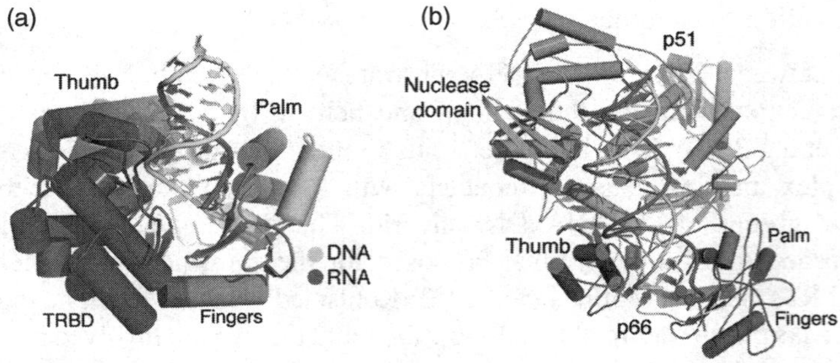

FIGURE 3.4 (a) Same as Figure 3.3a. (b) The structure of the HIV-1 RT bound to an RNA–DNA heteroduplex (PDB ID: 1RTD). (See the color version of this figure in Color Plates section.)

moieties of the modeled nucleotide. The presumed contacts would facilitate the binding of the nucleotide and help to place it in close apposition to the 3′-end of the incoming DNA primer. Supporting evidence for this nucleotide-binding model of TERT is provided by the partially occupied active site of TERT by the terminal DNA nucleotide G24 in the TERT–nucleic acid structure (Mitchell et al., 2010; Fig. 3.4b). The ribose group and to a lesser extent the guanosine base of G24 (which makes Watson–Crick pairing interactions with the rC1 located at the 5′-end of the RNA template) sit in a well-defined hydrophobic pocket formed by the side chains of the conserved Tyr256, Gln308, and Val342, while the α-phosphate is coordinated by the magnesium ion occupying the active site. The importance of Val342 in telomerase nucleotide binding and selectivity has been previously shown for the human telomerase holoenzyme (Drosopoulos and Prasad, 2007).

3.4.4 Mechanistic Similarities between TERT and HIV-1 RT

It is evident from the foregoing discussion that telomerase uses a mechanism of DNA synthesis that closely resembles retroviral RTs, at least with regard to substrate interactions and catalytic chemistry near the active site. To summarize, comparable mutations of equivalent residues in both groups of polymerases elicit comparable defects (Autexier and Lue, 2006). Moreover, structural comparison of the RNA–DNA bound TERT and HIV-1 RT shows a striking similarity in the relative disposition of the proteins and nucleic acids in the two structures (Fig. 3.4a and b; Sarafianos et al., 2001). Similar to HIV-1 RT, telomerase-dependent telomere DNA synthesis requires pairing of the template region with the incoming DNA primer and placement of the 3′-end of the DNA into the enzyme's active site for nucleotide addition. Moreover, TERT- or HIV-1 RT-nucleic acid associations are accompanied by domain rearrangements that facilitate the formation of a tight nucleoprotein assembly which places the DNA 3′-end at the active site of the enzyme for catalysis (Kohlstaedt et al., 1992; Rodgers et al., 1995; Steitz, 1997). Contacts between the protein and the RNA template region are specific and involve conserved motifs (motif 2 and B′ of the fingers and palm domain, respectively) that are shared between TERTs and RTs.

Contacts between the protein and the DNA substrate immediately upstream of the active site are mediated by the thumb domain. Despite the lack of sequence homology in this region between the two families of enzymes, the thumb helix of telomerase evidently performs a function that is analogous to helix H of the HIV-1 RT (Beese et al., 1993; Jacobo-Molina et al., 1993; Kohlstaedt et al., 1992). Placement of the DNA 3′-end at the active site of the enzyme is further facilitated by the primer grip region, another highly conserved motif shared between TERTs and RTs (Peng et al., 2001; Tantillo et al., 1994; Xie et al., 2010). The selective binding of nucleotides is guided by the same invariant Tyr and Gln amino acids in both classes of RTs as well (Tyr256 and Gln308 in *Tc*TERT) (Cases-Gonzalez et al., 2000; Huang et al., 1998). Thus, the mechanistic similarities between TERTs and RTs, hinted early on by sequence and mutagenesis analyses, are now supported by a wealth of shared features at the structural level.

3.4.5 TERT Domain Rearrangements upon Nucleic Acid Binding

Domain reorganization upon nucleic acid binding is a common feature of RNA and DNA polymerases, and retroviral RTs. Such rearrangements are geared towards the formation of a tight, catalytically poised protein–nucleic acid assembly that facilitates positioning of the DNA 3′-end at the active site of the enzyme for catalysis (Kohlstaedt et al., 1992; Rodgers et al., 1995; Steitz, 1997). Unlike the HIV-1 RT, telomerase appears to exist, at least in the absence of the full-length integral RNA component, in a closed-ring configuration, an arrangement mediated by extensive contacts between the TRBD and the thumb domains (Gillis et al., 2008). Comparison of the nucleic acid bound and substrate-free TERT structures suggest that TERT–nucleic acid associations induce subtle, rigid-body changes in the orientations of subunits of the enzyme that lead to a 3.5 Å decrease in the diameter of the interior cavity of the ring. The decrease in the diameter of the central cavity arises from a 6° inward rotation together with a 3.5 Å translation of the thumb domain with respect to the fingers and palm domains (Figs. 3.1b and 3.3a). Translation of the thumb domain towards the center of the ring is accompanied by a 3.5 Å shift of the TRBD towards the finger domain, creating a narrower RNA-binding pocket than the substrate free enzyme. The precise role of this subtle structural rearrangement is unclear at this point.

It is worth noting that in most polymerases, including the HIV-1 RT, the fingers and thumb domains undergo large conformational changes required for substrate binding and function (Ding et al., 1998; Steitz, 1997, 1999). For example, the fingers domain, which is known to bind and position the nucleotide at the active site of the enzyme, undergoes significant conformational changes between the "open" and "closed" states (Ding et al., 1998). By contrast, the interactions between the TRBD and the thumb domains of TERT may lock the fingers domain in place, thus preventing the conformational rearrangements observed in other polymerases. Telomerase may thus carry a preformed active site, as has been previously observed for the Hepatitis C viral RNA polymerase (NS5B) and the Y-family DNA polymerases (Bressanelli et al., 2002; Ling et al., 2001). On the other hand, if TERT does adopt an open conformation akin to that of HIV-1 RT, then this conformation is likely to entail very different interdomain interactions than those seen in the ring configuration.

3.5 TEMPLATE BOUNDARY DEFINITION

A unique property of TERT as an RT is its exclusive usage of a short segment within a large RNA molecule (TER) as the template. This property is essential for TERT's dedicated synthesis of telomeric repeats and can be attributed to the ability of the protein to recognize proper template boundaries within the TER molecule. Multiple mutations in TERT and TER have been shown to subvert this recognition, resulting in aberrant synthesis of nontelomeric DNA (Box et al., 2008; Moriarty et al., 2005a; Seto et al., 2003). The presence of nontelomeric DNA at the 3′ end of the DNA substrate is also expected to impair subsequent realignment and reduce enzyme processivity. Thus, recognition of template boundary by telomerase is crucial for multiple aspects of its function.

Interestingly, the structural barriers against aberrant "read through" appear to be different in different organisms (Drosopoulos and Prasad, 2009; Lai et al., 2001, 2002; Miller et al., 2000; Rouda and Skordalakes, 2007; also see Chapter 2). In budding yeast, a conserved stem (helix I) immediately 5′ of the template is required for proper boundary definition. A comparable stem in mammalian TER (the P1b helix in human TER) evidently provides the same function. The boundary remains intact with compensatory mutations that altered the sequences within the stem while preserving the base pairing, thus highlighting the role of structure rather than sequence. Notably, a role for protein–RNA interaction in specifying the template boundary has not been uncovered for either yeast or mammalian telomerase (Chen and Greider, 2003; Tzfati et al., 2000). By contrast, in ciliate telomerase, proper template boundary recognition requires both the TRBD domain, and an RNA segment (named TBE, located 5′ of the boundary) to which the TRBD binds (Lai et al., 2002). Mutations in both the TRBD domain of TERT and the TBE of TER from *Tetrahymena* have been shown to subvert proper recognition of template boundary. The recent high-resolution structures of TRBD domains provide hints for how this recognition can be accomplished at the molecular level.

The TBE in *Tetrahymena* TER consists of stem II and the flanking ssRNA regions and is located only a few nucleotides upstream of the RNA template. As described earlier, the T-pocket of TRBD represents a narrow hydrophobic cavity located on the surface of the protein that likely binds ssRNA (Gillis et al., 2008; Rouda and Skordalakes, 2007). If the T-pocket binds the ssRNA that connects stem I and stem II, this interaction may force stem II to act as a steric block, which would in turn restrict the TRBD domain to stay between the two stems. The TRBD domain may then act as a barrier that constrains the distance the RT domain can traverse and prevent it from moving beyond the boundaries of the RNA template. However, this structural explanation is most likely incomplete because the ciliate TRBD domain alone is not sufficient for template boundary definition and requires the action of the CP2 motif located to the N-terminal side of TRBD (Lai et al., 2002; Miller et al., 2000).

The precise mechanism by which the CP2 motif facilitates template boundary definition is unclear at this stage.

The foregoing hypothesis is also consistent with the structure of TERT in a complex with the putative RNA template region and telomeric DNA (Mitchell et al., 2010). In this structure, the RNA template does not directly engage the RNA-binding pocket of TRBD. The structure shows instead contacts between the 5′-end of the template and the entry of the RNA-binding pocket of TRBD (Fig. 3.3b). This arrangement would place the TBE present in most organisms or the short oligonucleotide overhang of rodent TER within the RNA-binding pocket of TRBD (Chen and Greider, 2003; Lai et al., 2002; Tzfati et al., 2000). The stable association of TER with the TRBD would force the enzyme to stall when reaching the nucleotide located at the 5′-end of the RNA template thus preventing replication beyond this point. Stalling of the enzyme for extended periods may result in dissociation of the RNA–DNA heteroduplex, leading to termination or another round of telomere repeat synthesis.

3.6 REPEAT ADDITION PROCESSIVITY

Despite the short template region in TER, telomerase has the ability to add multiple telomeric repeats to the DNA primer following a single binding event, a property referred to as RAP (Greider, 1991). Processive DNA synthesis is presumed to involve a "translocation" step in which the RNA template and the DNA substrate/product first becomes unpaired and then adopts an alternative alignment (Lue, 2004). The transient loss of the RNA–DNA hybrid in turn suggests the existence of protein–DNA interactions that allow the telomerase RNP to retain the DNA product during translocation. Indeed, the existence of a complex set of interactions involving contacts between multiple structural elements in TERT (and other telomerase subunits) and an extended segment of the DNA is now supported by many lines of investigation (Finger and Bryan, 2008; Wyatt et al., 2007). First, while a short DNA primer can support a low level of RAP, a much longer primer (~18 nt) is necessary for maximal processivity (Collins and Greider, 1993; Harrington and Greider, 1991; Lee and Blackburn, 1993). This observation has led to the notion of a telomerase "anchor site," which is physically distinct from the catalytic site and which mediates interaction with a 5′ region of the substrate DNA. In addition, mutations in many structural elements within the TEN and RT domains caused processivity defects (Autexier and Lue, 2006; Lue, 2004; Xie et al., 2010). In keeping with the potential importance of protein–DNA interaction, many of the structural elements implicated in processivity regulation are evidently involved directly or indirectly in DNA binding (Finger and Bryan, 2008; Hossain et al., 2002; Jacobs et al., 2006; Wyatt et al., 2007).

One DNA-binding domain that is strongly implicated in processivity control is the TEN domain. Specific mutations in this domain of *Tetrahymena*, *Saccharomyces*, and human telomerase have been shown to selectively impair multiple repeat addition *in vitro* and telomere maintenance *in vivo* (Lue, 2005; Lue and Li, 2007; Moriarty et al., 2005b; Xie et al., 2010; Zaug et al., 2008). Each of these domains from different species in isolation has also been reported to bind DNA, albeit with moderate to low affinity (Finger and Bryan, 2008; Wyatt et al., 2007; Xia et al., 2000). Cross-linking analysis suggests that in the context of the telomerase RNP, the TEN domain makes contact primarily with the upstream region (~10–15 nt away from the 3′ end) of the DNA primer (Jacobs et al., 2006; Lue, 2005). Thus, it has been argued that the TEN domain constitutes the classically defined "anchor site" of telomerase (Lee and Blackburn, 1993; Lue, 2004). Confirmation and elaboration of this proposal will require high-resolution structures of a telomerase–DNA complex. However, the existing structures of *Tc*TERT alone and in a complex with an RNA–DNA hybrid do provide tantalizing support for this notion. Specifically, although the TEN domain is missing from the *Tc*TERT, it is located on the N-terminal side of the TRBD domain in other TERTs, and hence naturally modeled on the "upstream" side of the TERT ring, in close proximity to the 5′ end of the DNA substrate (Sekaran et al., 2009). A detailed understanding of TEN domain–DNA interaction will likewise require high-resolution structural information. However, a combination of cross-linking and mutagenesis studies in *Tetrahymena* have revealed a potential DNA-binding groove on one face of the TEN domain near a highly conserved and functionally important Gln residue (Jacobs et al., 2006; Fig. 3.2d). This groove is also near a Trp187 that can be cross-linked to the DNA substrate, again supporting its potential in DNA binding (Romi et al., 2007). Further analyses will be required to unravel the mechanisms of this fascinating domain in telomerase processivity.

The other structures implicated in processivity control are located in the RT domains and evidently represent telomerase-specific elaborations of a canonical RT scaffold. They include motif 3, IFD, and the thumb domain. These structures or domains are not highly conserved at the sequence level. Hence only limited extrapolations can be made concerning their potential mechanisms from the available crystal structures of *Tc*TERT (Hossain et al., 2002; Xie et al., 2010). However, the evidence overall is consistent with a function for these structures in binding DNA or the DNA–RNA hybrid near the 3′ end. Motif 3 is discovered recently as an important structural element located between conserved RT motifs 2 and A (Xie et al., 2010). It consists of two α helices, a portion of which tracks closely the backbones of the nucleic acid substrate in the crystal structure of the complex, consistent with a role in nucleic acid interaction. Motif IFD is an ~70–150 amino acid region located between conserved RT motifs A and B′ (Lue et al., 2003). Only selected mutations in this region of yeast and human TERT caused preferential reduction in RAP, suggesting that it is not exclusively required for DNA binding during translocation (Lue et al., 2003; Xie et al., 2010). In the *Tc*TERT structure, IFD represents an extension of the palm domain and consists primarily of two antiparallel α helices that again does not make contact with nucleic acids (Gillis et al., 2008; Mitchell et al., 2010). However, the IFD regions in most other TERTs are considerably longer. The additional residues are likely to be inserted in the loop between the two α helices, resulting in a further extension of the palm domain that may contact nucleic acids directly (Fig. 3.2b).

Of the RT domain structures implicated in processivity control, the role of the thumb domain is perhaps most easily rationalized in light of the crystal structure of

*Tc*TERT and existing biochemical data. As described in detail earlier, the thumb helix within this domain sits in the minor groove of the RNA–DNA heteroduplex, making extensive contacts with the phosphodiester backbone and the ribose groups of the RNA–DNA hybrid (Mitchell et al., 2010; Fig. 3.3c). This mode of action is similar to that proposed for the equivalent helix (helix H) in retroviral RTs (Jacobo-Molina et al., 1993; Kohlstaedt et al., 1992). Another structure, the thumb loop, tracks closely the curvature of the DNA primer, and engages in a network of backbone and solvent-mediated interactions with the DNA (Gillis et al., 2008). Even though the RNA-DNA hybrid is disrupted during the translocation reaction, the thumb-RNA/DNA interactions observed in the crystal structure may still be relevant to RAP for two reasons. First, the unpaired DNA may retain some interaction with the thumb domain structures during translocation and this combined with the thumb domain's ability to adopt subtle rigid conformational changes could facilitate RAP. It is also possible that the thumb domain promotes the realignment reaction by virtue of its ability to bind and stabilize the RNA–DNA hybrid.

Altogether, the foregoing discussion paints a complex picture of processivity control involving multiple contacts between protein and nucleic acids during multiple steps of the translocation process. Indeed, TERT is not the only telomerase protein component that confers increased processivity. Other accessory or regulatory proteins have been shown to contribute. One notable example is *Tetrahymena* Teb1, which was isolated as a subunit of a highly processive subpopulation of telomerase (Min and Collins, 2009; Robart and Collins, 2010). In keeping with the theme of protein–nucleic acid interaction in processivity modulation, Teb1 has been shown to be an Rpa1-like protein that exhibits high-affinity binding to telomeric DNA (Min and Collins, 2009; Robart and Collins, 2010). Another notable factor that appears to be widely conserved is mammalian TPP1 and its potential homologue in yeast named Est3. TPP1 is a component of the shelterin complex that binds and protects mammalian telomeres (de Lange, 2009). It has also been shown to enhance the processivity of telomerase *in vitro* in cooperation with the single-strand telomere end-binding protein POT1 (Wang et al., 2007). TPP1 acts by reducing the dissociation rate of the telomerase RNP from telomeric DNA during active elongation, and appears to confer its effect at least in part by interacting with the TEN domain of TERT (Latrick and Cech, 2010; Zaug et al., 2010). Surprisingly, the yeast telomerase subunit Est3 is now believed to contain an OB fold domain that structurally resembles a comparable domain in TPP1 based on sequence and functional analyses (Lee et al., 2008; Yu et al., 2008). Moreover, a mutation in the TEN domain of yeast TERT reduced its association with Est3, hinting at a common mechanism of interaction between the TPP1–TERT and Est3–TERT protein pairs (Friedman et al., 2003; Friedman and Cech, 1999). TPP1/Est3-like OB folds are also known to be present in TEBPβ in ciliates and Tpz1 in fission yeast. Both TEBPβ and Tpz1 have been ascribed a function in telomerase regulation, though not specifically in processivity control (Miyoshi et al., 2008; Paeschke et al., 2008). The functional and mechanistic similarities and disparities between these apparently related OB fold proteins are interesting issues for future investigations.

3.7 hTERT MUTATIONS IN HUMAN DISEASES

Mutations in both hTERC and hTERT are associated with rare bone marrow failure syndromes, autosomal dominant dyskeratosis congenita (DC) and acquired aplastic anemia (AA) (Dokal and Vulliamy, 2003; Vulliamy et al., 2005; Yamaguchi et al., 2005). Studies of patients with acquired AA have revealed a number of mutations in the hTERC and hTERT, usually associated with reduced telomerase activity due either to haploinsufficiency or to dominant-negative effects of the mutations (Dokal and Vulliamy, 2003; Fogarty et al., 2003; Liang et al., 2006; Vulliamy et al., 2002, 2005; Yamaguchi et al., 2005). Current evidence suggests that sequence variations in telomerase gene components can operate by different mechanisms to cause enzymatic dysfunction that leads to telomere shortening and ultimately a reduced replicative potential of marrow stem cells. Consistent with this notion, the disease-associated TERT mutations are scattered throughout different domains of the protein (http://telomerase.asu.edu/diseases.html). The structure of *Tc*TERT provides substantial insights into the mechanistic bases of some of these mutations in telomerase dysfunction. As described below, by mapping the naturally occurring hTERT mutations on the *Tc*TERT structure, plausible rationales for how these mutations induce defects can be inferred (Table 3.1).

The naturally occurring mutant K570N is located at the tip of the T-motif, a highly conserved structural element implicated in telomerase RNP assembly and RNA template positioning near the active site of the enzyme (Fig. 3.2a). Single alanine substitutions of conserved residues in this motif of *T. thermophila* and yeast telomerase led to 80–100% loss of enzymatic activity (Bryan et al., 2000; Lai et al., 2002; Miller et al., 2000). The defect of the K570 mutation may thus be due to disruption of T-motif functions.

The G682D and V694M mutations alter two conserved residues located on the surface of the palm domain, at a significant distance from the active site of the enzyme. These characteristics suggest that the two mutations do not directly affect catalysis. Instead, because they form part of a conserved and shallow indentation on the surface of the protein, G682 and V694 may be involved in binding telomerase RNA or a regulatory protein. Disruption of the binding to the hypothetical target could explain the defects of these two mutants.

P721R and T726M are both located on the palm domain and form part of the IFD motif involved in the structural organization of the palm domain (Gillis et al., 2008)

TABLE 3.1 Naturally Occurring *h*TERT Mutants That can be Modeled on *Tc*TERT

hTERT	TcTERT	Reference
K570N	G143	Xin et al. (2007)
G682D	T226	Liang et al. (2006)
P721R	K260	Vulliamy et al. (2006)
T726M	C265	Liang et al. (2006)
K902N	K372	Armanios et al. (2005)
R979W	F433	Vulliamy et al. (2005); Xin et al. (2007)

and is implicated in telomerase processivity (Lue et al., 2003; Fig. 3.2b). Alteration of the IFD contacts with neighboring structural elements of the palm domain through the introduction of the much larger side chains of P721R and T726M would lead to the displacement or reorganization of the proximal structural elements from their current location, which would in turn affect substrate-binding and telomerase function.

K902N forms part of motif D, a conserved element of the palm domain, located beneath and in close proximity of the active site of TERT (Fig. 3.2b). Structural comparison of TERT with related enzymes such as HIV-1 RT (Huang et al., 1998) show that this residue is directly involved in nucleotide binding via hydrogen bonding of the nucleotide triphosphate. Mutation of this residue would affect nucleotide binding and catalysis, leading to loss of telomerase activity. The critical role of this residue in telomerase function is supported by the extremely severe phenotype of the K902N mutant, which includes onset of gray forelock, liver and lung fibrosis, and aplastic anemia (Armanios et al., 2005).

R979W is located on the thumb domain of hTERT and forms part of the thumb helix involved in RNA–DNA binding and telomerase processivity (Gillis et al., 2008; Fig. 3.2c). Deletion or mutation of the residues that form this helix in yeast and human TERT resulted in severe loss of telomerase processivity (Banik et al., 2002; Hossain et al., 2002; Huard et al., 2003). Replacing the basic side change of Arg979 with the large aromatic side chain of a Tryptophan could lead to displacement of this motif and disruption of nucleic acid binding.

Taken together, these examples of disease-related TERT mutations illustrate the myriad ways telomerase functions could be compromised, leading to impaired telomere maintenance and reduced replicative potentials in the affected cell populations. Future studies are likely to uncover even more mutations with pathological consequences. Conversely, in-depth investigations of the disease-related mutations could provide new insights on telomerase mechanisms and interaction partners.

3.8 CONCLUSIONS AND PROSPECTS

The discovery of the catalytic protein component of telomerase represents a milestone in the field (Lingner et al., 1997). By firmly establishing TERT as a distinct member of the RT family, it greatly catalyzed research on this special DNA polymerase of extraordinary medical relevance. Studies over the ensuing years have provided a rich illustration of how the basic RT scaffold can be elaborated to mediate a special reaction, that is, dedicated and repetitive synthesis of a short DNA sequence. By acquiring additional domains (the TRBD and TEN domain) and extra structural elements (motif 3 and IFD), TERT becomes adapt at performing a unique biological function in telomere maintenance. Herein lies undoubtedly a general lesson for other dedicated and specialized polymerases.

As evident from the discussions in this chapter, with the recent acquisition of high-resolution structural information for several TERT domains and full-length *Tc*TERT, much of the existing biochemical data can now be interpreted at the structural level.

However, because of some regions with limited sequence similarity between *Tc*TERT and TERTs in other model organisms, the precise workings of some critical structures such as motif 3 and IFD have remained uncertain and await further analyses. Perhaps the most challenging task in the future stems from the fact that telomerase is a stable RNP with mutually dependent and coevolving RNA and protein components (Cech, 2009). There is compelling evidence that TER does not simply play a passive role in providing the template, but rather actively participates in the polymerization reaction; regions of the RNA distant from the template are known to be essential for nucleotide addition and processivity (see Chapter 2 in this volume for more details). Though the structures of some of these essential TER regions are now available, their roles have remained largely obscure owing to the lack of cocrystal structures. Obtaining such structures and leveraging the insights from structures to molecular mechanisms would surely represent the next frontier in the ever expanding and ever fascinating exploration of this unique DNA polymerase.

ACKNOWLEDGMENT

The authors apologize to all colleagues whose primary work could not be cited due to space constraints. Works in the authors' laboratories have been supported by the NIH, The Ellison Medical, the V and The Emerald Foundations (Emanuel Skordalakes), and by NIH, ACS, Burroughs Wellcome Fund, and the STARR Cancer Consortium (Neal F. Lue).

REFERENCES

Arai, K., Masutomi, K., Khurts, S., Kaneko, S., Kobayashi, K., and Murakami, S. (2002) Two independent regions of human telomerase reverse transcriptase are important for its oligomerization and telomerase activity. *J. Biol. Chem.* **277**: 8538–8544.

Armanios, M., Chen, J.L., Chang, Y.P., Brodsky, R.A., Hawkins, A., Griffin, C.A., Eshleman, J.R., Cohen, A.R., Chakravarti, A., Hamosh, A., *et al.* (2005) Haploinsufficiency of telomerase reverse transcriptase leads to anticipation in autosomal dominant dyskeratosis congenita. *Proc. Natl. Acad. Sci. USA.* **102**: 15960–15964.

Autexier, C., and Lue, N.F. (2006) The structure and function of telomerase reverse transcriptase. *Annu. Rev. Biochem.* **75**: 493–517.

Bachand, F., and Autexier, C. (2001) Functional regions of human telomerase reverse transcriptase and human telomerase RNA required for telomerase activity and RNA–protein interactions. *Mol. Cell. Biol.* **21**: 1888–1897.

Banik, S.S., Guo, C., Smith, A.C., Margolis, S.S., Richardson, D.A., Tirado, C.A., and Counter, C.M. (2002) C-terminal regions of the human telomerase catalytic subunit essential for *in vivo* enzyme activity. *Mol. Cell. Biol.* **22**: 6234–6246.

Baran, N., Haviv, Y., Paul, B., and Manor, H. (2002) Studies on the minimal lengths required for DNA primers to be extended by the *Tetrahymena* telomerase: implications for primer positioning by the enzyme. *Nucleic Acids Res.* **30**: 5570–5578.

Beattie, T.L., Zhou, W., Robinson, M.O., and Harrington, L. (2000) Polymerization defects within human telomerase are distinct from telomerase RNA and TEP1 binding. *Mol. Biol. Cell.* **11**: 3329–3340.

Beese, L.S., Derbyshire, V., and Steitz, T.A. (1993) Structure of DNA polymerase I Klenow fragment bound to duplex DNA. *Science.* **260**: 352–355.

Blasco, M.A., Funk, W., Villeponteau, B., and Greider, C.W. (1995) Functional characterization and developmental regulation of mouse telomerase RNA. *Science.* **269**: 1267–1270.

Bosoy, D., and Lue, N. (2001) Functional analysis of conserved residues in the putative "finger" domain of telomerase reverse transcriptase. *J. Biol. Chem.* **276**: 46305–46312.

Bosoy, D., Peng, Y., Mian, I., and Lue, N. (2003) Conserved N-terminal motifs of telomerase reverse transcriptase required for ribonucleoprotein assembly *in vivo*. *J. Biol. Chem.* **278**: 3882–3890.

Box, J.A., Bunch, J.T., Zappulla, D.C., Glynn, E.F., and Baumann, P. (2008) A flexible template boundary element in the RNA subunit of fission yeast telomerase. *J. Biol. Chem.* **283**: 24224–24233.

Bressanelli, S., Tomei, L., Rey, F.A., and De Francesco, R. (2002) Structural analysis of the hepatitis C virus RNA polymerase in complex with ribonucleotides. *J. Virol.* **76**: 3482–3492.

Bryan, T.M., Goodrich, K.J., and Cech, T.R. (2000) Telomerase RNA bound by protein motifs specific to telomerase reverse transcriptase. *Mol. Cell.* **6**: 493–499.

Bryan, T.M., Sperger, J.M., Chapman, K.B., and Cech, T.R. (1998) Telomerase reverse transcriptase genes identified in *Tetrahymena thermophila* and *Oxytricha trifallax*. *Proc. Natl. Acad. Sci. USA.* **95**: 8479–8484.

Cases-Gonzalez, C.E., Gutierrez-Rivas, M., and Menendez-Arias, L. (2000) Coupling ribose selection to fidelity of DNA synthesis. The role of Tyr-115 of human immunodeficiency virus type 1 reverse transcriptase. *J. Biol. Chem.* **275**: 19759–19767.

Cech, T.R. (2009) Crawling out of the RNA world. *Cell.* **136**: 599–602.

Chen, J., and Greider, C. (2003) Template boundary definition in mammalian telomerase. *Genes Dev.* **17**: p2747–2752.

Collins K, Greider CW (1993) *Tetrahymena* telomerase catalyzes nucleolytic cleavage and nonprocessive elongation. *Genes Dev.* **7**: 1364–1376.

de Lange, T. (2009) How telomeres solve the end-protection problem. *Science.* **326**: 948–952.

Di Marco, S., Volpari, C., Tomei, L., Altamura, S., Harper, S., Narjes, F., Koch, U., Rowley, M., De Francesco, R., Migliaccio, G.,*et al.* (2005) Interdomain communication in hepatitis C virus polymerase abolished by small molecule inhibitors bound to a novel allosteric site. *J. Biol. Chem.* **280**: 29765–29770.

Ding, J., Das, K., Hsiou, Y., Sarafianos, S.G., Clark, A.D., Jr., Jacobo-Molina, A., Tantillo, C., Hughes, S.H., and Arnold, E. (1998) Structure and functional implications of the polymerase active site region in a complex of HIV-1 RT with a double-stranded DNA template-primer and an antibody Fab fragment at 2.8 A resolution. *J. Mol. Biol.* **284**: 1095–1111.

Dokal, I., and Vulliamy, T. (2003) Dyskeratosis congenita: its link to telomerase and aplastic anaemia. *Blood Rev.* **17**: 217–225.

Drosopoulos, W.C., and Prasad, V.R. (2007) The active site residue Valine 867 in human telomerase reverse transcriptase influences nucleotide incorporation and fidelity. *Nucl. Acids Res.* **35**: 1155–1168.

Drosopoulos, W.C., and Prasad, V.R. (2009) The telomerase-specific T motif is a restrictive determinant of repetitive reverse transcription by human telomerase. *Mol. Cell. Biol.* **30**: 447–459.

Figueiredo, L., Rocha, E., Mancio-Silva, L., Prevost, C., Hernandez-Verdun, D., and Scherf, A. (2005) The unusually large *Plasmodium* telomerase reverse-transcriptase localizes in a discrete compartment associated with the nucleolus. *Nucl. Acids Res.* **33**: p1111–1122.

Finger, S.N., and Bryan, T.M. (2008) Multiple DNA-binding sites in *Tetrahymena* telomerase. *Nucl. Acids Res.* **36**: 1260–1272.

Fogarty, P.F., Yamaguchi, H., Wiestner, A., Baerlocher, G.M., Sloand, E., Zeng, W.S., Read, E. J., Lansdorp, P.M., and Young, N.S. (2003) Late presentation of dyskeratosis congenita as apparently acquired aplastic anaemia due to mutations in telomerase RNA. *Lancet*. **362**: 1628–1630.

Forstemann, K., and Lingner, J. (2005) Telomerase limits the extent of base pairing between template RNA and telomeric DNA. *EMBO Rep*. **6**: p361–366.

Friedman, K., Heit, J., Long, D., and Cech, T. (2003) N-terminal domain of yeast telomerase reverse transcriptase: recruitment of Est3p to the telomerase complex. *Mol. Biol. Cell.* **14**: 1–13.

Friedman, K.L., and Cech, T.R. (1999) Essential functions of amino-terminal domains in the yeast telomerase catalytic subunit revealed by selection for viable mutants. *Genes Dev*. **13**: 2863–2874.

Gillis, A.J., Schuller, A.P., and Skordalakes, E. (2008) Structure of the *Tribolium castaneum* telomerase catalytic subunit TERT. *Nature*. **455**: 633–637.

Greider, C. (1991) Telomerase is processive. *Mol. Cell. Biol.* **11**: 4572–4580.

Haering, C.H., Nakamura, T.M., Baumann, P., and Cech, T.R. (2000) Analysis of telomerase catalytic subunit mutants *in vivo* and *in vitro* in *Schizosaccharomyces pombe*. *Proc. Natl. Acad. Sci. USA*. **97**: 6367–6372.

Hammond, P.W., and Cech, T.R. (1998) *Euplotes* telomerase: evidence for limited base-pairing during primer elongation and dGTP as an effector of translocation. *Biochemistry*. **37**: 5162–5172.

Harrington LA, Greider CW. (1991) Telomerase primer specificity and chromosome healing. *Nature*. **353**: 451–454.

Hossain, S., Singh, S., and Lue, N. (2002) Functional analysis of the c-terminal extension of telomerase reverse transcriptase. A putative "thumb" domain. *J. Biol. Chem.* **277**: 36174–36180.

Huang, H., Chopra, R., Verdine, G.L., and Harrison, S.C. (1998) Structure of a covalently trapped catalytic complex of HIV-1 reverse transcriptase: implications for drug resistance. *Science*. **282**: 1669–1675.

Huard, S., Moriarty, T., and Autexier, C. (2003) The C terminus of the human telomerase reverse transcriptase is a determinant of enzyme processivity. *Nucl. Acids Res.* **31**: 4059–4070.

Jacobo-Molina, A., Ding, J., Nanni, R.G., Clark, A.D., Jr., Lu, X., Tantillo, C., Williams, R.L., Kamer, G., Ferris, A.L., Clark, P.,*et al.* (1993) Crystal structure of human immunodeficiency virus type 1 reverse transcriptase complexed with double-stranded DNA at 3.0 A resolution shows bent DNA. *Proc. Natl. Acad. Sci. USA* **90**: 6320–6324.

Jacobs, S.A., Podell, E.R., and Cech, T.R. (2006) Crystal structure of the essential N-terminal domain of telomerase reverse transcriptase. *Nat. Struct. Mol. Biol.* **13**: 218–225.

Johansson, E., and Macneill, S.A. (2010) The eukaryotic replicative DNA polymerases take shape. *Trends Biochem. Sci.* **35**: 339–347.

Kohlstaedt, L.A., Wang, J., Friedman, J.M., Rice, P.A., and Steitz, T.A. (1992) Crystal structure at 3.5 A resolution of HIV-1 reverse transcriptase complexed with an inhibitor. *Science*. **256**: 1783–1790.

Lai, C., Miller, M., and Collins, K. (2002) Template boundary definition in *Tetrahymena* telomerase. *Genes Dev.* **16**: 415–420.

Lai, C.K., Mitchell, J.R., and Collins, K. (2001) RNA binding domain of telomerase reverse transcriptase. *Mol. Cell. Biol.* **21**: 990–1000.

Latrick, C.M., and Cech, T.R. (2010) POT1–TPP1 enhances telomerase processivity by slowing primer dissociation and aiding translocation. *EMBO J.* **29**: 924–933.

Lee, J., Mandell, E.K., Tucey, T.M., Morris, D.K., and Lundblad, V. (2008) The Est3 protein associates with yeast telomerase through an OB-fold domain. *Nat. Struct. Mol. Biol.* **15**: 990–997.

Lee, M.S., and Blackburn, E.H. (1993) Sequence-specific DNA primer effects on telomerase polymerization activity. *Mol. Cell. Biol.* **13**: 6586–6599.

Lee, S.R., Wong, J.M., and Collins, K. (2003) Human telomerase reverse transcriptase motifs required for elongation of a telomeric substrate. *J. Biol. Chem.* **278**: 52531–52536.

Lendvay, T.S., Morris, D.K., Sah, J., Balasubramanian, B., and Lundblad, V. (1996) Senescence mutants of *Saccharomyces cerevisiae* with a defect in telomere replication identify three additional EST genes. *Genetics*. **144**: 1399–1412.

Liang, J., Yagasaki, H., Kamachi, Y., Hama, A., Matsumoto, K., Kato, K., Kudo, K., and Kojima, S. (2006) Mutations in telomerase catalytic protein in Japanese children with aplastic anemia. *Haematologica*. **91**: 656–658.

Ling, H., Boudsocq, F., Woodgate, R., and Yang, W. (2001) Crystal structure of a Y-family DNA polymerase in action: a mechanism for error-prone and lesion-bypass replication. *Cell*. **107**: 91–102.

Lingner, J., and Cech, T.R. (1996) Purification of telomerase from *Euplotes aediculatis*: requirement of a primer 3′ overhang. *Proc. Natl. Acad. Sci. USA*. **93**: 10712–10717.

Lingner, J., Hughes, T.R., Shevchenko, A., Mann, M., Lundblad, V., and Cech, T.R. (1997) Reverse transcriptase motifs in the catalytic subunit of telomerase. *Science*. **276**: 561–567.

Lue, N. (2004) Adding to the ends: what makes telomerase processive and how important is it? *Bioessays*. **26**: 955–962.

Lue, N. (2005) A physical and functional constituent of telomerase anchor site. *J. Biol. Chem.* **280**: 26586–26591.

Lue, N., Lin, Y., and Mian, I. (2003) A conserved telomerase motif within the catalytic domain of telomerase reverse transcriptase is specifically required for repeat addition processivity. *Mol. Cell. Biol.* **23**: 8440–8449.

Lue, N.F., and Li, Z. (2007) Modeling and structure function analysis of the putative anchor site of yeast telomerase. *Nucl. Acids Res.* **35**: 5213–5222.

Malik, H.S., Burke, W.D., and Eickbush, T.H. (2000) Putative telomerase catalytic subunits from *Giardia lamblia* and *Caenorhabditis elegans*. *Gene*. **251**: 101–108.

Meier, B., Clejan, I., Liu, Y., Lowden, M., Gartner, A., Hodgkin, J., and Ahmed, S. (2006) trt-1 is the *Caenorhabditis elegans* catalytic subunit of telomerase. *PLoS Genet.* **2**: e18.

Miller, M.C., Liu, J.K., and Collins, K. (2000) Template definition by *Tetrahymena* telomerase reverse transcriptase. *EMBO J.* **19**: 4412–4422.

Min, B., and Collins, K. (2009) An RPA-related sequence-specific DNA-binding subunit of telomerase holoenzyme is required for elongation processivity and telomere maintenance. *Mol. Cell.* **36**: 609–619.

Mitchell, M., Gillis, A., Futahashi, M., Fujiwara, H., and Skordalakes, E. (2010) Structural basis for telomerase catalytic subunit TERT binding to RNA template and telomeric DNA. *Nat. Struct. Mol. Biol.* **17**: 513–518.

Miyoshi, T., Kanoh, J., Saito, M., and Ishikawa, F. (2008) Fission yeast Pot1–Tpp1 protects telomeres and regulates telomere length. *Science.* **320**: 1341–1344.

Moriarty, T., Huard, S., Dupuis, S., and Autexier, C. (2002) Functional multimerization of human telomerase requires an RNA interaction domain in the N terminus of the catalytic subunit. *Mol. Cell. Biol.* **22**: p1253–1265.

Moriarty, T.J., Marie-Egyptienne, D.T., and Autexier, C. (2005a). Regulation of 5′ template usage and incorporation of noncognate nucleotides by human telomerase. *RNA.* **11**: 1448–1460.

Moriarty, T.J., Ward, R.J., Taboski, M.A., and Autexier, C. (2005b). An anchor site-type defect in human telomerase that disrupts telomere length maintenance and cellular immortalization. *Mol. Biol. Cell.* **16**: 3152–3161.

O'Connor, C., Lai, C., and Collins, K. (2005) Two purified domains of telomerase reverse transcriptase reconstitute sequence-specific interactions with RNA. *J. Biol. Chem.* **280**: 17533–17539.

Osanai, M., Kojima, K.K., Futahashi, R., Yaguchi, S., and Fujiwara, H. (2006) Identification and characterization of the telomerase reverse transcriptase of *Bombyx mori* (silkworm) and *Tribolium castaneum* (flour beetle). *Gene.* **376**: 281–289.

Paeschke, K., Juranek, S., Simonsson, T., Hempel, A., Rhodes, D., and Lipps, H.J. (2008) Telomerase recruitment by the telomere end binding protein-beta facilitates G-quadruplex DNA unfolding in ciliates. *Nat. Struct. Mol. Biol.* **15**: 598–604.

Peng, Y., Mian, I.S., and Lue, N.F. (2001) Analysis of telomerase processivity: mechanistic similarity to HIV-1 reverse transcriptase and role in telomere maintenance. *Mol. Cell.* **7**: 1201–1211.

Podlevsky, J.D., Bley, C.J., Omana, R.V., Qi, X., and Chen, J.J. (2008) The telomerase database. *Nucl. Acids Res.* **36**: D339–343.

Robart, A.R., and Collins, K. (2010) Investigation of human telomerase holoenzyme assembly, activity, and processivity using disease-linked subunit variants. *J. Biol. Chem.* **285**: 4375–4386.

Rodgers, D.W., Gamblin, S.J., Harris, B.A., Ray, S., Culp, J.S., Hellmig, B., Woolf, D.J., Debouck, C., and Harrison, S.C. (1995) The structure of unliganded reverse transcriptase from the human immunodeficiency virus type 1. *Proc. Natl. Acad. Sci. USA.* **92**: 1222–1226.

Romi, E., Baran, N., Gantman, M., Shmoish, M., Min, B., Collins, K., and Manor, H. (2007) High-resolution physical and functional mapping of the template adjacent DNA binding site in catalytically active telomerase. *Proc. Natl. Acad. Sci. USA.* **104**: 8791–8796.

Rouda, S., and Skordalakes, E. (2007) Structure of the RNA-binding domain of telomerase: implications for RNA recognition and binding. *Structure.* **15**: 1403–1412.

Sarafianos, S.G., Clark, A.D., Jr., Das, K., Tuske, S., Birktoft, J.J., Ilankumaran, P., Ramesha, A. R., Sayer, J.M., Jerina, D.M., Boyer, P.L.,*et al.* (2002) Structures of HIV-1 reverse transcriptase with pre- and post-translocation AZTMP-terminated DNA. *EMBO J.* **21**: 6614–6624.

Sarafianos, S.G., Das, K., Tantillo, C., Clark, A.D., Jr., Ding, J., Whitcomb, J.M., Boyer, P.L., Hughes, S.H., and Arnold, E. (2001) Crystal structure of HIV-1 reverse transcriptase in complex with a polypurine tract RNA:DNA. *EMBO J.* **20**: 1449–1461.

Sealey, D.C., Zheng, L., Taboski, M.A., Cruickshank, J., Ikura, M., and Harrington, L.A. (2010) The N-terminus of hTERT contains a DNA-binding domain and is required for telomerase activity and cellular immortalization. *Nucl. Acids Res.* **38**: 2019–2035.

Sekaran, V.G., Soares, J., and Jarstfer, M.B. (2009) Structures of telomerase subunits provide functional insights. *Biochim. Biophys. Acta.* **1804**: 1190–1201.

Seto, A.G., Umansky, K., Tzfati, Y., Zaug, A.J., Blackburn, E.H., and Cech, T.R. (2003) A template-proximal RNA paired element contributes to *Saccharomyces cerevisiae* telomerase activity. *RNA.* **9**: 1323–1332.

Smith, R.A., Anderson, D.J., and Preston, B.D. (2006) Hypersusceptibility to substrate analogs conferred by mutations in human immunodeficiency virus type 1 reverse transcriptase. *J. Virol.* **80**: 7169–7178.

Steitz, T. (1999) DNA polymerases: structural diversity and common mechanisms. *J. Biol. Chem.* **274**: 17395–17398.

Steitz, T.A. (1997) DNA and RNA polymerases: structural diversity and common mechanisms. *Harvey Lect.* **93**: 75–93.

Tantillo, C., Ding, J., Jacobo-Molina, A., Nanni, R.G., Boyer, P.L., Hughes, S.H., Pauwels, R., Andries, K., Janssen, P.A., and Arnold, E. (1994) Locations of anti-AIDS drug binding sites and resistance mutations in the three-dimensional structure of HIV-1 reverse transcriptase. Implications for mechanisms of drug inhibition and resistance. *J. Mol. Biol.* **243**: 369–387.

Tzfati, Y., Fulton, T.B., Roy, J., and Blackburn, E.H. (2000) Template boundary in a yeast telomerase specified by RNA structure. *Science.* **288**: 863–867.

Vulliamy, T., Marrone, A., Dokal, I., and Mason, P.J. (2002) Association between aplastic anaemia and mutations in telomerase RNA. *Lancet.* **359**: 2168–2170.

Vulliamy, T.J., Walne, A., Baskaradas, A., Mason, P.J., Marrone, A., and Dokal, I. (2005) Mutations in the reverse transcriptase component of telomerase (TERT) in patients with bone marrow failure. *Blood Cells Mol. Dis.* **34**: 257–263.

Vulliamy, T.J., Marrone, A., Knight S.W., Walne, A., Mason, P.J., Dokal, I. (2006) Mutations in dyskeratosis congenita: their impact on telomere length and the diversity of clinical presentation. *Blood* **107**(7): 2680–2685.

Wang, F., Podell, E.R., Zaug, A.J., Yang, Y., Baciu, P., Cech, T.R., and Lei, M. (2007) The POT1–TPP1 telomere complex is a telomerase processivity factor. *Nature.* **445**: 506–510.

Wang, J., Sattar, A.K., Wang, C.C., Karam, J.D., Konigsberg, W.H., and Steitz, T.A. (1997) Crystal structure of a pol alpha family replication DNA polymerase from bacteriophage RB69. *Cell.* **89**: 1087–1099.

Wyatt, H.D., Lobb, D.A., and Beattie, T.L. (2007) Characterization of physical and functional anchor site interactions in human telomerase. *Mol. Cell. Biol.* **27**: 3226–3240.

Wyatt, H.D., Tsang, A.R., Lobb, D.A., and Beattie, T.L. (2009) Human telomerase reverse transcriptase (hTERT) Q169 is essential for telomerase function *in vitro* and *in vivo*. *PLoS One.* **4**: e7176.

Xia, J., Peng, Y., Mian, I.S., and Lue, N.F. (2000) Identification of functionally important domains in the N-terminal region of telomerase reverse transcriptase. *Mol. Cell. Biol.* **20**: 5196–5207.

Xie, M., Podlevsky, J.D., Qi, X., Bley, C.J., and Chen, J.J. (2010) A novel motif in telomerase reverse transcriptase regulates telomere repeat addition rate and processivity. *Nucl. Acids Res.* **38**: 1982–1996.

Xin, H., Liu, D., Wan, M., Safari, A., Kim, H., Sun, W., O'Connor, M.S., and Songyang, Z. (2007) TPP1 is a homologue of ciliate TEBP-beta and interacts with POT1 to recruit telomerase. *Nature*. **445**: 559–562.

Yamaguchi, H., Calado, R.T., Ly, H., Kajigaya, S., Baerlocher, G.M., Chanock, S.J., Lansdorp, P.M., and Young, N.S. (2005) Mutations in TERT, the gene for telomerase reverse transcriptase, in aplastic anemia. *N. Engl. J. Med.* **352**: 1413–1424.

Yu, E.Y., Wang, F., Lei, M., and Lue, N.F. (2008) A proposed OB-fold with a protein-interaction surface in *Candida albicans* telomerase protein Est3. *Nat. Struct. Mol. Biol.* **15**: 985–989.

Zaug, A.J., Podell, E.R., and Cech, T.R. (2008) Mutation in TERT separates processivity from anchor-site function. *Nat. Struct. Mol. Biol.* **15**: 870–872.

Zaug, A.J., Podell, E.R., Nandakumar, J., and Cech, T.R. (2010) Functional interaction between telomere protein TPP1 and telomerase. *Genes Dev*. **24**: 613–622.

4

TELOMERASE BIOGENESIS: RNA PROCESSING, TRAFFICKING, AND PROTEIN INTERACTIONS

TARA BEATTIE AND PASCAL CHARTRAND

4.1 INTRODUCTION

In vitro, the minimal catalytically active telomerase holoenzyme contains only the telomerase RNA (TER) and the catalytic subunit (TERT) (Beattie et al., 1998). However, as the size of the active telomerase complex *in vivo* has been estimated to vary between 0.6 to 2 MDa in humans (0.6 MDa in yeast), this suggests that other factors beside TER and TERT are present in this holoenzyme (Cohen et al., 2007; Lingner et al., 1997; Schnapp et al., 1998). Indeed, genetic and proteomic approaches have lead to the identification of several protein factors associated with telomerase (Fu and Collins, 2007; Stellwagen et al., 2003; Venteicher et al., 2008). Moreover, the composition of the telomerase ribonucleoprotein (RNP) has been shown to be particularly dynamic *in vivo*, especially during the various stages of the cell cycle (Collins, 2006). What are the functions of the telomerase-associated proteins? Where and when do they associate with the TER–TERT complex? How do these interactions regulate the activity of telomerase? These questions are still mostly unanswered and are the focus of active research.

In this chapter, we will present current ideas that link the processing and trafficking of the TER to the biogenesis of this holoenzyme in both yeasts and humans, organisms in which this trafficking has been most extensively studied. We will also introduce the currently known telomerase-associated proteins and describe their role in the

Telomerases: Chemistry, Biology, and Clinical Applications, First Edition.
Edited by Neal F. Lue and Chantal Autexier.

regulation of telomerase RNP biogenesis, assembly, and telomere recruitment, especially in human, ciliates, and yeast.

4.2 TELOMERASE RNA PROCESSING AND STABILITY

4.2.1 Yeasts Telomerase RNA

In the budding yeast *Saccharomyces cerevisiae*, the TER *TLC1* is a long transcript of ∼1250 nt transcribed by RNA polymerase II (pol II) (Fig. 4.1; Singer and Gottschling, 1994). Most of *TLC1* RNA accumulates as a nonpolyadenylated form, with only a minor fraction having a poly(A) tail (Chapon et al., 1997). Both poly(A) plus and minus forms are functional. It is not clear yet if the polyadenylated fraction is a precursor of the poly(A) minus form or if it is a different transcription termination product (Chapon et al., 1997). The poly(A) minus *TLC1* RNA has a Sm-binding site (5′-AAUUUUG-3′) at its 3′ end, which is recognized by Sm proteins (Seto et al., 1999). The Sm proteins form a ring structure around the Sm-binding site and are involved in 3′ end protection and stabilization of small nuclear RNAs (snRNAs) (Yong et al., 2004). Mutations in the *TLC1* RNA Sm site or in Sm proteins reduce the stability of this RNA, suggesting that Sm proteins also participate in the stabilization of the *TLC1* RNA (Seto et al., 1999). As a RNA pol II transcript, *TLC1* is synthesized with a monomethyl guanosine cap (m_7G), but it acquires a 2,2,7-trimethylguanosine (TMG) cap, like snRNAs (Seto et al., 1999). Modification of the monomethyl guanosine cap to a TMG cap occurs in the nucleolus and requires the trimethyl guanosine synthase Tgs1 (Franke et al., 2008; Gallardo et al., 2008).

In the fission yeast *Schizosaccharomyces pombe*, the TER *TER1* is a large ∼1200 nt transcript (Leonardi et al., 2008; Webb and Zakian, 2008). Similar to its budding yeast counterpart, *TER1* RNA exists mostly as a nonpolyadenylated form, with a minor fraction containing a poly(A) tail (Leonardi et al., 2008). *TER1* also has an Sm site at its 3′ end, which is essential for the stability of this RNA (Leonardi et al., 2008). Interestingly, recent data showed that 3′ end processing of the *TER1* RNA requires the splicing of an intron at the 3′ end of the unprocessed transcript (Box et al., 2008). In this specific case, the first cleavage step of the intron by the spliceosome generates the mature 3′ end of *TER1*, without promoting exon ligation. *S. pombe TER1* RNA also contains a 5′ TMG cap, which was used for its initial purification and identification (Webb and Zakian, 2008).

TER from other budding yeast species, like *Kluyveromyces* and *Candida*, have also been identified (Gunisova et al., 2009; McEarchern and Blackburn, 1995). They vary in size from 779 nt for *Candida guilliermondii* TER to 1817 nt for *Candida parapsilosis* TER. They all contain an Sm site at their 3′ end, indicating that stabilization of TER by Sm proteins is a conserved mechanism in yeasts (Gunisova et al., 2009). Surprisingly, several *Candida* TER genes contain consensus splicing elements at their 3′ end, suggesting that the 3′ end processing of these TER may be dependent on a partial splicing reaction, like for *S. pombe TER1* (Gunisova et al., 2009). In contrast, splicing has not been implicated in the processing of the *S. cerevisiae TLC1* RNA.

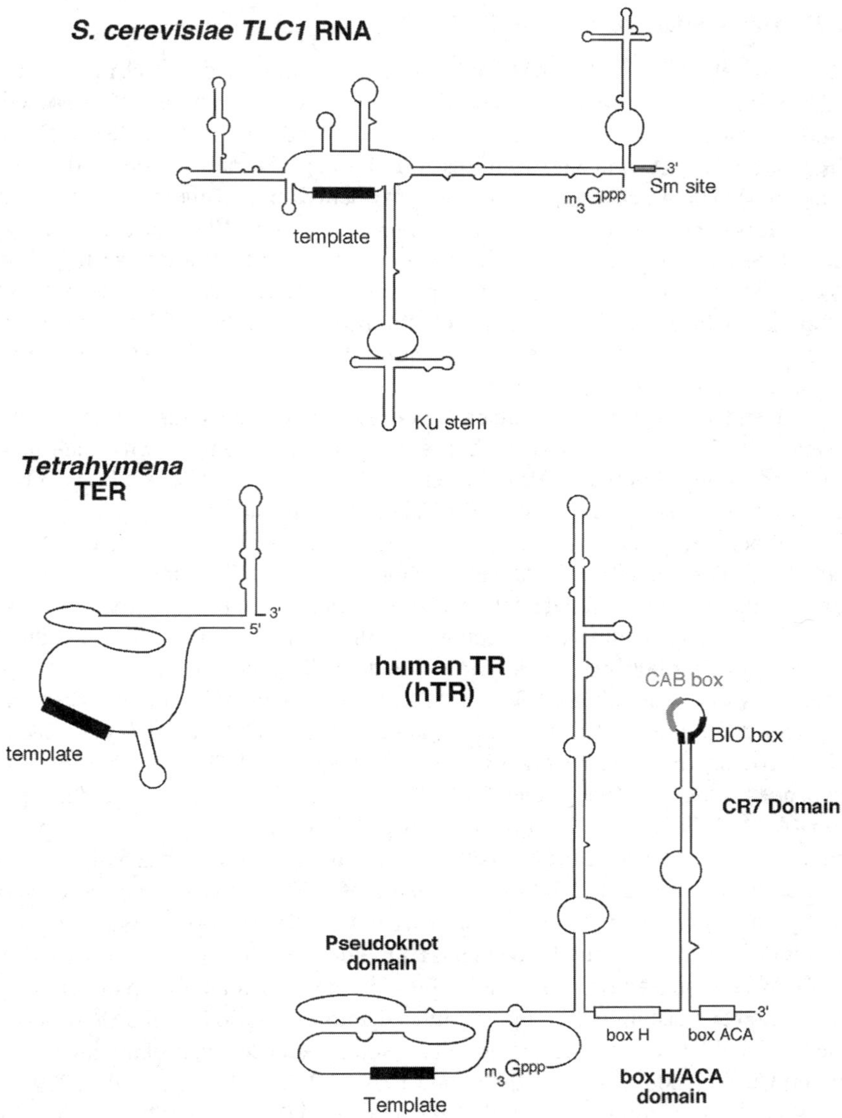

FIGURE 4.1 Structures and functional domains in telomerase RNA from yeast, ciliates, and vertebrate.

4.2.2 Ciliates Telomerase RNA

In ciliates, unlike other eukaryotes, the ~150 nucleotides TER (Fig. 4.1) is the product of RNA polymerase III (RNA pol III) transcription (Greider and Blackburn, 1989). Accordingly, it contains a poly(U) tail at its 3′ end, which is part of the transcription-termination process (Bogenhagen and Brown, 1981). No specific processing step has been reported for this RNA.

4.2.3 Mammalian Telomerase RNA

The human telomerase RNA (hTR) is ubiquitously transcribed by RNA pol II (Feng et al., 1995). As an RNA pol II transcript, hTR is synthesized with a monomethyl guanosine capped precursor and an unprocessed 3′ end. The mature, 451 nucleotide hTR has no poly(A) tail and acquires a TMG cap (Fu and Collins, 2006; Jady et al., 2004). The process leading to the hypermethylation of the hTR 5′ cap is still unclear. It was proposed that the TMG cap modification of hTR may occur in Cajal bodies (see Section 4.3.2), since an isoform of the methyltransferase hTgs1 was shown to be present in these bodies (Girard et al., 2008). However, an hTR mutated in its Cajal body localization element (or CAB box, see Section 4.2.2 below) still has a TMG cap (Fu and Collins, 2006), suggesting that cap methylation occurs before localization of hTR to Cajal bodies.

The 3′ end processing and cellular accumulation of hTR depends on two structures present at its 3′ end; the box H/ACA motif and the CR7 domain, which are highly conserved among vertebrate TRs (Chen et al., 2000; Fig. 4.1). The box H/ACA motif is constituted by a stem–H box–stem–ACA box structure and is recognized by four H/ACA associated proteins: Dyskerin/Cbf5, Nhp2, Nop10, and Gar1 (Mitchell et al., 1999; Pogacic et al., 2000; see Section 4.4 below). This motif is commonly found in a large family of small nucleolar (sno) and small Cajal body (sca) RNAs (Kiss et al., 2010). These small RNAs direct the modification of specific uridine residues to pseudouridine in ribosomal RNA (rRNAs) and small spliceosomal RNA (snRNAs), respectively (Yu et al., 2005). While hTR also contains an H/ACA motif, there is no evidence yet for a role of this RNA in pseudouridylation. Unlike vertebrate snoRNAs and scaRNAs, which are encoded in pre-mRNA introns, hTR is synthesized as an independent transcript (Feng et al., 1995). Even if snoRNA processing and assembly into H/ACA RNP complex requires splicing and exonucleolytic degradation of their 5′ and 3′ ends, evidence suggests that these steps occur cotranscriptionally (Darzacq et al., 2006; Richard et al., 2006). Processing of hTR may also occur cotranscriptionally since expression of this RNA from a RNA pol III promoter results in defective maturation and in the accumulation of its H/ACA domain alone (Mitchell et al., 1999). The H/ACA motif and its associated proteins are essential for mammalian TR stabilization and accumulation *in vivo*. Indeed, either mutations in the H/ACA motif or depletion of H/ACA-associated proteins result in reduced levels of mammalian TR (Fu and Collins, 2007; Martin-Rivera and Blasco, 2001; Mitchell et al., 1999).

A second processing and stability element in hTR is the CR7 domain, which overlaps with the H/ACA motif at the 3′ end (Fig. 4.1). This domain contains two different elements: a CAB box, necessary for hTR targeting to Cajal bodies (see Section 4.2.2) and a biogenesis box or BIO box, required for efficient 3′ end processing (Collins, 2006; Fu and Collins, 2003; Theimer et al., 2007). Both the CAB and BIO boxes are in the apical loop of the CR7 domain. While mutations in the CAB box have no effect on hTR accumulation, disruption of the BIO box strongly reduces hTR stability and telomerase activity *in vivo* (Theimer et al., 2007). It is still unclear why both the H/ACA motif and BIO box are required for hTR accumulation. Unlike the majority of H/ACA snoRNAs and scaRNAs, which contain only a box H/ACA motif at their 3′ end, hTR shares the dual H/ACA and BIO box motifs with at least one other RNA, the U64 snoRNA. Substitution of the CR7 domain of hTR by the U64 snoRNA terminal loop maintain hTR stability, while terminal loops from other H/ACA scaRNAs or snoRNAs cannot maintain proper hTR processing (Fu and Collins, 2003). This suggests that hTR 3′ end processing pathway is distinct from scaRNAs and snoRNAs, and occurs before its translocation to Cajal bodies (Theimer et al., 2007).

4.3 TELOMERASE RNA TRAFFICKING AND TELOMERASE BIOGENESIS

4.3.1 Nucleocytoplasmic Trafficking of the Yeast Telomerase RNA

In eukaryotes, small RNPs assembly frequently occurs in several subcellular compartments and requires the targeting of RNA and protein subunits to specific locations (Hopper, 2006). As a large RNP, telomerase assembly possibly follows a similar rule. Cytological studies of *S. cerevisiae* telomerase have been impeded due to the low level of endogenous *TLC1* RNA and Est proteins (Mozdy and Cech, 2006). The Est proteins (Est1, Est2, and Est3) were identified in a genetic screen and constitute the core components of the telomerase holoenzyme (Lendvay et al., 1996). Besides the Est proteins, the Ku70/80 heterodimer is another factor that directly interacts with the *TLC1* RNA (Stellwagen et al., 2003). The Ku70/80 heterodimer is a DNA-end binding complex that serves to recruit DNA repair machinery to site of DNA-double strand breaks. The first cytological studies on yeast telomerase used overexpressed *TLC1* RNA and Est proteins, and had little high-resolution information on the subnuclear localization of the telomerase components (Ferrezuelo et al., 2002; Teixeira et al., 2002). Nevertheless, overexpressed Est2 was found to accumulate in the nucleolus. When *TLC1* RNA was coexpressed, Est2 relocated to the nucleoplasm. In the same study, a heterokaryon assay was used to show that some *TLC1* RNA accumulated in all the nuclei of heterokarya derived from *TLC1* and *tlc1* cells, suggesting that this RNA may shuttle between the nucleus and the cytoplasm (Teixeira et al., 2002). Further evidence of a role for *TLC1* RNA trafficking in telomere homeostasis came from the finding that a mutation in the *MTR10* gene, which encodes for an importin β involved in nuclear import of mRNA-binding proteins, resulted in shorter telomeres (Ferrezuelo et al., 2002). In a *mtr10* mutant, overexpressed *TLC1* RNA accumulated in the cytoplasm.

In a recent study, the endogenous *TLC1* RNA was detected in yeast cells using fluorescent *in situ* hybridization (FISH) (Gallardo et al., 2008). With this approach, *TLC1* RNA was found to colocalize with the telomeres in the G1 and S phases of the cell cycle. These results confirmed chromatin immunoprecipitation experiments which showed that Est2 is present at the telomeres in G1–S in a *TLC1*-dependent manner (Taggart et al., 2002). Unexpectedly, deletions of any one of the *ESTs* or *YKUs* result in a cytoplasmic accumulation of *TLC1* RNA, raising the possibility that a stage of the biogenesis of telomerase occurs in the cytoplasm. However, an alternative explanation is that these deletions result in a defect in the nuclear retention of *TLC1*

RNA and induce its nuclear export. To determine if the endogenous *TLC1* RNA has a cytoplasmic phase in a *ESTs* and *YKUs* positive background, a heterokaryon-based nucleocytoplasmic shuttling assay was used and showed that this RNA indeed shuttles between the nucleus and the cytoplasm. Nuclear export of *TLC1* RNA was found to depend on the exportin Crm1, while nuclear import requires the β-importin Mtr10 and Kap122 (Gallardo et al., 2008). The fact that the *TLC1* RNA shuttles between the nucleus and the cytoplasm in the presence of Est1-3 and yKu factors, and that this shuttling occurs via a defined export–import pathway, strongly suggests that this trafficking has a specific biological function.

This nucleocytoplasmic shuttling is reminiscent of snRNPs assembly in metazoans, in which snRNAs are exported to the cytoplasm via a Crm1-dependent pathway, acquire a TMG cap and are bound by Sm proteins before their reimport into the nucleus (Matera et al., 2007). While *TLC1* RNA has a TMG cap and is bound by Sm proteins such as snRNAs, snRNPs nucleocytoplasmic shuttling is still controversial in yeast (Hopper, 2006; Olson and Siliciano, 2003). Moreover, the enzyme responsible for the hypermethylation of the guanosine cap of snRNAs and snoRNAs, the methyltransferase Tgs1, resides in the nucleolus in yeast (Mouaikel et al., 2002), suggesting that the *TLC1* RNA may have a nucleolar phase. Indeed, deletion of the *TGS1* gene inhibits the cap hypermethylation of *TLC1* RNA and results in the nucleolar accumulation of this RNA (Gallardo et al., 2008). This maturation event possibly occurs prior to *TLC1* RNA nuclear export since this RNA accumulates in the nucleus as TMG-capped in a Crm1-disrupted strain. Also, the *TLC1* RNA always has a TMG cap when it accumulates in the cytoplasm of *ESTs* and *YKUs* deleted strains (Gallardo et al., 2008). These results suggest that, unlike snRNAs in metazoans, the yeast TER acquires its TMG cap and Sm proteins before its nuclear export.

A model for telomerase biogenesis in yeast, which integrates the TMG cap modification of newly synthesized *TLC1* RNA in the nucleolus, its nucleocytoplasmic shuttling and the recruitment of the telomerase at the telomeres can be proposed (Fig. 4.2). Still, the results mentioned above raise several questions. For instance, what promotes the nuclear import of the *TLC1* RNA? No homologue of Snurportin, the nuclear import factor of metazoans TMG-capped snRNAs, has been identified in yeast (Mans et al., 2004). Moreover, nuclear accumulation of *TLC1* RNA requires the presence of all three Est proteins and the yKu70/80 heterodimer, suggesting that these factors play a direct role in the nuclear import and/or nuclear retention of this RNA. One model could be that a telomerase holoenzyme containing the *TLC1* RNA and the three Est proteins is assembled in the cytoplasm (where these proteins are synthesized), generating a complex which is competent for nuclear import (Figure 4.2). While the level of some Est proteins varies during the cell cycle, such as Est1 which is low in G1 and increases in S and G2 (Osterhage et al., 2006), no cell-cycle-dependent cytoplasmic accumulation of *TLC1* RNA has been observed (Gallardo and Chartrand, unpublished data), suggesting that a sufficient quantity of each Est protein is available to interact with newly synthesized *TLC1* RNA in the cytoplasm and promote its nuclear accumulation at every phase of the cell cycle. Once in the nucleus, the telomerase holoenzyme can be retained at the telomeres, possibly via the interaction between yKu70/80 and the *TLC1* RNA (Stellwagen et al., 2003).

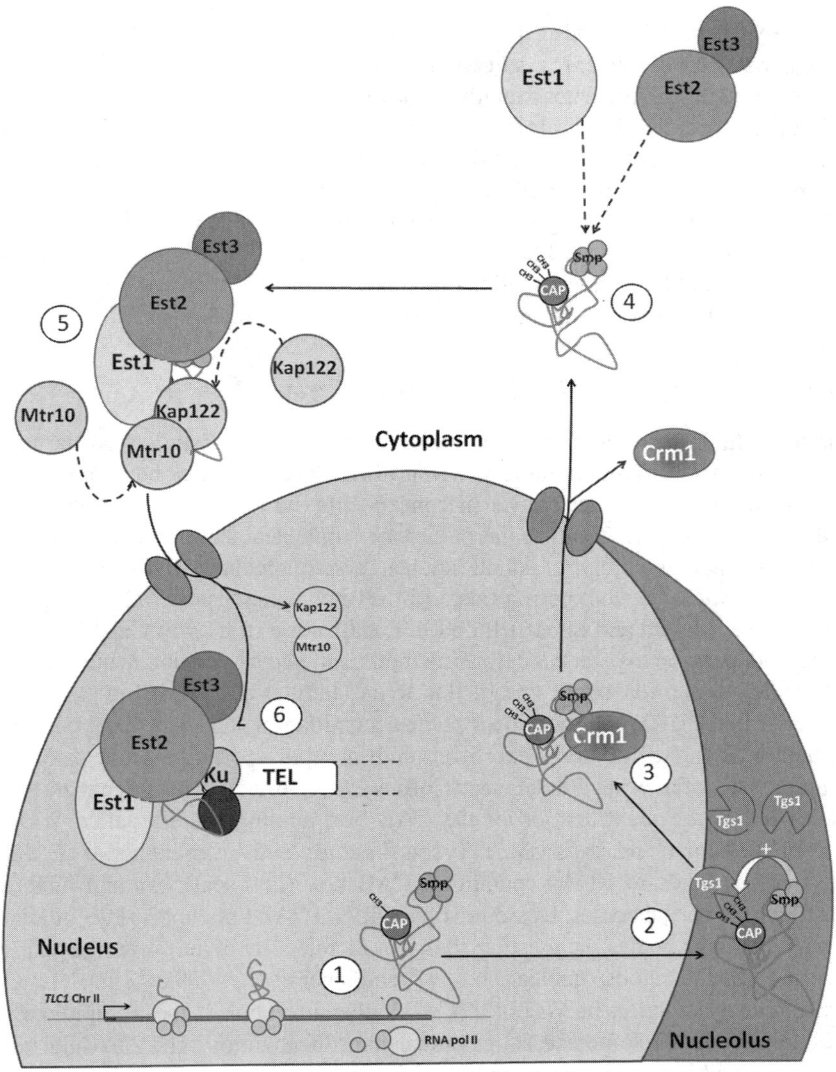

FIGURE 4.2 An integrated model of telomerase biogenesis in yeast. The *Saccharomyces cerevisiae* telomerase RNA *TLC1* is transcribed by the RNA polymerase II machinery (1) and targeted to the nucleolus where its 5′mono-methylguanosine cap is hypermethylated by Tgs1 (2). Following its 5′cap hypermethylation, the *TLC1* RNA is exported in the cytoplasm via the Crm1p-dependent pathway (3). In the cytoplasm, the *TLC1* RNA recruits the proteic components of the telomerase complex (4), assembles into a mature telomerase particle (5), and is imported back in the nucleus via a Mtr10/Kap122 pathway (5). Once in the nucleus, it can be recruited at the telomeres via the interaction between the *TLC1* RNA and the yKu heterodimer (6). (The telomerase holoenzyme at the telomere depicted in this figure would correspond to the one in S phase. As Est1 is actively degraded or not depending on the phase of the cell cycle, so the constitution of the telomerase recruited at the telomeres will vary accordingly). Taken from Gallardo and Chartrand (2008). ©Landes Bioscience. (See the color version of this figure in Color Plates section.)

Finally, these results raise questions concerning the function of a cytoplasmic assembly step for a nuclear resident complex such as telomerase. One possibility may be that this shuttling is part of a quality control pathway in the folding of *TLC1* RNA and in the assembly of the telomerase holoenzyme. Since the *TLC1* RNA acts as a scaffold that recruits the protein subunits of telomerase (Zappulla and Cech, 2004), the proper recruitment of the Est proteins may constitute a proofreading step which indicates that this RNA is properly folded. Since misfolded *TLC1* RNA or misassembled telomerase would have a negative function (i.e., the failure to elongate telomeres), the cytoplasmic retention of a misassembled telomerase would eliminate the damage that such an enzyme might produce in the nucleus.

4.3.2 Intranuclear Trafficking of the Human Telomerase RNA

Due to its high expression level in cancer cells, cytological studies on the human telomerase have progressed rapidly. Using FISH, several studies have reported that hTR accumulates in Cajal bodies in cancer cells, but not in primary cells (Jady et al., 2004; Zhu et al., 2004). Cajal bodies are subnuclear structures involved in the maturation and assembly of small nuclear and nucleolars RNPs (snRNPs and snoRNPs), which are factors implicated in mRNA splicing and rRNA biogenesis, respectively (Matera and Shpargel, 2006). Localization of hTR to Cajal bodies was shown to depend on a specific localization motif, called the CAB box, which is present in the CR7 domain at the 3′ end of this RNA (Jady et al., 2004; Fig. 4.1). While mutations in the CAB box motif disrupt the accumulation of hTR in Cajal bodies, the mutated hTR is still properly expressed and incorporated in an active telomerase complex (Cristofari et al., 2007; Fu and Collins, 2006). Interestingly, recent proteomic studies led to the identification of the CAB box binding protein called WDR79/TCAB1, in human and *Drosophila* (Tycowski et al., 2009; Venteicher et al., 2009). This protein binds to RNAs containing CAB box (i.e., scaRNAs and hTR) and accumulates in Cajal bodies. Depletion of WDR79/TCAB1 disrupts the accumulation of hTR in Cajal bodies, suggesting that this factor is involved in the targeting or retention of hTR in these nuclear bodies (Venteicher et al., 2009). As for hTR CAB box mutants, depletion of WDR79/TCAB1 does not abolish the accumulation of catalytically active telomerase, suggesting that the localization of hTR to Cajal bodies is not involved in the proper assembly of the active telomerase holoenzyme.

A better understanding of the role of Cajal bodies in telomerase function emerged from cell-cycle studies of hTR and hTERT trafficking. Unlike in yeast, where Est2 (yeast TERT) and *TLC1* RNA are associated with the telomeres in G1 and S phase (Gallardo et al., 2008; Taggart et al., 2002), there is no evidence that hTR and hTERT are associated with the telomeres during most phases of the cell cycle. Recent studies combined cytological techniques with cell-cycle synchronization of cancer cells to show that hTR and hTERT colocalize with telomeres only in S phase (Jady et al., 2006; Tomlinson et al., 2006). In G1 and early S phase, hTR is localized in Cajal bodies, but not with hTERT. In late S phase, when telomeres are elongated, a fraction of hTR and hTERT accumulate in foci that are distinct from but adjacent to Cajal bodies, and which colocalize with telomeres (Fig. 4.3). On average, hTR

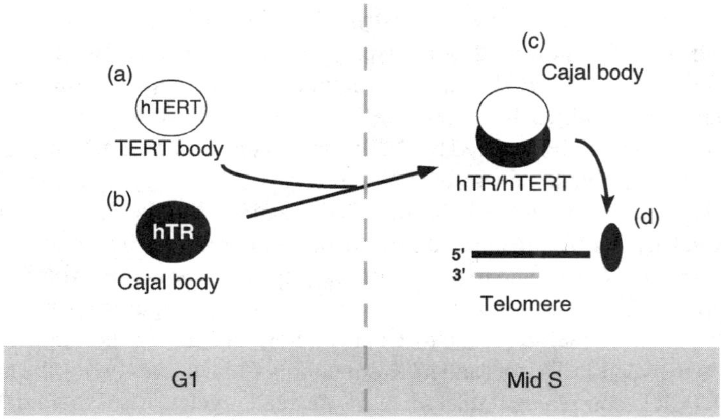

FIGURE 4.3 Cell-cycle-dependent trafficking of hTR and hTERT in the nucleoplasm. Model for cell-cycle-dependent nucleoplasmic trafficking of hTR and hTERT, based on fluorescent *in situ* hybridization and immunofluorescence data (Jady et al., 2006; Tomlinson et al., 2006). In G1, hTERT and hTR accumulate in separate subnuclear structures: TERT foci and Cajal bodies, respectively (a and b). In mid S phase, some TERT foci and Cajal bodies partially colocalize (c). The association between TERT foci and hTR-containing Cajal bodies may favor the formation of hTR–hTERT complex at this time of the cell cycle, leading to the recruitment of the active telomerase holoenzyme at short telomeres (d).

accumulation was observed at one or two telomeres per cell, with a maximum of six in a single cell, suggesting that only a subset of telomeres are elongated per cell cycle (Jady et al., 2006). Such data are in agreement with what was shown in yeast, where less than 10% of the telomeres are elongated per cell cycle (Teixeira et al., 2004). Since Cajal bodies are mobile structures (Platani et al., 2000), these data lead to a model in which the Cajal body itself delivers the telomerase to telomeres (Jady et al., 2006; Tomlinson et al., 2006). In support of this model, cancer cells expressing an hTR with a mutated CAB box displayed a reduced rate of telomere elongation (Cristofari et al., 2007). Depletion of WDR79/TCAB1 also impairs telomere elongation, but has no effect on telomerase activity (Venteicher et al., 2009). Altogether, these results support a role for Cajal bodies in the recruitment of telomerase to telomeres. However, the role of Cajal bodies in telomere homeostasis may vary depending on cell types, since coexpression of hTERT and a hTR CAB box mutant in a human X-linked dyskeratosis congenita (DC) primary fibroblasts still resulted in telomere elongation in these cells (Fu and Collins, 2007).

4.4 TERT TRAFFICKING AND TELOMERASE BIOGENESIS

4.4.1 Cell-Cycle Trafficking

While hTR is constitutively expressed in most cell types, the presence of hTERT only in germ cells, stem cells, and cancer cells limited the cytological studies of this protein factor to mostly cancer cells. Unlike hTR, which is enriched in Cajal bodies, hTERT exhibits a more diverse subnuclear distribution. Immunofluorescence analysis of endogenous hTERT showed that it displays a nucleolar enrichment (Tomlinson et al., 2006; Yang et al., 2002), but also accumulates in nucleoplasmic foci distinct from Cajal bodies, which have been called TERT foci (Tomlinson et al., 2006). Ectopic expression of GFP-tagged hTERT results in the accumulation of this factor in nucleoli of primary and cancer cells (Wong et al., 2002; Yang et al., 2002). More importantly, studies of cancer cells revealed that hTERT distribution is dynamic during the cell cycle (Tomlinson et al., 2006). In G1 phase, hTERT is mostly present in nucleoplasmic TERT foci and in the nucleolus, physically separated from hTR which accumulates in Cajal bodies. In early to mid-S phase, a significant proportion of hTERT is found in foci adjacent to Cajal bodies, raising the possibility that the association between TERT foci and hTR-containing Cajal bodies favors the formation of hTR–hTERT complex at this time of the cell cycle. Also in mid-S phase, colocalization between hTERT and telomeres is observed, with one to five colocalizations per cell, as for hTR (Fig. 4.3).

Altogether, the data presented above support a model in which hTR and hTERT are separated during most of the cell cycle and assembled at telomeres only in S phase. Physical separation of hTERT and hTR may be necessary in order to avoid unregulated telomerase activity. Since hTR and hTERT are both expressed in cells with active telomerase (i.e., lymphocytes; Liu et al., 1999), their separate sequestration may serve to participate in the regulation of enzyme activity. Human telomerase activity can be detected in all the phases of the cell cycle (Holt et al., 1997), suggesting the presence of a pool of preassembled hTR–hTERT complex in the cells. Another possibility is that the preparation of cell extracts caused disruption of cell compartments and resulted in the artifactual assembly of telomerase. Interestingly, accumulation of hTR in Cajal bodies depends on the presence of hTERT, even if this protein is not found in these bodies (Tomlinson et al., 2008). This suggests that, as in yeast, an hTR–hTERT interaction may be required for the nuclear or subnuclear localization of the hTR. It is possible that the cell-cycle-independent activity of telomerase may originate from this pool of hTR–hTERT complex.

4.4.2 Nucleocytoplasmic Trafficking of hTERT

Once synthesized in the cytoplasm, hTERT must translocate to the nucleus for its function in telomere maintenance. Even though no specific nuclear localization element has been identified in hTERT, evidence suggests that the nucleocytoplasmic trafficking of hTERT constitutes a level of post-transcriptional regulation. Indeed, a CRM1-dependent nuclear export signal (NES) has been mapped at the C-terminus of hTERT (amino acids 831–1132) (Seimiya et al., 2000). The protein 14-3-3, a known regulator of intracellular localization of signaling proteins (Beck and Hall, 1999), was shown to interact with telomerase in a yeast two-hybrid screen (Seimiya et al., 2000). In particular, the C-terminus of hTERT (amino acids 1030–1047) has been shown to interact with the C-terminus of the 14-3-3 protein, an interaction that is dispensable for catalytic activity. This 14-3-3 interaction domain on hTERT is just downstream of a putative hTERT NES (amino acids 970–981) which lies within the region of hTERT that also interacts with CRM1. If the interaction between hTERT and 14-3-3 is disrupted by mutation in either protein, there is no effect on the catalytic activity of the enzyme, however hTERT is unable to localize to the nucleus (Seimiya et al., 2000). These finding have lead to the hypothesis that binding of the 14-3-3 protein enhance the nuclear localization of hTERT by masking its NES and inhibiting CRM1 binding. Recent evidence suggests that nuclear export of hTERT has an important biological function. An hTERT mutant containing L980A and L987A substitutions disrupts the NES signal, and the resultant protein localized to the nucleus and did not accumulate in the cytoplasm. Furthermore, these mutations rendered hTERT unable to immortalize normal fibroblasts (Kovalenko et al., 2010). Cells expressing this mutant enter cellular senescence and display dysfunctional mitochondria, even if this NES-mutant retains telomerase activity *in vitro*.

Besides its accumulation in nucleoplasmic foci, hTERT can also be detected in the nucleolus of cancer and primary cells (Etheridge et al., 2002; Wong et al., 2002; Yang et al., 2002). Several nucleolar targeting signals (NTS) have been identified in hTERT, including two at its N-terminus and one at its C-terminus (Etheridge et al., 2002; Lin et al., 2008; Yang et al., 2002). Mutations in the two N-terminal NTS still result in partial nucleolar accumulation of hTERT, and only the disruption of the C-terminal NTS abolishes nucleolar localization (Lin et al., 2008). While it is still unclear which factor(s) bind these NTS motifs and mediate the nucleolar localization of hTERT, there are a number of known proteins that are thought to play a role in this process. One of these proteins is nucleolin, an abundant nucleolar phosphoprotein that is involved in ribosome biogenesis, chromatin structure, rDNA transcription, and nucleocytoplasmic transport (reviewed in Ginisty et al., 1999). Nucleolin interacts with hTERT in an hTR-dependent manner, and this interaction is critical for the nucleolar localization of telomerase (Khurts et al., 2004). The interaction between nucleolin and telomerase in the nucleolus may negatively regulate telomerase activity by sequestering the enzyme in the nucleolus until it needs to be delivered to its telomeric substrate at the correct time during the cell cycle. This is supported by the observation that in G1, hTERT is localized in part to the nucleolus and then is subsequently redistributed into the nucleoplasm during S-phase when telomeres are elongated by telomerase (Wong et al., 2002).

What could be the function for the nucleolar localization of hTERT? Evidence suggest that it is not involved in telomerase assembly, since the C-terminal NTS mutant of hTERT still maintains telomerase activity and counteracts telomere shortening (Lin et al., 2008). Another possibility is that this trafficking may participate in the regulation of telomerase activity. Indeed, the signals that regulate the distribution of hTERT in the nucleus are dependent on the cell cycle, cellular transformation, and DNA damage. Specifically, using confocal microscopy, Wong et al., demonstrated that hTERT is released from sequestration in the nucleoli into the nucleoplasm at a time in the cell cycle when telomeres are replicated (Wong et al., 2002). This study also showed a nucleolar exclusion of telomerase in transformed cells, and conversely, ionizing radiation induced DNA damage relocalized telomerase to the nucleoli (Wong et al., 2002). These findings demonstrate that telomere elongation is regulated by the shuttling of telomerase between the nucleolus and the nucleoplasm, and that transformation or DNA damage have opposite effects on this regulation.

Moreover, a potential tumor suppressor and a known resident of the nucleolus, PinX1, can act as a repressor of telomerase activity (Zhou and Lu, 2001). While PinX1 was first identified as a TRF1-binding protein (Zhou and Lu, 2001), it also interacts with the telomerase subunits hTERT and hTR (Banik and Counter, 2004). PinX1 therefore colocalizes to both the telomeres, where TRF1 is bound, and to the nucleolus, where the telomerase subunits are located. PinX1 is an endogenous inhibitor of telomerase activity and a negative regulator of telomere function (Zhou and Lu, 2001). The yeast orthologue of PinX1 (Gno1) was found to compete with *TLC1* RNA for binding free Est2 and inhibit telomerase activity, possibly by sequestering uncomplexed Est2 in the nucleolus (Lin and Blackburn, 2004). Conversely, in humans, PinX1 also binds hTERT, but the PinX1–hTERT complex does not preclude the binding of hTR to hTERT, which leads to the hypothesis that human PinX1 represses telomerase activity by binding the preassembled hTERT–hTR complex (Banik and Counter, 2004). Since PinX1 can also bind TRF1, it remains possible that telomerase inhibition by PinX1 in human cells may occur at the telomere. Together these data indicate a differential regulatory role for the PinX1 protein in different species. Paradoxically, a recent study suggests that PinX1 can also promote cancer-cell proliferation and possesses a stimulatory role for telomerase, as knockdown of PinX1 reduces telomere length and proliferation in telomerase-positive cancer cells (Zhang et al., 2009). One possible explanation of the conflicting findings is that PinX1 functions as a tumor suppressor by sequestration of telomerase in the nucleolus, and that when this regulatory function is compromised, telomerase shuttles out of the nucleolus into the nucleoplasm where it can interact with and elongate telomeres (Zhang et al., 2009).

4.5 TELOMERASE ASSOCIATED PROTEINS THAT REGULATE RNP BIOGENESIS, ASSEMBLY, AND TELOMERE RECRUITMENT

The biogenesis of human telomerase and the assembly of hTERT and hTR into an active complex involves a number of important regulatory steps. One of these critical steps is the assembly of an active telomerase holoenzyme. Human telomerase associates either directly or indirectly with a variety of proteins that have important roles in RNP biogenesis, and enzyme assembly. These proteins vary greatly between species and the relevance of many of these proteins to telomerase function remains to be determined. In this section, we will present the proteins that are known to play integral roles in the regulation and assembly of the telomerase holoenzyme.

4.5.1 Mammalian Telomerase RNP Assembly

4.5.1.1 Role of H/ACA Binding Proteins As previously described, hTR contains an evolutionarily conserved box H/ACA domain at the 3′ end of the molecule that is required for *in vivo* hTR accumulation, stability, and end-processing

(Mitchell et al., 1999). The H/ACA motif is a conserved structure that guides the pseudouridylation of a variety of small noncoding RNA molecules, including scaRNAs and snoRNAs (Meier, 2005). The H/ACA motif of these small RNAs interacts with a set of core proteins which include dyskerin, NOP10, and NHP2 and then subsequently GAR1 (Matera et al., 2007). As with the scaRNAs and snoRNAs, hTR also interacts with these core H/ACA motif binding proteins (Fu and Collins, 2003).

Dyskerin is a conserved nucleolar protein that contains the active site for pseudouridinylation of target RNA molecules. It was first shown to associate with the hTR in 1999, and it was demonstrated that this interaction is critical for the stability and accumulation of hTR *in vivo* (Mitchell et al., 1999). More recently, Cohen et al., demonstrated that dyskerin is a core component of an active telomerase complex purified from human cell extracts (Cohen et al., 2007).

Each of the core H/ACA motif binding proteins has the ability to directly interact with the TER. However, *in vivo*, they do not assemble independently (Fu and Collins, 2003, 2007). The proteins dyskerin, NOP10, and NHP2 are assembled cotranscriptionally onto the TER (Darzacq et al., 2006). There is a direct interaction between dyskerin and NOP10, with NHP2 being recruited to the complex through its binding to NOP10 (Darzacq et al., 2006; Fu and Collins, 2003). An additional RNA chaperone protein, NAF1, is required for the interaction of this trimeric complex with the newly transcribed RNA (Hoareau-Aveilla et al., 2006). This is followed by a remodeling of the complex, resulting in the release of NAF1 and the recruitment and binding of GAR1 (Darzacq et al., 2006). This interaction is important for the assembly and localization of the mature RNP (Fig. 4.4).

Mutations in the telomerase holoenzyme genes *TRT, TERC, DKC2, NOP10*, and *NHP2* have been identified in patients with dyskeratosis congenita (Trahan and Dragon, 2009; Trahan et al., 2010; Vulliamy et al., 2008; Walne et al., 2007). This disease is associated with a deficiency in the renewal of highly proliferative tissues, and progressive bone marrow failure. Although the cause and progression of this disease is not yet fully understood, it is believed to be caused by the inability of the cells to maintain telomere lengths. Mutations in each of the genes encoding for *DKC1, NOP10,* and *NHP2* results in a decreased accumulation of hTR levels, which consequently results in a decrease in telomerase activity. Similarly, mutations in *TERC* that disrupt the sequence and structure of the box H/ACA motif also prevents its cellular accumulation (Mitchell et al., 1999). This decrease in telomerase activity leads to accelerated or premature telomere shortening, which may contribute to the limited replicative potential of highly proliferative tissues. The role of telomerase deficiency in human diseases has been discussed in more detail in Chapters 2, 3, and 9.

4.5.1.2 Role of Chaperone Proteins Although it is clear that the H/ACA motif binding proteins play a role in the biogenesis of hTR, the *in vivo* assembly of hTERT with this hTR-containing RNP is not well defined. It is possible that the interaction of hTR with the core H/ACA binding proteins induces a conformational rearrangement that facilitate hTERT binding, similar to what is observed in *Tetrahymena*

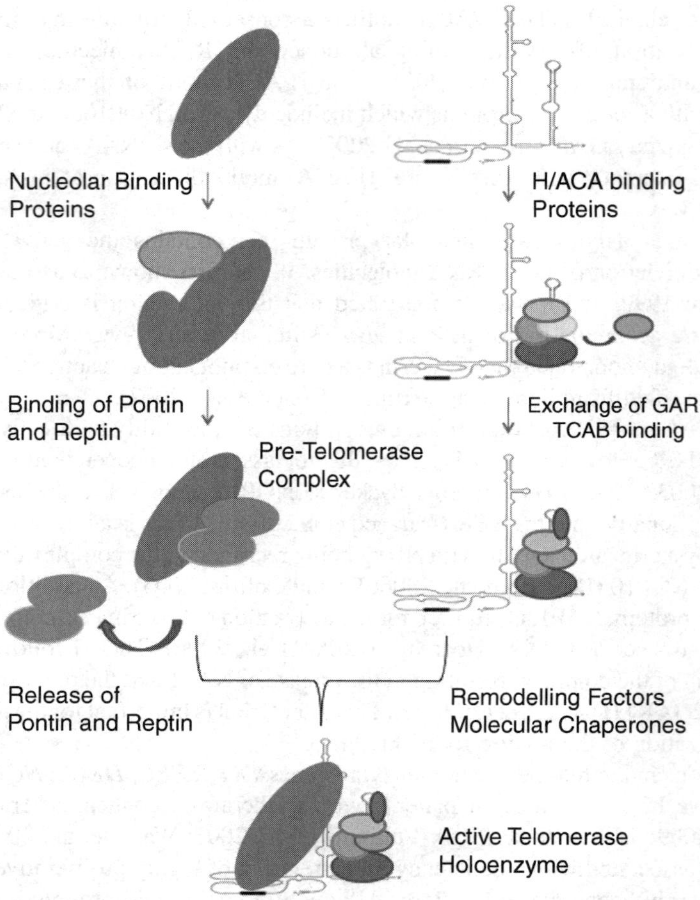

FIGURE 4.4 Assembly of the human telomerase complex. After the transcription and processing of the human telomerase RNA, the H/ACA proteins (dyskerin—blue, Nop10—green, NHP2—orange, and NAF1—yellow) bind to the 3′ end of the telomerase RNA. Subsequently, NAF1 is exchanged for GAR1 (burgundy), and TCAB1 (purple) binds to the hTR. After TERT (red) is localized to the nucleolus, mediated in part by interactions with 14-3-3 (gray) and nucleolin (pink) it is assembled with the ATPases pontin and reptin (shown in green), to form a pretelomerase complex. During S-phase, pontin and reptin are released from hTERT and the complex is remodeled or assembled with the help of additional factors (such as the molecular chaperones Nat10, GNL3L, heat shock proteins, and SMN) with hTR to form an active telomerase holoenzyme. (See the color version of this figure in Color Plates section.)

thermophila (see below). It is also likely that additional factors could function to promote hTERT/hTR interactions and the assembly of an active telomerase complex.

In rabbit reticulocyte lysates (RRL), hTERT and hTR are sufficient to reconstitute an active telomerase enzyme (Beattie et al., 1998; Weinrich et al., 1997). However, this reconstitution is dependent on molecular chaperones found in the RRL (Holt et al., 1999). More specifically, two factors that have been shown to be involved in the assembly of active telomerase in vertebrates are the molecular chaperones p23 and hsp90 (Holt et al., 1999). Immunodepletion of hps90 or p23, or inhibition of hsp90 activity using geldanamycin, results in a decrease in telomerase catalytic activity in cell extracts (Forsythe et al., 2001; Holt et al., 1999). This may in part be due to the inability of hTERT and hTR to assemble into an active complex (Forsythe et al., 2001; Holt et al., 1999), and/or a requirement to keep hTERT in a conformation that is competent for interaction with telomeric DNA. Additional studies have demonstrated a decrease in hTERT levels in human cells treated with geldanamycin due to the ubiquitination and subsequent proteosomal degradation of uncomplexed hTERT (Kim et al., 2005).

The survival of motor neuron (SMN) protein is another molecular chaperone that is involved in the biogenesis of cellular RNPs and has been reported to specifically interact with catalytically active telomerase, both *in vitro* and *in vivo* (Bachand et al., 2002). SMN is found both in the cytoplasm and the nucleus, in particular in Cajal bodies, gems, and the nucleolus. It is believed that SMN helps to localize hTERT to the nucleoli and Cajal bodies, where its assembly with hTR into an active complex can occur (Bachand et al., 2002). It has been postulated, based on additional studies, that SMN may interact with GAR1 and/or dyskerin, suggesting that it is involved in telomerase biogenesis in conjunction with the core H/ACA binding motif proteins (Jones et al., 2001; Pellizzoni et al., 2001).

Dual-affinity purification of hTERT complexes from HeLa cell extracts coupled with mass spectrometric analysis identified the AAA + ATPases pontin and reptin as telomerase associated proteins (Venteicher et al., 2008). Pontin and reptin are two closely related proteins that are frequently found in a number of chromatin remodeling complexes and are thought to play a role in transcription, epigenetic regulation, nuclear complex assembly, and the repair of DNA damage (reviewed in Gallant, 2007). Biochemical analysis revealed that pontin interacts directly with both hTERT and dyskerin, while reptin associates with the telomerase complex through its interaction with pontin (Venteicher et al., 2008). The interaction of hTERT with pontin and reptin peaks during S-phase, suggesting that pontin and reptin are involved in the cell-cycle regulation of telomerase assembly. Reduction in either pontin or reptin protein levels in human cells results in a significant decrease in the levels of hTR and dyskerin *in vivo*, coupled with a reduction of telomerase activity. Telomerase activity and the accumulation of dyskerin and hTR are dependent on the ATPase function of these proteins. The current hypothesis is that pontin and reptin act as a scaffold required for the accumulation of an hTR/dyskerin RNP in an ATP-dependent manner (Venteicher et al., 2008). The interaction of pontin and reptin with hTERT may aid in the interaction of the hTR/dyskerin RNP with hTERT, or may serve to induce a conformational change in the hTR/dyskerin/hTERT complex to form an active telomerase enzyme. Once the active RNP is assembled, pontin and reptin are thought to dissociate from the complex resulting in a catalytically competent enzyme (Venteicher et al., 2008; Fig. 4.4).

Other proteins known to regulate hTERT/hTR interactions are the nucleolar acetyltransferase NAT10 and the nucleolar GTPase GNL3L (Fu and Collins, 2007). The functions of these two proteins are still largely uncharacterized. They were shown to interact with the telomerase holoenzyme via an interaction with hTERT by affinity purification of endogenous telomerase from HeLa cells (Fu and Collins, 2007). Results from this purification differ from those reported by Cohen et al., which identified only hTERT and dyskerin as proteins that purified with telomerase, presumably due to the increased stringency and complexity of the purification steps (Cohen et al., 2007). Transient knockdown of either protein by shRNA did not impact telomerase activity or hTR accumulation. Overexpression of either GNL3L or NAT10, resulted in a progressive decrease in telomerase length and increased telomerase activity in cell extracts (Fu and Collins, 2007). Therefore, GNL3L and NAT10 can negatively regulate telomere length without decreasing telomerase enzymatic activity (Fu and Collins, 2007). It is postulated therefore that these proteins regulate telomere length by affecting the assembly and localization of telomerase subunits, possibly by sequestration.

4.5.1.3 Proteins Involved in Recruitment of Telomerase to Telomeres in Mammalian Systems Another class of proteins implicated in telomere length maintenance is the hnRNPs. Members of this family are involved in a variety of RNA-related processes, such as alternative splicing, mRNA maturation/turnover, mRNA transport, and telomere and telomerase regulation (Ford et al., 2002). Telomere-associated hnRNPs include hnRNPA1, A2-B1, D, and E. In fact, hnRNPA1, C1/C2, and D have been shown to be bound to the telomerase enzyme itself (Ford et al., 2002). It has been proposed that hnRNPA1 and C help to recruit telomerase to the telomeres, possibly by binding simultaneously to both telomeric DNA and the TER (Ford et al., 2000, 2002; LaBranche et al., 1998). More recently, it was demonstrated that hnRNPA1 contributes to telomere elongation and stimulates telomerase activity by unwinding G–G hairpins of G-quadruplexes that may form after telomerase translocation (Zhang et al., 2006). In the purification scheme performed by the Collins lab, hnRNPC and hnRNPU were also identified as telomerase-associated proteins (Fu and Collins, 2007). In this study, it was proposed that the nuclear localization of hnRNP C and hnRNP U might function to retain telomerase in the nucleus through an interaction with hTR.

4.5.2 Biogenesis and Assembly of the Tetrahymena Telomerase Holoenzyme

Unlike the human telomerase holoenzyme, ciliates appear to have a unique RNP biogenesis pathway. While vertebrates utilize proteins that are important for a number of different cellular processes, ciliates primarily utilize telomerase-specific proteins. Much of what is known about telomerase biogenesis in *T. thermophila* comes from the affinity purification of endogenously expressed, epitope tagged *Tetrahymena* TERT (Witkin and Collins, 2004). Using this purification scheme, several telomerase holoenzyme subunits were identified. These proteins include p75, p65, p45, and p20.

p65 is a protein that contains a La motif, a domain that is found in a number of RNA-binding proteins, including those that are important for processing of RNA pol III transcripts (Wolin and Cedervall, 2002). Interestingly, an antisense-oligonucleotide affinity purification of telomerase from *Euplotes aediculatus* identified a 43-kDa protein that copurified with telomerase activity (Lingner and Cech, 1996). This protein is also a La homologue. In *Euplotes*, p43 binds to the TER and interacts with telomerase *in vivo*. Although it contains homology to the human La autoantigen, p43 does not bind to RNA pol III precursor transcripts (Witkin and Collins, 2004), suggesting that p43 plays a distinct role in telomerase biogenesis, rather than the traditional role of the La autoantigen in the maturation of RNA pol III transcripts (Aigner et al., 2003). Similar to the p43 protein from *Euplotes*, p65 from *T. thermophila* specifically interacts with the *Tetrahymena* TER (Witkin and Collins, 2004). When p65 is genetically depleted, telomere length decreases and there is reduced accumulation of the TER, suggesting a role for p65 in RNA stability. The knockdown of p65, however, does not appear to affect the stability or accumulation of other snRNA molecules (Witkin and Collins, 2004). This is different from what is observed with dyskerin in human cells, suggesting that the function of p65 is specific to telomerase. Subsequent studies using recombinant factors have shown that the interaction between p65 and TER not only stabilizes the RNA, but also enhances the interaction between TER and TERT by inducing a conformational change in TER that is favorable for TERT binding and telomerase function (O'Connor and Collins, 2006; Prathapam et al., 2005; Stone et al., 2007). It has therefore been suggested that in *T. thermophila*, the early stage of telomerase biogenesis is a stepwise, hierarchical pathway. Whether these findings can be extended to the biogenesis of telomerase from other species remains to be determined.

Both the p45 and the p75 proteins that were identified in the affinity purification scheme from *Tetrahymena* have also been shown to play a role in the biogenesis of the telomerase holoenzyme (Witkin and Collins, 2004; Witkin et al., 2007). Both proteins are unique and lack any sequence homology to other known proteins. Similar to p65, genetic depletion of p45 and p75 results in shortened telomeres, however no change in TER or TERT accumulation was observed *in vivo* (Witkin and Collins, 2004; Witkin et al., 2007). These findings suggest that p45 and p75 likely act downstream of p65–TER–TERT assembly, with p45 and p75 incorporated into the holoenzyme after the formation of a catalytically active complex to allow telomerase to correctly function at chromosome ends (Witkin and Collins, 2004; Witkin et al., 2007). For example, telomerase subunits may be required to induce conformational changes within telomerase for the recruitment and utilization of a native telomeric substrate. Another possibility is that these telomerase subunits may be required for the correct subcellular localization of the assembled holoenzyme. Telomere shortening, coupled with the lack of TER accumulation defect with depletion of p45 and p75 is very similar to what is observed in Est1 and Est3 null strains in *S. cerevisiae* (see below). It remains to be determined if the *Tetrahymena* proteins p45 and p75 are in fact functional analogues of Est1 or Est3, and what role they play in telomere homeostasis.

The p20 protein was recently reported to be the *Tetrahymena* Skp1 orthologue (Witkin et al., 2007). Skp1 is a multifunctional F-box protein that is the essential substrate recognition component of the SCF (Skp1/cullin-1/F-box protein) ubiquitin ligase complex (Cardozo and Pagano, 2004; Willems et al., 2004). In contrast to what was observed with the depletion of p65, p45, and p75, the depletion of p20 resulted in "over-elongated" telomeres, suggesting that it is a negative regulator of telomere length (Witkin et al., 2007). While the affinity purification of p65, p45, and p75 all enriched for active telomerase, the affinity purification of p20 did not. Rather, it primarily purified proteins of the SCF ubiquitin ligase complex (Witkin et al., 2007). It is possible that *Tetrahymena* Skp1 functions in the regulation and assembly of the telomerase complex, much like the yeast Skp1 protein functions in the centromere DNA-binding protein CBF3 assembly and checkpoint control. Alternatively, Skp1 could function as part of the SCF complex to target either TERT or other telomere maintenance/structural proteins for ubiquitin-mediated degradation (Witkin et al., 2007).

4.5.3 Yeast Telomerase Biogenesis and Recruitment to Telomeres

Once the yeast telomerase holoenzyme is assembled, it needs to be recruited to the telomeres. A genetic screen identified *est1* and *est3* as genes that when mutated resulted in shortened telomeres (Lendvay et al., 1996). Yeast strains that lack these proteins exhibit telomere shortening and a decrease in cell viability without any effect on telomerase catalytic activity (Lingner et al., 1997). It has also been demonstrated that both Est1 and Est3 can coimmunoprecipitate telomerase activity, suggesting that they are in fact telomerase holoenzyme proteins (Hughes et al., 2000). Although these proteins are not required for the enzymatic activity of telomerase *per se* (Lingner et al., 1997), they are critical for the *in vivo* function of telomerase. Est1 is an 82 kDa basic protein that can bind directly to *TLC1* (the yeast TER, Seto et al., 2002). In addition, Est1 can also bind to the yeast DNA-end binding protein Cdc13 and single-stranded DNA (Virta-Pearlman et al., 1996). The main role for Est1 is to recruit telomerase to telomeres via protein/protein interactions with Cdc13 (Evans and Lundblad, 2002).

As mentioned previously, the function of telomerase at chromosome ends is regulated in part by the recruitment of the enzyme to the telomere. In yeast, there is evidence for at least two independent mechanisms that serve to recruit telomerase to telomeres. Chromatin immunoprecipitation experiments revealed that Est2 (the catalytic subunit and TERT homologue) is constitutively associated with chromatin, although telomere elongation occurs during S-phase only (Taggart et al., 2002). The association of Est2 with telomeres increases in the G1 phase of the cell cycle, decreases during early S phase, and then increase again in late S-phase (Taggart et al., 2002). The association of telomerase with telomeric chromatin in G1 is dependent on the interaction of *TLC1* RNA with the Ku70/80 heterodimer (Fisher et al., 2004). If the Ku/*TLC1* interaction is disrupted, telomere length decreases as does chromosome healing at double-strand breaks (Stellwagen et al., 2003).

Conversely, the increase in Est2 at chromatin in late S-phase is dependent on Cdc13. There are a number of steps that are required at this point for telomerase to correctly function at the telomere. It was demonstrated that Est1 associates with telomeric DNA only in late S-phase, when the levels of this protein increase (Taggart et al., 2002). Est1 may then interact with Est2/*TLC1* that is associated with telomeric DNA (Taggart et al., 2002). This complex could then interact with Cdc13 to allow for recruitment of active telomerase to telomere ends. Alternatively, it has been argued that the Cdc13–Est1 interaction changes the state of the bound Est2, possibly by inducing a conformational change of the complex that activates the enzyme (Taggart et al., 2002).

Yeast Est3 is a unique 19 kDa protein that interacts with Est2 (Hughes et al., 2000). Although the exact role that Est3 plays in telomerase function is unclear, its association with the yeast telomerase complex is dependent on Est1 (Friedman et al., 2003; Osterhage et al., 2006). Est3 might contribute to the activation of telomerase whereas the role of Est1 is to recruit Est3 (Hsu et al., 2007). Further studies will be required to fully elucidate the molecular mechanisms of Est3 in telomere length maintenance.

4.6 CONCLUDING REMARKS

While hTERT transcription and expression is limiting for the catalytic activity of telomerase in cells, there are many additional regulatory processes that are critical for the assembly of an active telomerase complex and for the correct cellular trafficking of this complex. The biogenesis of telomerase enzymes and the assembly of individual subunits into an active complex involves a number of important regulatory steps and a number of factors that associate either directly or indirectly with the telomerase core components, with a variety of these proteins having important roles in RNP biogenesis, enzyme assembly, localization, catalytic activity, and telomere recruitment. The identities and the roles of many of these proteins are becoming clearer. Although much is known about the protein subunits that contribute to telomerase biogenesis, there is still much about this process that remains to be determined.

REFERENCES

Aigner S, Postberg J, Lipps HJ, Cech TR. (2003) The *Euplotes* La motif protein p43 has properties of a telomerase-specific subunit. *Biochemistry*. **42**(19): 5736–5747.

Bachand F, Boisvert FM, Cote J, Richard S, Autexier C. (2002) The product of the survival of motor neuron (SMN) gene is a human telomerase-associated protein. *Mol. Biol. Cell.* **13**(9): 3192–3202.

Banik SSR, Counter CM. (2004) Characterization of interactions between PinX1 and human telomerase subunits hTERT and hTR. *J. Biolo. Chem.* **279**(50): 51745–51748.

Beattie TL, Zhou W, Robinson MO, Harrington L. (1998) Reconstitution of human telomerase activity *in vitro*. *Curr. Biol.* **8**: 177–180.

Beck T, Hall MN. (1999) The TOR signalling pathway controls nuclear localization of nutrient-regulated transcription factors. *Nature*. **402**(6762): 689–692.

Bogenhagen DF, Brown DD. (1981) Nucleotide sequences in *Xenopus* 5S DNA required for transcription termination. *Cell*. **24**(1): 261–270.

Box JA, Bunch JT, Tang W, Baumann P. (2008) Spliceosomal cleavage generates the 3′ end of telomerase RNA. *Nature*. **456**(7224): 910–914.

Cardozo T, Pagano M. (2004) The SCF ubiquitin ligase: insights into a molecular machine. *Nat. Rev. Mol. Cell. Biol.* **5**(9): 739–751.

Chapon C, Cech TR, Zaug AJ. (1997) Polyadenylation of telomerase RNA in budding yeast. *RNA*. **3**(11): 1337–1351.

Chen J-L, Blasco MA, Greider CW. (2000) Secondary structure of vertebrate telomerase RNA. *Cell*. **100**(5): 503–514.

Cohen SB, Graham ME, Lovrecz GO, Bache N, Robinson PJ, Reddel RR. (2007) Protein composition of catalytically active human telomerase from immortal cells. *Science*. **315** (5820): 1850–1853.

Collins K. (2006) The biogenesis and regulation of telomerase holoenzymes. *Nat. Rev. Mol. Cell. Biol.* **7**(7): 484–494.

Cristofari G, Adolf E, Reichenbach P, Sikora K, Terns RM, Terns MP, Lingner J. (2007) Human telomerase RNA accumulation in Cajal bodies facilitates telomerase recruitment to telomeres and telomere elongation. *Molec. Cell.* **27**(6): 882–889.

Darzacq X, Kittur N, Roy S, Shav-Tal Y, Singer RH, Meier UT. (2006) Stepwise RNP assembly at the site of H/ACA RNA transcription in human cells. *J. Cell Biol.* **173**(2): 207–218.

Etheridge KT, Banik SSR, Armbruster BN, Zhu Y, Terns RM, Terns MP, Counter CM. (2002) The nucleolar localization domain of the catalytic subunit of human telomerase. *J. Biol. Chem.* **277**(27): 24764–24770.

Evans SK, Lundblad V. (2002) The Est1 subunit of *Saccharomyces cerevisiae* telomerase makes multiple contributions to telomere length maintenance. *Genetics*. **162**(3): 1101–1115.

Feng J, Funk WD, Wang SS, Weinrich SL, Avilion AA, Chiu CP, Adams RR, Chang E, Allsopp RC, Yu J,al e. (1995) The RNA component of human telomerase. *Science*. **269**(5228): 1236–1241.

Ferrezuelo F, Steiner B, Aldea M, Futcher B. (2002) Biogenesis of yeast telomerase depends on the importin Mtr10. *Mol. Cell. Biol.* **22**(17): 6046–6055.

Fisher TS, Taggart AK, Zakian VA. (2004) Cell cycle-dependent regulation of yeast telomerase by Ku. *Nat. Struct. Mol. Biol.* **11**(12): 1198–1205.

Ford LP, Suh JM, Wright WE, Shay JW. (2000) Heterogeneous nuclear ribonucleoproteins C1 and C2 associate with the RNA component of human telomerase. *Mol. Cell Biol.* **20**(23): 9084–9091.

Ford LP, Wright WE, Shay JW. (2002) A model for heterogeneous nuclear ribonucleoproteins in telomere and telomerase regulation. *Oncogene*. **21**(4): 580–583.

Forsythe HL, Jarvis JL, Turner JW, Elmore LW, Holt SE. (2001) Stable association of hsp90 and p23, but Not hsp70, with active human telomerase. *J. Biol. Chem.* **276**(19): 15571–15574.

Franke J, Gehlen J, Ehrenhofer-Murray AE. (2008) Hypermethylation of yeast telomerase RNA by the snRNA and snoRNA methyltransferase Tgs1. *J. Cell. Sci.* **121**(21): 3553–3560.

Friedman KL, Heit JJ, Long DM, Cech TR. (2003) N-terminal domain of yeast telomerase reverse transcriptase: recruitment of Est3p to the telomerase complex. *Mol. Biol. Cell.* **14**(1): 1–13.

Fu D, Collins K. (2003) Distinct biogenesis pathways for human telomerase RNA and H/ACA small nucleolar RNAs. *Mol. Cell.* **11**(5): 1361–1372.

Fu D, Collins K. (2006) Human telomerase and Cajal body ribonucleoproteins share a unique specificity of Sm protein association. *Genes Dev*. **20**(5): 531–536.

Fu D, Collins K. (2007) Purification of human telomerase complexes identifies factors involved in telomerase biogenesis and telomere length regulation. *Mol. Cell.* **28**(5): 773–785.

Gallant P. (2007) Control of transcription by pontin and reptin. *Trends Cell. Biol.* **17**(4): 187–192.

Gallardo F, Chartrand P. (2008) Telomerase biogenesis: the long road before getting to the end. *RNA Biol.* **5:** 212–215.

Gallardo F, Olivier C, Dandjinou AT, Wellinger RJ, Chartrand P. (2008) TLC1 RNA nucleo-cytoplasmic trafficking links telomerase biogenesis to its recruitment to telomeres. *EMBO J.* **27:** 748–757.

Ginisty H, Sicard H, Roger B, Bouvet P. (1999) Structure and functions of nucleolin. *J. Cell Sci.* **112**(Pt 6): 761–772.

Girard C, Verheggen C, Neel H, Cammas A, Vagner S, Soret J, Bertrand E, Bordonne R. (2008) Characterization of a short isoform of human Tgs1 hypermethylase associating with small nucleolar ribonucleoprotein core proteins and produced by limited proteolytic processing. *J. Biol. Chem.* **283**(4): 2060–2069.

Greider CW, Blackburn EH. (1989) A telomeric sequence in the RNA of *Tetrahymena* telomerase required for telomere repeat synthesis. *Nature.* **337**(6205): 331–337.

Gunisova S, Elboher E, Nosek J, Gorkovoy V, Brown Y, Lucier J-F, Laterreur N, Wellinger RJ, Tzfati Y, Tomaska L. (2009) Identification and comparative analysis of telomerase RNAs from *Candida* species reveal conservation of functional elements. *RNA.* **15:** 546–559.

Hoareau-Aveilla C, Bonoli M, Caizergues-Ferrer M, Henry Y. (2006) hNaf1 is required for accumulation of human box H/ACA snoRNPs, scaRNPs, and telomerase. *RNA.* **12**(5): 832–840.

Holt SE, Aisner DL, Baur J, Tesmer VM, Dy M, Ouellette M, Trager JB, Morin GB, Toft DO, Shay JW, Wright WE, White MA. (1999) Functional requirement of p23 and Hsp90 in telomerase complexes. *Genes Dev.* **13**(7): 817–826.

Holt SE, Aisner DL, Shay JW, Wright WE. (1997) Lack of cell cycle regulation of telomerase activity in human cells. *Proc. Natl. Acad. Sci. USA* **94**(20): 10687–10692.

Hopper AK. (2006) Cellular dynamics of small RNAs. *Crit. Rev. Biochem. Mol. Biol.* **41**(1): 3–19.

Hsu M, Yu EY, Singh SM, Lue NF. (2007) Mutual dependence of *Candida albicans* Est1p and Est3p in telomerase assembly and activation. *Eukaryot. Cell.* **6**(8): 1330–1338.

Hughes TR, Evans SK, Weilbaecher RG, Lundblad V. (2000) The Est3 protein is a subunit of yeast telomerase. *Curr. Biol.* **10**(13): 809–812.

Jady BE, Bertrand E, Kiss T. (2004) Human telomerase RNA and box H/ACA scaRNAs share a common Cajal body-specific localization signal. *J. Cell. Biol.* **164**(5): 647–652.

Jady BE, Richard P, Bertrand E, Kiss T. (2006) Cell cycle-dependent recruitment of telomerase RNA and Cajal bodies to human telomeres. *Mol. Biol. Cell.* **17**(2): 944–954.

Jones KW, Gorzynski K, Hales CM, Fischer U, Badbanchi F, Terns RM, Terns MP. (2001) Direct interaction of the spinal muscular atrophy disease protein SMN with the small nucleolar RNA-associated protein fibrillarin. *J. Biol. Chem.* **276**(42): 38645–38651.

Khurts S, Masutomi K, Delgermaa L, Arai K, Oishi N, Mizuno H, Hayashi N, Hahn WC, Murakami S. (2004) Nucleolin interacts with telomerase. *J. Biol. Chem.* **279**(49): 51508–51515.

Kim JH, Park SM, Kang MR, Oh SY, Lee TH, Muller MT, Chung IK. (2005) Ubiquitin ligase MKRN1 modulates telomere length homeostasis through a proteolysis of hTERT. *Genes Dev.* **19**(7): 776–781.

Kiss T, Fayet-Lebaron E, Jady B. (2010) Box H/ACA small ribonucleoproteins. *Mol. Cell.* **37**(5): 597–606.

Kovalenko OA, Caron M, J., Ulema P, Medrano C, Thomas A, P., Kimura M, Bonini M, G., Herbig U, Santos Janine H. (2010) A mutant telomerase defective in nuclear–cytoplasmic shuttling fails to immortalize cells and is associated with mitochondrial dysfunction. *Aging Cell.* **9**(2): 203–219.

LaBranche H, Dupuis S, Ben-David Y, Bani MR, Wellinger RJ, Chabot B. (1998) Telomere elongation by hnRNP A1 and a derivative that interacts with telomeric repeats and telomerase. *Nat. Genet.* **19**(2): 199–202.

Lendvay TS, Morris DK, Sah J, Balasubramanian B, Lundblad V. (1996) Senescence mutants of *Saccharomyces cerevisiae* with a defect in telomere replication identify three additional EST genes. *Genetics.* **144**(4): 1399–1412.

Leonardi J, Box JA, Bunch JT, Baumann P. (2008) TER1, the RNA subunit of fission yeast telomerase. *Nat. Struct. Mol. Biol.* **15**(1): 26–33.

Lin J, Blackburn EH. (2004) Nucleolar protein PinX1p regulates telomerase by sequestering its protein catalytic subunit in an inactive complex lacking telomerase RNA. *Genes Dev.* **18**(4): 387–396.

Lin J, Jin R, Zhang B, Chen H, Bai YX, Yang PX, Han SW, Xie YH, Huang PT, Huang C, Huang JJ. (2008) Nucleolar localization of TERT is unrelated to telomerase function in human cells. *J. Cell. Sci.* **121**(13): 2169–2176.

Lingner J, Cech TR. (1996) Purification of telomerase from *Euplotes aediculatus*: requirement of a primer 3′ overhang. *Proc. Natl. Acad. Sci. USA.* **93**(20): 10712–10717.

Lingner J, Cech TR, Hughes TR, Lundblad V. (1997) Three ever shorter telomere (EST) genes are dispensable for in vitro yeast telomerase activity. *Proc. Natl. Acad. Sci. USA* 94 (21): 11190–11195.

Liu K, Schoonmaker MM, Levine BL, June CH, Hodes RJ, Weng N-p. (1999) Constitutive and regulated expression of telomerase reverse transcriptase (hTERT) in human lymphocytes. *Proc. Natl. Acad. Sci. USA* 96 (9): 5147–5152.

Mans B, Anantharaman V, Aravind L, Koonin EV. (2004) Comparative genomics, evolution and origins of the nuclear envelope and nuclear pore complex. *Cell Cycle.* **3**: 1612–1637.

Martin-Rivera L, Blasco MA. (2001) Identification of functional domains and dominant negative mutations in vertebrate telomerase RNA using an *in vivo* reconstitution system. *J. Biol. Chem.* **276:** 5856–5665.

Matera AG, Shpargel KB. (2006) Pumping RNA: nuclear bodybuilding along the RNP pipeline. *Curr. Opin. Cell Biol.* **18**(3): 317–324.

Matera AG, Terns RM, Terns MP. (2007) Non-coding RNAs: lessons from the small nuclear and small nucleolar RNAs. *Nat. Rev. Mol. Cell. Biol.* **8**(3): 209–220.

McEarchern MJ, Blackburn EH. (1995) Runaway telomere elongation caused by telomerase RNA gene mutations. *Nature.* **376:** 403–409.

Meier UT. (2005) The many facets of H/ACA ribonucleoproteins. *Chromosoma.* **114**(1): 1–14.

Mitchell JR, Cheng J, Collins K. (1999) A Box H/ACA small nucleolar RNA-like domain at the human telomerase RNA 3′ end. *Mol. Cell. Biol.* **19**(1): 567–576.

Mouaikel J, Verheggen C, Bertrand E, Tazi J, Bordonne R. (2002) Hypermethylation of the cap structure of both yeast snRNAs and snoRNAs requires a conserved methyltransferase that is localized to the nucleolus. *Mol. Cell.* **9**(4): 891–901.

Mozdy AD, Cech TR. (2006) Low abundance of telomerase in yeast: Implications for telomerase haploinsufficiency. *RNA*. **12**(9): 1721–1737.

O'Connor CM, Collins K. (2006) A novel RNA binding domain in *Tetrahymena* telomerase p65 initiates hierarchical assembly of telomerase holoenzyme. *Mol. Cell. Biol.* **26**(6): 2029–2036.

Olson BL, Siliciano PG. (2003) A diverse set of nuclear RNAs transfer between nuclei of yeast heterokaryons. *Yeast*. **20**(10): 893–903.

Osterhage JL, Talley JM, Friedman KL. (2006) Proteasome-dependent degradation of Est1p regulates the cell cycle-restricted assembly of telomerase in *Saccharomyces cerevisiae*. *Nat. Struct. Mol. Biol.* **13**(8): 720–728.

Pellizzoni L, Baccon J, Charroux B, Dreyfuss G. (2001) The survival of motor neurons (SMN) protein interacts with the snoRNP proteins fibrillarin and GAR1. *Curr. Biol.* **11**(14): 1079–1088.

Platani M, Golberg I, Swedlow JR, Lamond AI. (2000) In vivo analysis of Cajal body movement, separation, and joining in live human cells. *J. Cell. Biol.* **151:** 1561–1574.

Pogacic V, Dragon F, Filipowicz W. (2000) Human H/ACA small nucleolar RNPs and telomerase share evolutionarily conserved proteins NHP2 and NOP10. *Mol. Cell. Biol.* **20**(23): 9028–9040.

Prathapam R, Witkin KL, O'Connor CM, Collins K. (2005) A telomerase holoenzyme protein enhances telomerase RNA assembly with telomerase reverse transcriptase. *Nat. Struct. Mol. Biol.* **12**(3): 252–257.

Richard P, Kiss AM, Darzacq X, Kiss T. (2006) Cotranscriptional recognition of human intronic box H/ACA snoRNAs occurs in a splicing-independent manner. *Mol. Cell. Biol.* **26**(7): 2540–2549.

Schnapp G, Rodi HP, Rettig WJ, Schnapp A, Damm K. (1998) One-step affinity purification protocol for human telomerase. *Nucl. Acids Res.* **26**(13): 3311–3313.

Seimiya H, Sawada H, Muramatsu Y, Shimizu M, Ohko K, Yamane K, Tsuruo T. (2000) Involvement of 14-3-3 proteins in nuclear localization of telomerase. *EMBO J.* **19**(11): 2652–2661.

Seto AG, Livengood AJ, Tzfati Y, Blackburn EH, Cech TR. (2002) A bulged stem tethers Est1p to telomerase RNA in budding yeast. *Genes Dev.* **16**(21): 2800–2812.

Seto AG, Zaug AJ, Sobel SG, Wolin SL, Cech TR. (1999) *Saccharomyces cerevisiae* telomerase is an Sm small nuclear ribonucleoprotein particle. *Nature*. **401**(6749): 177–180.

Singer MS, Gottschling DE. (1994) TLC1: template RNA component of *Saccharomyces cerevisiae* telomerase. *Science*. **266**(5184): 404–409.

Stellwagen AE, Haimberger ZW, Veatch JR, Gottschling DE. (2003) Ku interacts with telomerase RNA to promote telomere addition at native and broken chromosome ends. *Genes Dev.* **17**(19): 2384–2395.

Stone MD, Mihalusova M, O'Connor CM, Prathapam R, Collins K, Zhuang X. (2007) Stepwise protein-mediated RNA folding directs assembly of telomerase ribonucleoprotein. *Nature*. **446**(7134): 458–461.

Taggart AKP, Teng S-C, Zakian VA. (2002) Est1p as a cell cycle-regulated activator of telomere-bound telomerase. *Science*. **297**(5583): 1023–1026.

Teixeira MT, Arneric M, Sperisen P, Lingner J. (2004) Telomere length homeostasis is achieved via a switch between telomerase-extendible and -nonextendible states. *Cell*. **117**(3): 323–335.

Teixeira TM, Förstemann K, Gasser SM, Lingner J. (2002) Intracellular trafficking of yeast telomerase components. *EMBO Rep.* **3**(7): 652–659.

Theimer CA, Jady BE, Chim N, Richard P, Breece KE, Kiss T, Feigon J. (2007) Structural and functional characterization of human telomerase RNA processing and Cajal body localization signals. *Mol. Cell.* **27**(6): 869–881.

Tomlinson RL, Abreu EB, Ziegler T, Ly H, Counter CM, Terns RM, Terns MP. (2008) Telomerase reverse transcriptase is required for the localization of telomerase RNA to Cajal bodies and telomeres in human cancer cells. *Mol. Biol. Cell.* **19**(9): 3793–3800.

Tomlinson RL, Ziegler TD, Supakorndej T, Terns RM, Terns MP. (2006) Cell cycle-regulated trafficking of human telomerase to telomeres. *Mol. Biol. Cell.* **17**(2): 955–965.

Trahan C, Dragon F. (2009) Dyskeratosis congenita mutations in the H/ACA domain of human telomerase RNA affect its assembly into a pre-RNP. *RNA*. **15**(2): 235–243.

Trahan C, Martel C, Dragon F. (2010) Effects of dyskeratosis congenita mutations in dyskerin, NHP2 and NOP10 on assembly of H/ACA pre-RNPs. *Hum. Mol. Genet.* **19**(5): 825–836.

Tycowski KT, Shu M-D, Kukoyi A, Steitz JA. (2009) A conserved WD40 protein binds the Cajal body localization signal of scaRNP particles. *Mol. Cell.* **34**(1): 47–57.

Venteicher AS, Abreu EB, Meng Z, McCann KE, Terns RM, Veenstra TD, Terns MP, Artandi SE. (2009) A human telomerase holoenzyme protein required for Cajal body localization and telomere synthesis. *Science*. **323**(5914): 644–648.

Venteicher AS, Meng Z, Mason PJ, Veenstra TD, Artandi SE. (2008) Identification of ATPases pontin and reptin as telomerase components essential for holoenzyme assembly. **132**(6): 945–957.

Virta-Pearlman V, Morris DK, Lundblad V. (1996) Est1 has the properties of a single-stranded telomere end-binding protein. *Genes Dev.* **10**(24): 3094–3104.

Vulliamy T, Beswick R, Kirwan M, Marrone A, Digweed M, Walne A, Dokal I. (2008) Mutations in the telomerase component NHP2 cause the premature ageing syndrome dyskeratosis congenita. *Proc. Natl. Acad. Sci. USA*. **105**(23): 8073–8078.

Walne AJ, Vulliamy T, Marrone A, Beswick R, Kirwan M, Masunari Y, Al-Qurashi FH, Aljurf M, Dokal I. (2007) Genetic heterogeneity in autosomal recessive dyskeratosis congenita with one subtype due to mutations in the telomerase-associated protein NOP10. *Hum. Mol. Genet.* **16**(13): 1619–1629.

Webb CJ, Zakian VA. (2008) Identification and characterization of the *Schizosaccharomyces pombe* TER1 telomerase RNA. *Nat. Struct. Mol. Biol.* **15**(1): 34–42.

Weinrich SL, Pruzan R, Ma L, Ouellette M, Tesmer VM, Holt SE, Bodnar AG, Lichtsteiner S, Kim NW, Trager JB, Taylor RD, Carlos R, Andrews WH, Wright WE, Shay JW, Harley CB, Morin GB. (1997) Reconstitution of human telomerase with the template RNA component hTR and the catalytic protein subunit hTRT. *Nat. Genet.* **17**(4): 498–502.

Willems AR, Schwab M, Tyers M. (2004) A hitchhiker's guide to the cullin ubiquitin ligases: SCF and its kin. *Biochim. Biophys. Acta.* **1695**(1–3): 133–170.

Witkin KL, Collins K. (2004) Holoenzyme proteins required for the physiological assembly and activity of telomerase. *Genes Dev.* **18**(10): 1107–1118.

Witkin KL, Prathapam R, Collins K. (2007) Positive and negative regulation of *Tetrahymena* telomerase holoenzyme. *Mol. Cell. Biol.* **27**(6): 2074–2083.

Wolin SL, Cedervall T. (2002) The La protein. *Annu. Rev. Biochem.* **71**: 375–403.

Wong JMY, Kusdra L, Collins K. (2002) Subnuclear shuttling of human telomerase induced by transformation and DNA damage. *Nat. Cell. Biol.* **4**(9): 731–736.

Yang Y, Chen Y, Zhang C, Huang H, Weissman SM. (2002) Nucleolar localization of hTERT protein is associated with telomerase function. *Exp. Cell Res.* **277**(2): 201–209.

Yong J, Wan L, Dreyfuss G. (2004) Why do cells need an assembly machine for RNA–protein complexes? *Trends Cell Biol.* **14**(5): 226–232.

Yu YT, Terns RM, Terns MP. (2005) Mechanisms and functions of RNA-guided RNA modification, New York: Springer-Verlag

Zappulla DC, Cech TR. (2004) Yeast telomerase RNA: a flexible scaffold for protein subunits. *Proc. Natl. Acad. Sci. USA* 101 (27): 10024–10029.

Zhang B, Bai YX, Ma HH, Feng F, Jin R, Wang ZL, Lin J, Sun SP, Yang P, Wang XX, Huang PT, Huang CF, Peng Y, Chen YC, Kung H-f, Huang JJ. (2009) Silencing PinX1 compromises telomere length maintenance as well as tumorigenicity in telomerase-positive human cancer cells. *Cancer Res.* **69**(1): 75–83.

Zhang QS, Manche L, Xu RM, Krainer AR. (2006) hnRNP A1 associates with telomere ends and stimulates telomerase activity. *RNA.* **12**(6): 1116–1128.

Zhou XZ, Lu KP. (2001) The Pin2/TRF1-interacting protein PinX1 is a potent telomerase inhibitor. *Cell.* **107**(3): 347–359.

Zhu Y, Tomlinson RL, Lukowiak AA, Terns RM, Terns MP. (2004) Telomerase RNA accumulates in Cajal bodies in human cancer cells. *Mol. Biol. Cell.* **15**(1): 81–90.

5

TRANSCRIPTIONAL REGULATION OF HUMAN TELOMERASE

ANTONELLA FARSETTI AND YU-SHENG CONG

Telomerase activity has been assayed in a wide variety of normal human tissues and in a whole spectrum of human tumors (Shay and Bacchetti, 1997). Enzymatic activity is repressed during embryonic differentiation and is indeed absent in most types of normal human somatic cells with limited proliferative capacity. The enzyme however remains active in some tissues, such as male germ cells, activated lymphocytes, and certain types of stem-cell populations, and is activated in over 90% of cancerous cells and *in vitro*-immortalized cells (Cong et al., 2002; Kim et al., 1994). Telomerase activation and/or telomere maintenance may be one of the six key events common to cancer (Hanahan, 2000; Hanahan and Weinberg, 2000). The differential activity of the hTERT promoter in normal and cancer cells has attracted considerable attention to its potential usefulness in therapeutic applications (Gu et al., 2000; Kyo et al., 2008).

The regulation of telomerase activity occurs at various levels, including transcription of hTR and hTERT, post-translational modification of hTERT and assembly of active telomerase ribonucleoprotein complexes. In most cases, hTERT expression is closely correlated with telomerase activity being transcriptionally repressed in many normal cells and activated or upregulated during cellular immortalization and tumorigenesis. The transcriptional regulation of the hTERT gene represents the primary and rate-limiting step in the activation of the enzyme in most cells.

In this chapter, we will summarize our current understanding of the transcriptional regulation of telomerase, focusing primarily on the regulation of hTERT but including also that of hTR. In addition we will illustrate how hTERT may be situated at the cross point of several regulatory pathways mediated by estrogens, nitric oxide (NO), and

Telomerases: Chemistry, Biology, and Clinical Applications, First Edition.
Edited by Neal F. Lue and Chantal Autexier.

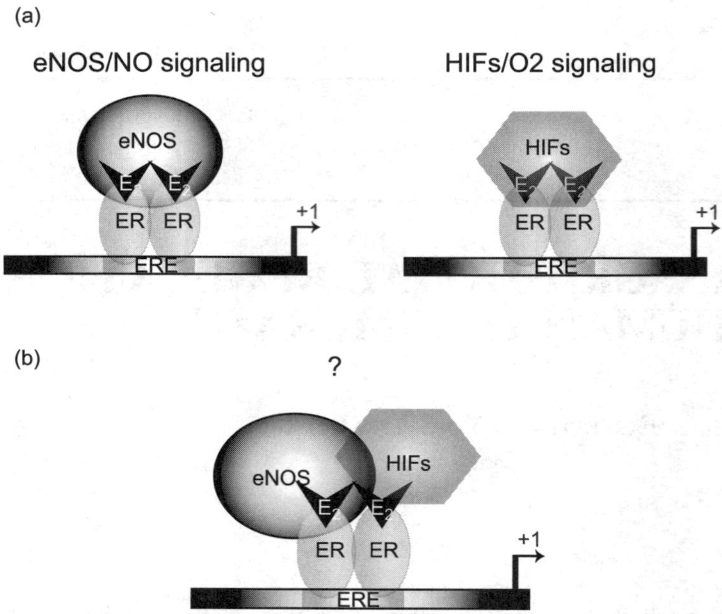

FIGURE 5.1 Cartoons illustrating the functional cooperation between the ERs, eNOS and HIFs pathways in the regulation of hTERT. (a) In primary cultures of human endothelium and in prostate cancer cell lines, ligand-activated ER and eNOS form a combinatorial complex on the estrogen response element (ERE) within the hTERT gene promoter (left panel). In prostate cancer cell lines with a constitutive hypoxic phenotype, ER/HIF-1α or ER/HIF-2α complexes are recruited upon estrogen treatment onto the hTERT–ERE (right panel). All these events lead to increased hTERT gene transcription and telomerase activity. (b) Speculative model of formation of a ER/eNOS/HIF trimeric complex. Since eNOS, ERs, and HIFs play a key role in prostate cancer progression, it is conceivable that they may cooperate in the tumor microenviroment by coregulating their transcriptional targets. We propose that in the presence of estrogen and of reduced O2 availability (hypoxia), these factors may form a trimeric complex recruited by the ERE. This event may induce a local chromatin remodeling significantly affecting the transcription of target genes. (See the color version of this figure in Color Plates section.)

hypoxia-induced signaling, all essentially involved in the cell adaptation to the microenvironment (see Section 5.5 and Fig. 5.1).

5.1 THE hTERT GENE AND THE hTERT PROMOTER

5.1.1 Localization and Organization of hTERT Gene

In human diploid cells, the hTERT gene is present as a single copy on chromosome band 5p15.33, about 2 Mb distance from the telomere (Bryce et al., 2000). The subtelomeric localization of the gene suggests that telomere positional effect may contribute to the repression of hTERT expression in normal human cells

(Baur et al., 2001). Although there is no available evidence in support of this hypothesis, it would be interesting to investigate whether critically short telomeres in normal cells at the end of their replicative capacity may result in a telomeric heterochromatin environment that is more competent for transcriptional activation of the hTERT gene during cellular immortalization and tumorigenesis.

The hTERT gene consists of 16 exons and 15 introns and extends over 40 kb; a similar gene organization has been reported for the TERT gene of other mammals (Cong et al., 1999). The hTERT gene is subject to post-transcriptional regulation and over 10 differentially spliced transcripts have been detected in human cells (Kilian et al., 1997; Sykorova and Fajkus, 2009). Abundant transcripts include the full-length, the α spliced transcript with a 36 nucleotide deletion from the 5′ end of exon 6, and the β spliced transcript with deletion of exons 7 and 8. The product of the α transcript is enzymatically inactive and appears to be a dominant inhibitor of telomerase when overexpressed (Colgin et al., 2000; Yi et al., 2000). Several hTERT transcripts are expressed during human development in a tissue- and gestational age-dependent manner, but only the full-length transcript is associated with telomerase activity (Ulaner et al., 1998, 2000). The specific expression pattern of splice variants during development indicates that splicing events are not random and could have physiological functions in cells. Recent studies indicate that the hTERT protein has roles in several essential cellular processes independent of telomere maintenance (Cong and Shay, 2008). Whether the differentially spliced hTERT transcripts are implicated in nontelomeric functions of telomerase merits further investigation.

Additional copies of the hTERT gene have been detected in human tumors and tumor-derived cell lines (Bryce et al., 2000; Zhang et al., 2000) and are due primarily to increased numbers of chromosomes carrying a single copy of hTERT. This suggests that amplification of the gene at a single locus may not be a major mechanism accounting for increased hTERT expression. However, chromosome rearrangements associated with telomere dysfunction or other forms of DNA damage may release the hTERT gene from a repressive chromatin environment.

5.1.2 Features of the hTERT Promoter

Transcriptional regulation of the hTERT gene is the primary mechanism of telomerase regulation in human cells. We and others have cloned and examined the regulatory activity of the 5′-upstream sequences of the hTERT gene (Cong et al., 1999; Horikawa et al., 1999; Takakura et al., 1999). In transient transfections of a luciferase reporter under control of a 4.0 kb hTERT promoter fragment, we showed that this promoter is inactive in normal and preimmortal transformed cells but, similar to telomerase activity, is activated in immortalized and tumor-derived cells. This indicates that, in this cellular context, a potential telomere position effect on hTERT expression triggered by short telomeres is prevented or counteracted by other factors.

Deletion analysis of the promoter suggests that the minimum sequence requirements for transcription are contained within the 330 bp upstream of the hTERT ATG. However, sequences upstream or downstream of this minimal promoter may be necessary for the transcriptional regulation of the hTERT gene (Cong et al., 1999).

Sequence analysis revealed that the hTERT promoter has no TATA or CAAT boxes but is highly GC-rich, suggesting epigenetic regulation of the gene (see Section 5.3). The transcription initiation site maps 60–120 bp upstream of the translational start site (Horikawa et al., 1999; Takakura et al., 1999). The 4 kb hTERT promoter contains binding sites for many transcription factors (TRs) including two E-boxes for the Myc/Mad network, five GC boxes for Sp1 located within the minimal promoter, and binding sites for AP1 (activator protein 1), AP2 (activator protein 2), WT1 (Wilms tumor protein), ATF (activating transcription factor), CREB (cAMP-responsive element binding protein), PR (progesterone receptor), ER (estrogen receptor), and USF (upstream stimulating factor) (Cong et al., 1999). The abundance of these sites suggests a complex mechanism of transcriptional modulation of hTERT expression that may involve specific factors in different cell types and/or microenvironments.

5.2 TRANSCRIPTIONAL REGULATION OF hTERT

All of the factors affecting hTERT transcription are summarized in Table 5.1.

5.2.1 Regulation of hTERT by Transcription Factors

The transcriptional regulation of the hTERT gene is the major mechanism that negatively and positively controls telomerase activity in human normal and cancer cells, respectively.

5.2.1.1 Nonhormonal Transcription Factors Among a multitude of transcription factors (TFs) critical for hTERT basal expression emerges the proto-oncogene c-Myc, whose pleiotropic effects on tumorigenesis are likely to be mediated by its target genes. A known c-Myc transcriptional target is indeed hTERT. Wang et al. (1998) first reported that c-Myc induces *de novo* hTERT mRNA and telomerase activity in normal human mammary epithelial cell and primary fibroblasts. Sequence analysis indicates that the hTERT promoter comprises two E-boxes upstream of the ATG (see Section 5.1.2) and using electrophoretic mobility shift assays, promoter reporter gene assays and chromatin immunoprecipitations (ChIPs) have confirmed the functionality of these binding sites (Cong et al., 1999; Greenberg et al., 1999; Horikawa et al., 1999; Takakura et al., 1999; Wick et al., 1999; Wu et al., 1999). Despite a large number of studies, however, the mechanisms by which c-Myc induces hTERT expression in tumorigenesis remain unclear. Deregulation of the complex between c-Myc and its antagonist Mad can only partially account for the activation of hTERT in human cancers, and a fine balance between c-Myc and other molecular partners has been proposed to play a major role in cancer-cell metabolism and proliferation. Several studies (Gordan et al., 2007a,b; Koshiji et al., 2004; Zhang et al., 2007) for instance, have described the existence of a crosstalk between c-Myc and the hypoxia inducible factors (HIFs). In this regard, the interplay between c-Myc and HIF may contribute to the fine tuning of the response of tumor cells growing in low oxygen conditions (Gordan et al., 2007a,b).

TABLE 5.1 Factors Involved in the Regulation of hTERT Transcription

Factor	Type	Effect	References
AR	HR	Repression	Meeker et al. (1996);Culig et al. (2002);Teske et al. (2002); Moehren et al. (2008)
bFGF (FGF2)	GF	Upregulation	Kurz et al. (2003)
c-Myc	TF	Turn on/basal transcription	Wang et al. (1998); Greenberg et al. (1999); Wu et al. (1999)
EBV LMP-1	VP	Upregulation	Counter et al. (1994); Yang et al. (2004); Mei et al. (2006); Terrin et al. (2008)
EGF	GF	Upregulation	Maida et al. (2002)
eNOS	E	Upregulation	Narducci et al. (2007); Misiti et al. (2000); Nanni et al. (2002); Grasselli et al. (2008); Farsetti et al. (2009); Nanni et al. (2009)
ER	HR	Upregulation	Kyo et al. (1999); Misiti et al. (2000); Nanni et al. (2002); Grasselli et al. (2008); Farsetti et al. (2009); Nanni et al. (2009)
Ets	TF	Upregulation/repression	Maida et al. (2002); Xiao et al. (2003); Dwyer et al. (2007, 2010); Xu et al. (2008)
EWS/ETS	VP	Upregulation	Takahashi et al. (2003)
HBV	V	Upregulation	Horikawa and Barrett (2001); Ferber et al. (2003), Paterlini-Brechot et al. (2003)
HBV HBx, preS2	VP	Upregulation	Qu et al. (2005); Liu et al. (2007a,b); Luan et al. (2009)
HER2/Neu	Onc	Upregulation	Goueli and Janknecht (2004)
HIFs	TF	Upregulation	Nishi et al. (2004); Yatabe et al. (2004)
HPVE6	VP	Upregulation	Klingelhutz et al. (1996); Gewin et al. (2004); Liu et al. (2005, 2009)
IGF1	GF	Upregulation	Tu et al. (1999); Wetterau et al. (2003)
IL2	CK	Upregulation	Kang et al. (1999); Akiyama et al. (2002)
INF α and γ	CK	Repression	Lindkvist et al. (2006)
Mad1	TF	Repression	Gunes et al. (2000); Oh et al. (2000); Xu et al. (2001)

(Continued)

TABLE 5.1 (*Continued*)

Factor	Type	Effect	References
MENIN	TS	Repression	Lin and Elledge (2003)
NFX1	TF	Repression	Gewin et al. (2004)
PR	HR	Upregulation/repression	Wang et al. (2000)
p53	TF,TS	Repression	Roos et al. (1998); Kusumoto et al. (1999); Nair et al. (2000);Shats et al. (2004)
Raf	Onc	Upregulation	Goueli and Janknecht (2004)
Ras	Onc	Upregulation	Goueli and Janknecht (2004)
RARα, RXR	HR	Repression	Pendino et al. (2003); Phipps et al. (2009)
Sp1	TF	Turn on/basal transcription	Kyo et al. (2000); Deng et al. (2007)
TGFβ	GF	Repression/upregulation	Katakura et al. (1999); Yang et al. (2001); Lindkvist et al. (2005); Li et al. (2006a,b); Li and Liu (2007); Lacerte et al. (2008)
VEGF	GF	Upregulation	Zaccagnini et al. (2005)
WT1	TF, TS	Repression	Oh et al. (1999)

Note: CK, cytokine; E, enzyme; GF, growth factor; HR, hormone receptor; Onc, oncoprotein; TF, transcription factor; TS, tumor suppressor; V, Virus; VP, viral protein.

Regulation of hTERT expression is *per se* highly sensitive to variations in the intracellular oxygen levels. Two original reports (Minamino et al., 2001; Seimiya et al., 1999) have shown that hypoxia induces telomerase activity, proposing this effect as protective against hypoxia-dependent genetic stress in tumor cells. The novel mechanism underlying this phenomenon has been deciphered in subsequent studies (Nishi et al., 2004; Yatabe et al., 2004) reporting that under hypoxia HIF-1α binds directly to hypoxia response elements (HREs) in the hTERT core promoter, acting as a potent transactivator. It is necessary, however, to mention that alternative indirect mechanisms have been proposed by Koshiji et al. (2004) whose study demonstrated that neither HIF-1a transcriptional activity nor its DNA binding to the consensus sequences are required for activation of less "traditional" hypoxia-responsive genes, for example, $p21^{cip1}$. In the latter case, HIF-1a appears to induce cell-cycle arrest by functionally counteracting c-Myc, thereby derepressing $p21^{cip1}$. In this scenario, HIF-1α displaces c-Myc from the $p21^{cip1}$ promoter (it should be noted that their specific target sequences, HREs and E boxes largely overlap). In conclusion, hypoxia induces telomerase via transactivation of hTERT, possibly via multiple hypoxia-induced mechanisms of action operating on this process, among which the involvement of the MAP kinase signaling pathway in colon and ovarian cancer cells as well as phosphorylation of the hTERT protein in vascular smooth muscle cells (Minamino et al., 2001).

Among the general TFs regulating hTERT expression, the role of Sp1 whose binding sites are highly represented (at least five GC-boxes) within the hTERT core promoter (Kyo et al., 2000) should be emphasized. These sites are located between two E boxes forming a critical regulatory region that encompasses the major transcription start site. Importantly, since the hTERT promoter lacks a TATA box, the Sp1-binding sites, as shown by mutation analysis (Cong and Bacchetti, 2000; Kyo et al., 2000), ensure hTERT transcription. For the basal transcription a cooperation between several TFs including c-Myc, Sp1 and, more recently, the activating enhancer binding protein 2 (AP-2) has been postulated (Deng et al., 2007)

Several Ets (see Section 5.2.2.1) TFs have been identified as regulators of telomerase both directly or through other partners such as c-Myc (summarized in Dwyer et al. 2010)(Dwyer and Liu), sometimes functioning as coactivators with the transcriptional initiation complex, as in the case for EWS–E1AF, or as positive or negative transcriptional regulators via Ets-binding sites in the hTERT promoter (Maida et al., 2002; see Section 5.2.2.1).

5.2.1.2 Nuclear Hormone Receptors A significant part of hTERT transcriptional regulation is mediated by members of the nuclear hormone receptor super family, mostly the ERs, ERα and ERβ and, to a lesser extent, the androgen, the progesterone, and the retinoic acid receptors. The original findings revealing that hTERT is a direct target of hormone action were reported in two studies from Kyo et al. (1999) and Misiti et al. (2000). It was demonstrated that estrogen-induced activation of hTERT transcription is mediated by the interaction between the ligand-activated ER and the estrogen response elements (EREs) identified within the hTERT 5′ flanking genomic region. The interaction requires an intact DNA-binding domain evolutionarily conserved among the nuclear hormone receptors and specifically causes accumulation of hTERT mRNA in normal human ovary epithelial as well as in breast cancer cells. Furthermore *in vivo* DNA footprinting revealed that distinct and cell type-specific chromatin remodeling takes place over the hTERT–EREs in ERα-positive but not in ERβ-negative cells, demonstrating the requirement for the expression of ERα. This is in agreement with the finding that the ER antagonist tamoxifen, inhibits telomerase activity in MCF7 breast cancer cells (Aldous et al., 1999).

However, the most convincing evidence in favor of a physiopathological role of telomerase in the senescence and malignant conversion of hormone-dependent cells and tissues has been provided by the human prostate cancer (PCa) model (Nanni et al., 2002). Although sex steroid receptors, androgens and estrogens, are well-recognized hormonal effectors in prostate tumorigenesis (Abate-Shen and Shen, 2000), the report by Nanni et al. (2002) established an essential involvement of ER signaling in the malignant conversion of human prostate epithelium through activation and/or modulation of telomerase. In particular, a prompt induction of hTERT mRNA and of telomerase activity, but not of hTR, was detected in normal and transformed human prostate epithelial cells upon E2 treatment, suggesting a mechanism acting primarily at the transcription level. In the same study, evidence was also provided in favor of a relevant role of ERs as therapeutic targets in PCa.

In contrast to the above and surprisingly, androgen ablation in rats led to induction of telomerase activity in the prostate (Culig et al., 2002; Meeker et al., 1996; Teske et al., 2002) whereas androgen treatment of castrated animals reverses this effect, thus silencing telomerase (Moyzis et al., 1988). In agreement with the *in vivo* androgen ablation experiments, Moehren et al. (2008) showed that, in the presence of agonists, the androgen receptor (AR) represses the expression of hTERT and telomerase activity in PCa cell lines by being recruited to specific sites within 4 kb from the ATG in the hTERT promoter. However, as reported by Guo et al. (2003), androgens restored basal telomerase activity in PCa cells in which the enzyme had been downregulated by serum starvation. These seemingly contradictory results are likely attributable to the use of cell lines that differ in their androgen dependency and/or harbor a mutated AR.

Some studies have also demonstrated modulation of hTERT expression by the traditional antagonist of estrogen, progesterone. Wang et al. (2000) first demonstrated that progesterone acts in a dual and time-dependent manner in regulating hTERT transcription in breast and endometrial cancer cells expressing the PR: in the short term (3 h of treatment) augmenting hTERT transcription while causing its repression upon prolonged exposure. The underlying molecular mechanisms of this negative regulation appears to involve the MAP kinase pathway, although a binding site for the PR in the hTERT regulatory regions is functional by reporter gene assay (Ducrest et al., 2002).

A strong synergistic repression of hTERT transcription mediated by retinoic-acid receptor α, RARα and retinoid-X receptor, RXR in the presence of their specific ligands, has been reported by Pendino et al. (2003) in maturation-resistant acute promyelocytic leukemia cells, suggesting the possible development of more effective receptor-specific retinoids for tumor treatments leading to cell death through telomerase repression. Consistent with these findings, a report (Phipps et al., 2009) has shown a rapid decrease in histone H3–lysine 9 acetylation at the hTERT promoter as an important mechanism by which all trans retinoic acid (ATRA) represses telomerase and mediates its antitumor effects in ER-negative breast cancer cells. Most recently it has been reported that ATRA treatment cooperates with methylation in the repression of the hTERT promoter (Azouz *et al.* 2010 and Section 5.3; Azouz et al.).

5.2.2 Regulation of hTERT by Cellular and Viral Oncoproteins and by Tumor Suppressors

Unlike the case for most oncogenes, genomic amplification or rearrangements of the hTERT locus have not been linked to the up-regulation of hTERT transcription in cancer cells. In the past decade, much attention has been focused on the identification of the factors and pathways that control hTERT transcription in normal somatic cells (repression) and cancer cells (up-regulation). Besides factors whose binding sites are present in the hTERT promoter and those discussed in Section 5.1.2, a number of oncogenes and tumor suppressors that directly or indirectly regulate hTERT transcription have been identified.

5.2.2.1 Cellular Oncoproteins Regulate hTERT Transcription The hTERT promoter contains two Myc/Mad-binding sites (E-boxes) and several groups have demonstrated that c-Myc, a well-characterized cellular oncogene frequently activated in human cancers, binds to these sites and activates hTERT transcription (Greenberg et al., 1999; Wu et al., 1999) whereas Mad 1, a c-Myc antagonist, binds and suppresses hTERT expression (Gunes et al., 2000; Oh et al., 2000). However, the *in vivo* contribution of the c-Myc oncoprotein to telomerase activation during tumorigenesis remains unclear. In some cancer cells, the expression level of hTERT does not correlate with the amount of the endogenous c-Myc protein (Gewin and Galloway, 2001; Horikawa et al., 2002). Thus, overexpression of c-Myc may not be a significant mechanism contributing to telomerase activation in all tumors.

The Ets TFs and related Id proteins have been reported to participate in the regulation of telomerase via formation of a Ets–Id2–DNA complex (Dwyer et al., 2007; Maida et al., 2002; Xiao et al., 2002, 2003; Xu et al., 2008). The Ets/Id family of TFs interact with several signaling pathways, including those of MAP kinases, Erk1/2, p38/JNK, and PI3K, that connect cytoplasmic signals to the control of gene expression. This interaction in response to diverse cellular stresses merits future investigation. In addition, the Ewing's sarcoma fusion oncoprotein EWS/ETS is involved in stimulating hTERT transcription (Takahashi et al., 2003). Ewing's sarcoma is an aggressive bone neoplasia that is characterized by specific chromosomal translocations wherein the EWS gene on chromosome 22 is fused to one of five members of the ETS gene family (FLI1, ERG, ETV1/ER81, E1AF/PEA3, and FEV). These translocations produce five EWS/ETS chimeric proteins that contain the NH2-terminal transactivation domain of EWS and the COOH-terminal DNA-binding domain of ETS family TFs (Arvand and Denny, 2001). Both

EWS/FLI1 and EWS/E1AF activate telomerase by up-regulating hTERT transcription (Takahashi et al., 2003).

Recent studies have shown that three oncoproteins, HER2/Neu, Ras, and Raf, all stimulate hTERT transcription via the Ets TF ER81 (Goueli and Janknecht, 2004). Accordingly, hTERT expression is increased in HER2/Neu-positive breast tumors and breast tumor cell lines. These findings suggest that these oncoproteins may facilitate tumor formation by activating hTERT transcription and telomerase activity.

5.2.2.2 Viral Oncoproteins Regulate hTERT Transcription Human tumor viruses have evolved multiples strategies to circumvent immune defenses and the growth suppressive and proapoptotic functions of tumor suppressors in order to persist and amplify in host cells. One of these strategies may be the activation of telomerase by up-regulation of hTERT transcription. There are at least six human viruses, Epstein–Barr (EBV), hepatitis B (HBV), hepatitis C (HCV), herpersvirus 8/Kaposi sarcoma-associated herpersvirus (HHV-8/KSHV), human papilloma (HPV), and T-cell lymphotropic virus (HTLV-1) that encode viral oncoproteins capable of regulating hTERT transcription (Bellon and Nicot, 2008).

High-risk HPV are the etiologic agents of cervical cancer. The viral E6 and E7 proteins cooperate in the transformation of infected cells by targeting respectively the p53 and the pRb proteins for proteasomal degradation (zur Hausen, 2002). Early studies found that elevated telomerase activity is associated with E6 expression in precrisis human cervical keratinocytes (Klingelhutz et al., 1996). Further studies showed that E6 regulates telomerase activity through interaction with a cellular E6-associated protein E6AP (Gewin et al., 2004; Liu et al., 2005, 2009). It was proposed that E6AP, an ubiquitin-ligase, may target a cellular inhibitor of telomerase for ubiquitin degradation. Additionally, a transcriptional repressor, NFX1, that interacts with E6/E6AP was identified by a yeast two-hybrid screen. Silencing of NFX1 expression by siRNA results in hTERT transcription and telomerase activity in primary human epithelial cells, suggesting that the E6/E6AP complex induces hTERT transcription by ubiquitin degradation of the hTERT transcriptional repressor (Gewin et al., 2004).

EBV is a human gammaherpesvirus with a potent transforming ability and is implicated in the pathogenesis of human malignancies, but the mechanism of its oncogenesis remains largely unknown. Both EBV-positive nasopharyngeal carcinoma and EBV-immortalized B-lymphoblastoid cell lines have elevated telomerase activity, suggesting telomerase as a target of EBV (Counter et al., 1994; Kataoka et al., 1997). The EBV-encoded latent membrane protein-1 (LMP-1) is expressed during viral latency and has analogous properties to those of the constitutively active tumor necrosis factor receptor.Several studies indicate that LMP-1 is able to transactivate the hTERT gene by simultaneously modulating multiple signaling pathways including NF-kB and the MAPK kinases (Mei et al., 2006; Terrin et al., 2008; Yang et al., 2004). Activation of telomerase by the viral oncoprotein LMP-1 may be one of the mechanisms accounting for EBV oncogenicity.

HBV infection causes chronic liver diseases and malignant transformation. A number of reports have shown that integration of the viral genome and viral enhancer elements is associated with elevated telomerase activity in liver hepatocellular carcinoma (Ferber et al., 2003; Horikawa and Barrett, 2001; Paterlini-Brechot et al., 2003). Besides *cis*-activation of the hTERT promoter through viral integration, HBV encodes two viral oncoproteins, the HBx and preS2 with transcriptional activator function, both of which have been shown to transactivate the hTERT gene (Liu et al., 2007b; Luan et al., 2009; Qu et al., 2005).

Although the mechanisms of hTERT transactivation by viral oncoproteins are not fully elucidated, targeting tumor suppressors and TFs (such as p53, c-Myc, or Sp1) and signaling pathways (such as NF-kB, MAP kinases, and Phosphoinositide 3-kinase (PI3K) appear to be common strategies deployed by these tumor viruses.

5.2.2.3 Tumor Suppressors Regulate hTERT Transcription Inactivation of tumor suppressor and activation of telomerase are common features in cancer cells, suggesting a link between these two pathways. Absence of telomerase activity in normal human somatic cells is due to transcriptional repression of the hTERT gene; loss of this repression results in activation of hTERT expression and telomerase activity, which is critical for cellular immortality and tumorigenesis. Therefore transcriptional repressors of the hTERT gene may have tumor suppressor functions. Cell fusions between normal somatic cells and telomerase-positive immortal cells and microcell-mediated transfer of specific human chromosomes into cancer cells both result in repression of hTERT expression and downregulation of telomerase activity, indicating normal cells express functional transcriptional repressors of hTERT which may be inactivated in cancer cells (Ducrest et al., 2002).

The tumor suppressor protein p53 is a sequence-specific TF that regulates many genes involved in cell cycle, differentiation, senescence, and apoptosis. The wild-type protein, which inhibits tumorigenesis by inducing cell-cycle arrest or apoptosis in response to cellular damage, is inactivated by mutations in more than 50% of all human cancers (Levine and Oren, 2009). Inactivation of p53 tumor-suppressor function correlates well with up-regulation of hTERT expression and telomerase activity in several cancers (Kusumoto et al., 1999; Nair et al., 2000; Roos et al., 1998; Shats et al., 2004). Overexpression of wild-type p53 down-regulates telomerase activity through transcriptional repression of the hTERT gene, an effect independent from those of p53 on cell-cycle arrest and apoptosis induction (Kusumoto et al., 1999). However, there are no p53-binding sites in the hTERT promoter (Cong et al., 1999) and the mechanisms of p53-mediated hTERT repression remain to be elucidated. p53 may interact with Sp1 and thereby prevent Sp1 binding to the hTERT promoter. It is also possible that p53 recruits histone deacetylases onto the hTERT promoter (Kanaya et al., 2000; Shats et al., 2004; Xu et al., 2000). The ability of p53 to activate or repress the transcription of a large number of genes involved in growth and apoptosis is critical for its tumor suppressor functions (Levine and Oren, 2009). Repression of hTERT expression represents another important tumor suppression function that prevents uncontrolled cell proliferation and oncogenic transformation. The telomerase and p53 double-knockout mice display increased genomic instability and susceptibility to oncogenic transformation (Chin et al., 1999), supporting the potential existence of a functional interaction between the two proteins.

A screen of a cDNA library from normal human kidney cells for the candidate hTERT transcriptional repressor has lead to the identification of the Wilms' tumor 1 protein WT1 and the c-Myc antagonist E-box binding factor Mad1 (Oh et al., 1999, 2000). WT1 is a tumor suppressor involved in a common pediatric cancer. The hTERT core promoter has a WT1-binding site and WT1 represses hTERT transcription by direct interaction with these sequences. Overexpression of WT1 significantly reduces hTERT mRNA expression and telomerase activity, whereas mutation in the WT1 DNA-binding site increases hTERT promoter activity in 293 cells but not in HeLa cells (Oh et al., 1999) indicating that, consistent with its tissue-specific expression (kidney, gonad, and spleen), the role of WT1 may be cell-type specific. The second transcriptional repressor, Mad1 (Oh et al., 2000), competes with c-Myc for the binding to E-boxes with their common interacting partner Max. c-Myc/Max heterodimers bound to E-boxes activate E-box-containing promoters, whereas Mad/Max heterodimers compete for this binding and repress these promoters. The antagonistic effect of endogenous Mad1 and c-Myc on the hTERT promoter activity was demonstrated during the differentiation of HL60 cells by ChIP assays (Xu et al., 2001). In exponentially proliferating HL60 cells expressing hTERT and telomerase activity, the hTERT promoter E-boxes are occupied primarily by c-Myc/Max heterodimers. In contrast, in differentiated HL60 cells, the Mad1 protein is induced and binds to the E-boxes. Thus the switch from c-Myc/Max to Mad1/Max may determine activation or repression of hTERT transcription. This is consistent with the observation that Mad1 expression is either lost or is too low to be detected in tumor samples as compared to the level in matched normal tissues (Oh et al., 2000), whereas c-Myc expression is often up-regulated in telomerase-positive immortal cell lines and tumor cells.

The autocrine transforming growth factor β, TGF-β, can act either as tumor suppressor or oncogene with inhibitory and stimulatory effects on cell proliferation, respectively (Siegel and Massague, 2003). In normal development, TGF-β inhibits proliferation, promotes cell differentiation, and suppresses carcinogenesis; whereas during tumorigenesis, its autocrine activity is inhibited or attenuated. TGF-β exerts its function through serine/threonine kinase receptors that recruit downstream a Smad complex containing coactivator or corepressors and regulating gene transcription in a cell-type and tissue-dependent manner(Siegel and Massague, 2003). Several reports have shown that TGF-β inhibits telomerase activity through transcriptional repression of hTERT (Lacerte et al., 2008; Li and Liu, 2007; Yang et al., 2001). The precise mechanism of this repression remains unclear, but it has been proposed that TGF-β may act by suppressing c-Myc through Smad3 (Li et al., 2006a, b; Yang et al., 2001) or SIP1, a transcriptional target of the TGF-β pathway (Lin and Elledge, 2003). Recent reports also suggest the involvement of the E2F-1 TF (Lacerte et al., 2008).

Menin is the product of the tumor suppressor gene MEN1 (multiple endocrine neoplasia type 1; Agarwal et al., 2009). It is known to act as negative regulator of gene transcription through interaction with AP1 or NF-kB, whose binding sites are present in the hTERT promoter. Indeed, menin can physically interact with the hTERT promoter (Lin and Elledge, 2003). In telomerase-positive cancer cells overexpression of Menin represses hTERT transcription, whereas in telomerase-negative normal human fibroblasts menin down-regulation by siRNA leads to reactivation of hTERT expression. These results indicate that Menin is an hTERT transcriptional repressor. Moreover, menin-deficient cells are more susceptible than menin-positive cells to transformation by SV40 large T and small T antigens and oncogenic Ras, which supports the hypothesis that activation of telomerase may be one of the six critical events leading to oncogenic transformation (Hanahan and Weinberg, 2000).

5.2.3 Regulation of hTERT by Growth Factors and Cytokines

Telomerase activity is regulated directly or indirectly by specific growth factors (Haik et al., 2000; Hiyama et al., 1995; Tu et al., 1999; Zaccagnini et al., 2005) and by some cytokines (Akiyama et al., 1999). In this regulation, a prevalent role is played by several kinase cascades among which are PI3K, AKT, and MAPK as well as members of the SMAD family directly shuttling from the cytoplasm to the nucleus.

Insulin-like growth factor-1, IGF-1, a hormone with a molecular structure similar to insulin, acts by binding to its specific receptor, IGF1R, present on many cell types. In this way it modulates cell growth, proliferation, and transformation (Baserga et al., 1993; Rubin and Baserga, 1995). Peripheral blood T and B lymphocytes express detectable levels of telomerase (Igarashi and Sakaguchi, 1997), and Tu et al. (1999) have shown that, in human cord blood mononuclear cells (MNC) stimulated by Phytohaemagglutinin (PHA), IGF-1 is able to increase telomerase activity, and expression of hTERT and of the telomerase-associated protein TP1. IGF-1 on its own has no effect on these molecules but potentiates the increase in telomerase activity induced by PHA (Tu et al., 1999). In cultured PCa cell lines, IGF-1 induces hTERT expression, an effect blocked by the Akt inhibitor wortmanin but not by MAP kinase inhibitors (Wetterau et al., 2003). Thus, IGF-1 may exert its effect on the hTERT expression through its membrane receptor and PI3K/Akt signaling pathway (Wetterau et al., 2003).

In normal tissues, the basic fibroblast growth factor bFGF or FGF-2, is present in basement membranes and in the subendothelial extracellular matrix of blood vessels. In wound healing and tumor development, activation of FGF-2 mediates the formation of new blood vessels, a process known as angiogenesis, in which vascular endothelial growth factor, VEGF, is also involved. Telomerase activity was measured in cultured HUVECs maintained in exponential growth or induced to undergo quiescence by growth factor withdrawal or contact inhibition. The levels of hTERT mRNA were considerably reduced in quiescent cells and were stimulated together with those of Sp1 and telomerase activity by addition of FGF-2 but not VEGF (Kurz et al., 2003). On the other hand, in proliferating HUVEC cells selectively deprived of VEGF, addition of the growth factor induces telomerase activity (Zaccagnini et al., 2005).

The activity of VEGF is not confined to endothelial tissue, but depends on the presence of its receptors in other tissues such as muscle fibers and skeletal muscle satellite cells (Rissanen et al., 2002; Vale et al., 2001) in which VEGF is involved in postischemic tissue regeneration (Vale et al., 2001). In a rat model of hind-limb ischemia Zaccagnini et al. (2005) demonstrated that the VEGF-dependent activation of telomerase occurs through the NO pathway. Specifically, VEGF via the PI3K/AKT signaling (Zachary, 2003) regulates production of NO which is important for hTERT expression (see Section 5.5).

The epidermal growth factor, EGF, plays an important role in the regulation of cell growth, proliferation and differentiation by binding to its receptor EGFR, and inducing its dimerization followed by activation of an intrinsic tyrosine kinase which in turn leads to receptor autophosphorylation and induction of gene expression (Carpenter, 1992; Ullrich and Schlessinger, 1990; van der Geer et al., 1994). The effects of EGF on telomerase activity and the pathways involved in this process were studied using specific inhibitors for MEK, PI3K, and p38 pathways (Maida et al., 2002). The results showed that EGF activates telomerase and that this process is abrogated only by MEK inhibitors, suggesting the involvement of the MAP kinase pathway, which also affects the ETs family of proteins. Since the hTERT promoter contains ETs-binding motifs (Wasylyk et al., 1990), it seems possible that these factors transduce the EGF signals on this promoter.

Tumor growth factor β, TGF β, has opposing effects in different cellular contexts. During normal development, it promotes cell differentiation and inhibits cell proliferation, whereas during carcinogenesis it behaves like an oncogenic factor (Siegel and Massague, 2003). TGF-β binds to a type II receptor (TRII), which recruits and phosphorylates a type I receptor (TRI) which in turn phosphorylates a member of a receptor-regulated family of proteins known as SMADs (Fanayan et al., 2002). Within this family, Smad3 specifically mediates TGF-β action by binding to a specific site on the hTERT promoter leading to inhibition of hTERT transcription (Lacerte et al., 2008; Li and Liu, 2007; Li et al., 2006a,b; Luke and Lingner, 2009). Since a binding site for Smad3 is a CAGA box near an E box, it has been proposed that Smad3 may inhibit hTERT expression by counteracting the stimulatory effect of c-myc by either binding to DNA in its proximity (Li et al., 2006a,b) or through interaction with the myc protein (Feng et al., 2002). Inhibition of telomerase by TGF-β has been observed in lung adenocarcinoma, colon carcinoma, and breast cancer cells (Katakura et al., 1999; Yang et al., 2001); recent studies have shown that in thyroid carcinoma cell lines, TGF-β1 can activate or inhibit telomerase (Lindkvist et al., 2005). These contrasting effects are correlated with the presence, respectively, of mutant or wild-type p53 (Kanaya et al., 2000; Kusumoto et al., 1999; Rahman et al., 2005; Xu et al., 2000).

Among cytokines, Interferon α and γ (IFN-α and IFN-γ) sensitize multiple myeloma (MM) cells to Fas-mediated apoptosis, most probably via hTERT regulation (Lindkvist et al., 2006). In fact, stimulation of MM cells by either IFN-α and IFN-γresults in decreased hTERT mRNA expression with different time kinetics, an effect potentially mediated by c-myc (Akiyama et al., 1999).

Interleukin-2 (IL-2) induces telomerase activity in NK cells, a subpopulation of lymphoid cells, via both a transcriptional and a post-translational mechanism. In the first case, the binding of IL-2 to its receptor mediates the activation of the MAPK, PI3K, and Ras pathways. PI3Ks play an important role in cell survival through the activation of phosphatidylinositol-dependent kinase 1/2 which activates Akt kinase. (Kawauchi et al., 2005). The latter has been shown to regulate telomerase activity through phosphorylation of hTERT in melanoma (Kang et al., 1999) and myeloma cells (Akiyama et al., 2002). Post-translational regulation of hTERT by IL-2 appears to occur through a multiprotein complex that includes Akt, Hsp90, mTOR (mammalian target of rapamycin), and S6 kinase (Kawauchi et al., 2005).

5.3 EPIGENETIC REGULATION OF hTERT

Epigenetics generates an additional regulatory layer for the modulation of hTERT expression in normal, transformed, and senescent cells. DNA methylation, an epigenetic process involved in embryonic development, differentiation and aging, typically results in gene silencing during carcinogenesis, especially in the case of tumor-suppressor genes. The CpG island proximal to the transcription start site in the hTERT promoter (see Section 5.1.2) was initially thought to be associated with hTERT gene silencing (Liu et al., 2007a; Shin et al., 2003). However, no significant correlation between hTERT expression and methylation status either overall or at a specific site was detected (Devereux et al., 1999). Furthermore, increased promoter methylation in some cancer cells and lack of methylation in normal hTERT-negative cells was reported (Guilleret and Benhattar, 2004; Renaud et al., 2007). Zinn et al. (2007) using bisulfite sequencing, first identified that all telomerase-positive cancer cell lines examined retained hTERT alleles with little or no methylation around the transcription start site despite being densely methylated in more upstream regions. These findings suggested that a low methylation level around the transcription start site could be important for hTERT basal transcription whereas more distal regions, heavily methylated, may be involved in rapid silencing of the hTERT locus during development or in pathophysiological conditions. Expression of hTERT is regulated by a number of signaling pathways that are responsible for the activation or silencing of genes involved in cell proliferation, differentiation, apoptosis, and senescence. In this regard, the modification of the local chromatin structure at the hTERT locus represents a further regulatory layer at which hTERT expression and function is controlled. ChIPs indicated that both active and inactive chromatin marks are present within the hTERT promoter, including histone H3 Lysine 9 acetylation (H3K9Ac) and lysine 4 di-methylation (H3K4me2) as well as Lysine 9 trimethylation (H3K9me3) or Lysine 27 trimethylation (H3K27me3) (Zinn et al., 2007). Most of the active chromatin marks were found at the transcription start site in a region which is generally hypomethylated on CpG residues, further indicating that the DNA methylation pattern of the hTERT promoter is consistent with the dynamics of gene expression.

Modification of nucleosome histones, including acetylation/deacetylation and methylation, is known to regulate chromatin structure and thereby to affect gene transcription. Histone-modification-mediated chromatin remodeling in the regulation of hTERT transcription have been well-characterized (Kyo et al., 2008; Liu et al., 2004). Several epigenetic enzymes seem important in this process, but histone acetylases (HATs) and deacetylases (HDACs) are certainly the best characterized. There are additional enzymes that carry acetyl-transferase activity however, because of their lower sequence similarities, they cannot be clearly grouped. Auto-acetylation is a further functionally relevant feature of HATs. HDACs are a family of 18 molecules divided in four subclasses (I–II–III–IV) defined according to structural similarities (Hildmann et al., 2007). Their functional role is variegated, ranging from regulation of chromatin structure, repression or activation of gene expression, histone and nonhistone protein modification, and the regulation of cell metabolism (Hildmann et al., 2007). Interestingly, transient expression assays revealed that the HDAC inhibitor trichostatin A (TSA) activates the hTERT promoter and that Sp1 overexpression enhanced TSA responsiveness indicating that HDAC inhibitors activate hTERT promoter in an Sp1-dependent manner (Cong and Bacchetti, 2000; Takakura et al., 2001). It is possible that endogenous Sp1 interacts with HDAC and recruits it to the hTERT promoter resulting in the deacetylation of nucleosomes and gene silencing (Doetzlhofer et al,. 1999). Further evidence suggests however, that Sp1 interacts with the p300 HAT coactivator (Suzuki et al., 2000). Therefore, Sp1 may interact with various factors that have HAT or HDAC activity explaining the different actions of Sp1 on the hTERT promoter in normal and tumor cells.

The E-box binding activator c-Myc and repressor Mad1 (Xu et al., 2001) which compete with each other for the common binding partner Max, are also involved in histone-modification-mediated chromatin remodeling of the hTERT promoter. The endogenous c-Myc/Max complex on the hTERT promoter in fact was found to be associated with acetylated histones, resulting in enhanced hTERT expression. In contrast, the Mad1/Max complex is associated with an increase in deacetylated histones and hTERT repression (Xu et al., 2001). Recently, a role for histone methylation in hTERT regulation has also been demonstrated. Atkinson et al. (2005) observed that highly H3K4me3 was associated with the actively transcribed hTERT gene in telomerase-proficient tumor cells.

Additional work is required to understand the epigenetic regulation of hTERT the complexity of which is mediated either by the expression of numerous intra/extra cellular signals converging on this gene or of the involvement of hTERT in multiple biological processes. In this regard, the recent work of Blasco and coworkers clearly show that during the induced staminalization of adult cells the epigenetic reactivation of hTERT and the elongation of telomeres are among the earliest requirements (Marion and Blasco) further providing evidence about the important contribution of hTERT and its regulatory signals to the fundamental physiology of an organism.

5.4 TRANSCRIPTIONAL REGULATION OF hTR

In the past decade, studies on telomerase regulation have focused mainly on the transcription of hTERT, because its expression is strictly associated with and rate limiting for enzymatic activity, whereas hTR is ubiquitously expressed in cells regardless of telomerase activity. However, expression of hTR is also subject to transcriptional regulation (Downey et al., 2001; Feng et al., 1995; Nakano et al., 1998; Soder et al., 1998). Interestingly, recent studies have shown that overexpression of both hTERT and hTR induces telomerase activity and maintains telomere length to a higher extent than overexpression of either hTERT or hTR alone in both cancer cell lines and primary human fibroblasts (Cristofari and Lingner, 2006), suggesting that both components control telomerase activity and telomere length homeostasis *in vivo*.

The gene encoding hTR is a single-copy gene located on chromosome 3 at 3q26.3 (Soder et al., 1997), with a 5′ flanking region containing a CCAAT box, a TATA box, and a number of TF-binding sites including Sp1, Sp3, AP1, HIF-1, NF-Y (Zhao et al., 1998). Physical and functional interaction of Sp1, Sp3, and NF-Y TFs with the hTR promoter were revealed by electrophoretic mobility gel shift assays *in vitro*, by ChIP assays *in vivo,* and by promoter reporter assays (Zhao et al., 1998, 2003).

A number of different signaling pathways have been implicated in hTR transcriptional regulation. The JNK inhibitor SP600125 was shown to induce hTR promoter activity and hTR expression in a dose-dependent manner. In contrast, a constitutively active kinase domain of WEKK1, upstream of the JNK pathway, strongly represses promoter activity, suggesting that the mitogen-activated protein kinase (MEKK1)/JNK pathway is involved in hTR transcription (Bilsland et al., 2006). Further investigations indicated that the JNK pathway functions in cooperation with Sp1 and Sp3, where Sp1 activates hTR transcription and Sp3 represses it (Bilsland et al., 2006). The precise roles of Sp1 and Sp3 in hTR transcriptional regulation need to be elucidated. Similar to the hTERT promoter (Nishi et al., 2004), the hTR core promoter contains a hypoxia response element (HRE) binding site and is regulated by hypoxic stress signals (Anderson et al., 2006). Thus, transcriptional activation of both hTERT and hTR may contribute to induction of telomerase activity under hypoxia.

5.5 CELLULAR MICROENVIRONMENT AND TELOMERASE REGULATION

The experimental evidence described above is compatible with a model in which hTERT is a critical target of multiple signaling pathways in different cellular microenvironments. In this scenario, recent studies indicate that three key molecules appear to significantly affect telomerase regulation in the endothelium and prostate microenvironments: hormones (in particular estrogens (E2), NO, and low intracellular oxygen (hypoxia)).

Estrogens are implicated in many physiological and pathological processes and in particular they exert a relevant ateroprotective effect on the cardiovascular system. Of relevance in this context is the role that estrogens play in the apoptotic process and the migration of endothelial cells through the regulation of target genes such as endothelial NO synthase, eNOS (Coulet et al., 2003). Intriguingly, it has been shown that both estrogen and NO signaling can strongly counteract endothelial senescence through a common effector, hTERT. In keeping with these observations, Narducci et al. (2007) obtained evidence of higher telomerase activity in polymorphonuclear neutrophils (PMN) in the coronary plaques of patients with unstable angina (UA) as compared to their peripheral blood. This finding is in agreement with delayed PMN apoptosis in UA patients (Garlichs et al., 2004) and suggests local extended lifespan and prolonged activity of these inflammatory cells in the early phases of UA.

This conclusion is supported by Hayashi et al. (2006) who demonstrated that NO prevents senescence in cooperation with the antisenile effect mediated by estrogen treatment, suggesting a direct involvement of the eNOS/NO and ligand-activated ER-signaling pathways in delaying endothelial senescence

Activation of telomerase is mediated in part by NO production due to estrogen-dependent activation of eNOS through genomic and nongenomic mechanisms. Misiti et al. (2000) and Nanni et al. (2002) also provided evidence for a direct role of E2-activated ERs in the transcriptional regulation of hTERT, and consequently, in the activation of the enzyme through a specific estrogen-dependent chromatin remodeling of the hTERT genomic sequences at specific EREs.

The contribution to endothelial function of both estrogen and NO signaling and reports indicating that activated eNOS can translocate into the nucleus where it regulates gene transcription(Feng et al., 1999; Gobeil et al., 2006; Goetz et al., 1999; Klinz et al., 2005), provide the rationale for postulating a functional cooperation between the ERs and eNOS pathways. Recent studies have indeed provided insights on the molecular mechanisms underlying the eNOS nuclear function in two different experimental models, primary cultures of human endothelium and PCa cell lines (Farsetti et al., 2009; Grasselli et al., 2008; Nanni et al., 2009). In both systems, transcriptional regulation of telomerase was dependent on the presence of both ERs and eNOS bound to the chromatin region encompassing specific hTERT-EREs. Coincident with their presence, an epigenetic modification consisting of H3 Lysine 79 di-methylation (H3K79me2), typically associated with a transcription permissive configuration (Ng, 2003), was detected at these sites (Grasselli et al., 2008).

Further, it has been recently demonstrated that: (i) the NO/eNOS nuclear signaling is a key pathway in the progression of PCa; and (ii) the existence of critical combinatorial complexes, the eNOS/ERβ (specifically active in the prostate tumor microenvironment) and the eNOS/ERα (specifically involved in the maintenance of vascular homeostasis), determines localized remodeling of the chromatin leading to transcriptional regulation of previously identified prognostic target genes (Nanni et al., 2006), including hTERT, a gene extremely sensitive to estrogen stimulation and to variations in the intracellular levels of oxygen and NO.

Other key molecules, on which a number of studies in solid tumors, such as hormone-dependent breast, prostate, and ovarian tumors, have focused in the last decade, are the hypoxia-inducible factors, HIF-1α and HIF2 α. The regulation of transcription by HIF-1α represents the most important mechanism mediating adaptative response to a O_2-reduced environment or hypoxia (Semenza, 2003). Recently we and others reported the induction of genes involved in the activation of cellular response to hypoxia, in particular of the two members of the family, HIF-1a and HIF-2a, whose overexpression is associated with early invasion (Pouyssegur et al., 2006) and tumor progression (Keith and Simon, 2007; Pouyssegur et al., 2006). Since hTERT and several other genes belonging to a prognostic transcriptional signature (Nanni et al., 2006) respond to both estrogens or hypoxia, the potential cooperation of eNOS, ERs, and HIFs, in regulating hTERT transcription was investigated in PCa (Nanni et al., 2009).

The results by ChIPs and re-ChIPs assays revealed the dynamic occupancy by ERβ, eNOS, and HIF-1α/HIF-2α at specific estrogen and HREs within the hTERT regulatory sequences as well as the existence and prognostic role of functional combinatorial complexes among all factors, exclusively in the cells associated with aggressive behavior (Nanni et al., 2009).

Mechanistically, the novel role of nuclearized eNOS in both normal and tumor cells, resides in its ability to form combinatorial complexes with ERα (in the human endothelium) (Farsetti et al., 2009; Grasselli et al., 2008; Zeng and Xu, 2008) or with ERβ or HIF-1α/HIF-2α (in the prostate tumor epithelium) (Nanni et al., 2009 and Aiello *et al.* unpublished data). Therefore, the formation of active eNOS/ERs or eNOS/HIFs complexes along the hTERT genomic sequences, reflecting specific modifications of the cell microenvironment (e.g., imbalance of androgen/estrogen ratio in favor of estrogen or activation of hypoxia-response and downstream effectors), is capable of initiating a transcriptional program with major consequences on endothelial cell function and on the malignant conversion of human prostate epithelial cells.

5.6 CONCLUSIONS AND PERSPECTIVES

The biological importance of telomerase in aging and cancer has attracted enormous interests and significant progresses have been made in the past decade. It is clear that telomerase is subjected to complex regulation by many TFs involved in several intracellular and extracellular signaling pathways depending on cell types, tissue environments, and physiopathological conditions. Elucidation of the molecular mechanisms underlying telomerase regulation should have considerable impact on our understanding of the relevance of telomerase in aging and cancer and on the ability to manipulate telomerase for prophylactics and therapeutic applications.

In light of the experimental evidence summarized in this review, it may be anticipated that the two following aspects will be of increasing interest in the field of telomerase regulation:

(i) the novel role of eNOS as an essential cofactor of ERs and HIFs in hTERT transcriptional regulation.
(ii) the understanding that telomerase is regulated at the epigenetic level, which amplifies the number of possible regulatory factors involved in the fine expression and functional tuning of this very important molecule.

There are however additional players emerging from the rapidly evolving field of non-coding RNA (ncRNA). Telomerase, with its dual components, structural and messenger RNAs, is a potential target for short RNA sequences (see for review Grillari and Grillari-Voglauer, 2010) including those miRNAs recently reported in the case of thyroid cancer and hepatoma (Mitomo et al., 2008; Miura et al., 2009). Similarly, the long noncoding transcripts TERRA (telomeric repeat-containing RNA) appear to be involved in regulating not only telomere length and telomerase but also chromatin stability (Luke and Lingner, 2009). Limited information is presently available about telomerase control by these ncRNAs in physiological and pathological contexts. This situation is likely not to persist and substantial advances in this direction are envisaged in the near future.

ACKNOWLEDGMENTS

The authors are indebted to Silvia Bacchetti and Carlo Gaetano for constructive discussions and critical reading of the manuscript, and thank Annalisa Grasselli, Simona Nanni, and Valentina Pantisano for their help with literature search and valuable suggestions. Support from Associazione Italiana Ricerca sul Cancro (AIRC) and Italian Minister of Education, University and Research (PRIN 2008NY72SJ) to Antonella Farsetti1, and from the "973" Project of the Ministry of Science and Technology (2007CB507402, 2007CB914402) to Yu-Sheng Cong is acknowledged.

REFERENCES

Abate-Shen C, Shen MM. (2000) Molecular genetics of prostate cancer. *Genes Dev*. **14**: 2410–2434.

Agarwal SK, Ozawa A, Mateo CM, Marx SJ. (2009) The MEN1 gene and pituitary tumours. *Horm. Res.* **71**: Suppl 2: 131–138.

Akiyama M, Hideshima T, Hayashi T, Tai YT, Mitsiades CS, Mitsiades N, Chauhan D, Richardson P, Munshi NC, Anderson KC. (2002) Cytokines modulate telomerase activity in a human multiple myeloma cell line. *Cancer Res*. **62**: 3876–3882.

Akiyama M, Iwase S, Horiguchi-Yamada J, Saito S, Furukawa Y, Yamada O, Mizoguchi H, Ohno T, Yamada H, (1999) Interferon-alpha repressed telomerase along with G1-accumulation of Daudi cells. *Cancer Lett*. **142**: 23–30.

Aldous WK, Marean AJ, DeHart MJ, Matej LA, Moore KH. (1999) Effects of tamoxifen on telomerase activity in breast carcinoma cell lines. *Cancer*. **85**: 1523–1529.

Anderson CJ, Hoare SF, Ashcroft M, Bilsland AE, Keith WN. (2006) Hypoxic regulation of telomerase gene expression by transcriptional and post-transcriptional mechanisms. *Oncogene*. **25**: 61–69.

Arvand A, Denny CT. (2001) Biology of EWS/ETS fusions in Ewing's family tumors. *Oncogene*. **20**: 5747–5754.

Atkinson SP, Hoare SF, Glasspool RM, Keith WN. (2005) Lack of telomerase gene expression in alternative lengthening of telomere cells is associated with chromatin remodeling of the hTR and hTERT gene promoters. *Cancer Res*. **65**: 7585–7590.

Azouz A, Wu YL, Hillion J, Tarkanyi I, Karniguian A, Aradi J, Lanotte M, Chen GQ, Chehna M, Segal-Bendirdjianx E. (2010) Epigenetic plasticity of hTERT gene promoter determines retinoid capacity to repress telomerase in maturation-resistant acute promyelocytic leukemia cells. *Leukemia*. **24**: 613–622.

Baserga R, Porcu P, Sell C. (1993) Oncogenes, growth factors and control of the cell cycle. *Cancer Surv*. **16**: 201–213.

Baur JA, Zou Y, Shay JW, Wright WE. (2001) Telomere position effect in human cells. *Science*. **292**: 2075–2077.

Bellon M, Nicot C. (2008) Regulation of telomerase and telomeres: human tumor viruses take control. *J. Natl. Cancer Inst.* **100**: 98–108.

Bilsland AE, Stevenson K, Atkinson S, Kolch W, Keith WN. (2006) Transcriptional repression of telomerase RNA gene expression by c-Jun-NH2-kinase and Sp1/Sp3. *Cancer Res.* **66**: 1363–1370.

Bryce LA, Morrison N, Hoare SF, Muir S, Keith WN. (2000) Mapping of the gene for the human telomerase reverse transcriptase, hTERT, to chromosome 5p15.33 by fluorescence *in situ* hybridization. *Neoplasia.* **2**: 197–201.

Carpenter G. (1992) Receptor tyrosine kinase substrates: src homology domains and signal transduction. *FASEB J.* **6**: 3283–3289.

Chin L, Artandi SE, Shen Q, Tam A, Lee SL, Gottlieb GJ, Greider CW, DePinho RA. (1999) p53 deficiency rescues the adverse effects of telomere loss and cooperates with telomere dysfunction to accelerate carcinogenesis. *Cell.* **97**: 527–538.

Colgin LM, Wilkinson C, Englezou A, Kilian A, Robinson MO, Reddel RR. (2000) The hTERTalpha splice variant is a dominant negative inhibitor of telomerase activity. *Neoplasia.* **2**: 426–432.

Cong Y, Shay JW. (2008) Actions of human telomerase beyond telomeres. *Cell Res.* **18**: 725–732.

Cong YS, Bacchetti S. (2000) Histone deacetylation is involved in the transcriptional repression of hTERT in normal human cells. *J. Biol. Chem.* **275**: 35665–35668.

Cong YS, Wen J, Bacchetti S. (1999) The human telomerase catalytic subunit hTERT: organization of the gene and characterization of the promoter. *Hum. Mol. Genet.* **8**: 137–142.

Cong YS, Wright WE, Shay JW. (2002) Human telomerase and its regulation. *Microbiol. Mol. Biol. Rev.* **66**: 407–425 table of contents.

Coulet F, Nadaud S, Agrapart M, Soubrier F. (2003) Identification of hypoxia-response element in the human endothelial nitric-oxide synthase gene promoter. *J. Biol. Chem.* **278**: 46230–46240.

Counter CM, Botelho FM, Wang P, Harley CB, Bacchetti S. (1994) Stabilization of short telomeres and telomerase activity accompany immortalization of Epstein–Barr virus-transformed human B lymphocytes. *J. Virol.* **68**: 3410–3414.

Cristofari G, Lingner J. (2006) Telomere length homeostasis requires that telomerase levels are limiting. *EMBO J.* **25**: 565–574.

Culig Z, Klocker H, Bartsch G, Hobisch A. (2002) Androgen receptors in prostate cancer. *Endocr. Relat Cancer.* **9**: 155–170.

Deng WG, Jayachandran G, Wu G, Xu K, Roth JA, Ji L. (2007) Tumor-specific activation of human telomerase reverses transcriptase promoter activity by activating enhancer-binding protein-2beta in human lung cancer cells. *J. Biol. Chem.* **282**: 26460–26470.

Devereux TR, Horikawa I, Anna CH, Annab LA, Afshari CA, Barrett JC. (1999) DNA methylation analysis of the promoter region of the human telomerase reverse transcriptase (hTERT) gene. *Cancer Res.* **59**: 6087–6090.

Doetzlhofer A, Rotheneder H, Lagger G, Koranda M, Kurtev V, Brosch G, Wintersberger E, Seiser C. (1999) Histone deacetylase 1 can repress transcription by binding to Sp1. *Mol Cell Biol.* **19**(8): 5504–5511.

Downey MG, Going JJ, Stuart RC, Keith WN. (2001) Expression of telomerase RNA in oesophageal and oral cancer. *J. Oral Pathol. Med.* **30**: 577–581.

Ducrest AL, Szutorisz H, Lingner J, Nabholz M. (2002) Regulation of the human telomerase reverse transcriptase gene. *Oncogene*. **21**: 541–552.

Dwyer J, Li H, Xu D, Liu JP. (2007) Transcriptional regulation of telomerase activity: roles of the Ets transcription factor family. *Ann. N Y Acad. Sci.* **1114**: 36–47.

Dwyer JM, Liu JP. (2010) Ets2 transcription factor, telomerase activity and breast cancer. *Clin. Exp. Pharmacol. Physiol.* **37**: 83–87.

Fanayan S, Firth SM, Baxter RC. (2002) Signaling through the Smad pathway by insulin-like growth factor-binding protein-3 in breast cancer cells. Relationship to transforming growth factor-beta 1 signaling. *J. Biol. Chem.* **277**: 7255–7261.

Farsetti A, Grasselli A, Bacchetti S, Gaetano C, Capogrossi MC. (2009) The telomerase tale in vascular aging: regulation by estrogens and nitric oxide signaling. *J. Appl. Physiol.* **106**: 333–337.

Feng J, Funk WD, Wang SS, Weinrich SL, Avilion AA, Chiu CP, Adams RR, Chang E, Allsopp RC, Yu J, et al. (1995) The RNA component of human telomerase. *Science*. **269**: 1236–1241.

Feng XH, Liang YY, Liang M, Zhai W, Lin X. (2002) Direct interaction of c-Myc with Smad2 and Smad3 to inhibit TGF-beta-mediated induction of the CDK inhibitor p15(Ink4B). *Mol Cell*. **9**: 133–143.

Feng Y, Venema VJ, Venema RC, Tsai N, Caldwell RB. (1999) VEGF induces nuclear translocation of Flk-1/KDR, endothelial nitric oxide synthase, and caveolin-1 in vascular endothelial cells. *Biochem. Biophys. Res. Commun.* **256**: 192–197.

Ferber MJ, Montoya DP, Yu C, Aderca I, McGee A, Thorland EC, Nagorney DM, Gostout BS, Burgart LJ, Boix L, Bruix J, McMahon BJ, Cheung TH, Chung TK, Wong YF, Smith DI, Roberts LR. (2003) Integrations of the hepatitis B virus (HBV) and human papillomavirus (HPV) into the human telomerase reverse transcriptase (hTERT) gene in liver and cervical cancers. *Oncogene*. **22**: 3813–3820.

Garlichs CD, Eskafi S, Cicha I, Schmeisser A, Walzog B, Raaz D, Stumpf C, Yilmaz A, Bremer J, Ludwig J, Daniel WG. (2004) Delay of neutrophil apoptosis in acute coronary syndromes. *J. Leukoc. Biol.* **75**: 828–835.

Gewin L, Galloway DA. (2001) E box-dependent activation of telomerase by human papillomavirus type 16 E6 does not require induction of c-myc. *J. Virol.* **75**: 7198–7201.

Gewin L, Myers H, Kiyono T, Galloway DA. (2004) Identification of a novel telomerase repressor that interacts with the human papillomavirus type-16 E6/E6-AP complex. *Genes Dev*. **18**: 2269–2282.

Gobeil F Jr., Zhu T, Brault S, Geha A, Vazquez-Tello A, Fortier A, Barbaz D, Checchin D, Hou X, Nader M, Bkaily G, Gratton JP, Heveker N, Ribeiro-da-Silva A, Peri K, Bard H, Chorvatova A, D'Orleans-Juste P, Goetzl EJ, Chemtob S. (2006) Nitric oxide signaling via nuclearized endothelial nitric-oxide synthase modulates expression of the immediate early genes iNOS and mPGES-1. *J. Biol. Chem.* **281**: 16058–16067.

Goetz RM, Thatte HS, Prabhakar P, Cho MR, Michel T, Golan DE. (1999) Estradiol induces the calcium-dependent translocation of endothelial nitric oxide synthase. *Proc. Natl. Acad. Sci. USA*. **96**: 2788–2793.

Gordan JD, Bertout JA, Hu CJ, Diehl JA, Simon MC. (2007a) HIF-2alpha promotes hypoxic cell proliferation by enhancing c-myc transcriptional activity. *Cancer Cell*. **11**: 335–347.

Gordan JD, Thompson CB, Simon MC. (2007b) HIF and c-Myc: sibling rivals for control of cancer cell metabolism and proliferation. *Cancer Cell*. **12**: 108–113.

Goueli BS, Janknecht R. (2004) Upregulation of the catalytic telomerase subunit by the transcription factor ER81 and oncogenic HER2/Neu, Ras, or Raf. *Mol. Cell Biol.* **24**: 25–35.

Grasselli A, Nanni S, Colussi C, Aiello A, Benvenuti V, Ragone G, Moretti F, Sacchi A, Bacchetti S, Gaetano C, Capogrossi MC, Pontecorvi A, Farsetti A. (2008) Estrogen receptor-alpha and endothelial nitric oxide synthase nuclear complex regulates transcription of human telomerase. *Circ. Res.* **103**: 34–42.

Greenberg RA, O'Hagan RC, Deng H, Xiao Q, Hann SR, Adams RR, Lichtsteiner S, Chin L, Morin GB, DePinho RA. (1999) Telomerase reverse transcriptase gene is a direct target of c-Myc but is not functionally equivalent in cellular transformation. *Oncogene.* **18**: 1219–1226.

Grillari J, Grillari-Voglauer R. (2010) Novel modulators of senescence, aging, and longevity: small non-coding RNAs enter the stage. *Exp. Gerontol.* **45**: 302–311.

Gu J, Kagawa S, Takakura M, Kyo S, Inoue M, Roth JA, Fang B. (2000) Tumor-specific transgene expression from the human telomerase reverse transcriptase promoter enables targeting of the therapeutic effects of the Bax gene to cancers. *Cancer Res.* **60**: 5359–5364.

Guilleret I, Benhattar J. (2004) Unusual distribution of DNA methylation within the hTERT CpG island in tissues and cell lines. *Biochem. Biophys. Res. Commun.* **325**: 1037–1043.

Gunes C, Lichtsteiner S, Vasserot AP, Englert C. (2000) Expression of the hTERT gene is regulated at the level of transcriptional initiation and repressed by Mad1. *Cancer Res.* **60**: 2116–2121.

Guo C, Armbruster BN, Price DT, Counter CM. (2003) *In vivo* regulation of hTERT expression and telomerase activity by androgen. *J. Urol.* **170**: 615–618.

Haik S, Gauthier LR, Granotier C, Peyrin JM, Lages CS, Dormont D, Boussin FD. (2000) Fibroblast growth factor 2 up regulates telomerase activity in neural precursor cells. *Oncogene.* **19**: 2957–2966.

Hanahan D. (2000) Benefits of bad telomeres. *Nature.* **406**: 573–574.

Hanahan D, Weinberg RA. (2000) The hallmarks of cancer. *Cell.* **100**: 57–70.

Hayashi T, Matsui-Hirai H, Miyazaki-Akita A, Fukatsu A, Funami J, Ding QF, Kamalanathan S, Hattori Y, Ignarro LJ, Iguchi A. (2006) Endothelial cellular senescence is inhibited by nitric oxide: implications in atherosclerosis associated with menopause and diabetes. *Proc. Natl. Acad. Sci. USA.* **103**: 17018–17023.

Hildmann C, Riester D, Schwienhorst A. (2007) Histone deacetylases—an important class of cellular regulators with a variety of functions. *Appl. Microbiol. Biotechnol.* **75**: 487–497.

Hiyama K, Hirai Y, Kyoizumi S, Akiyama M, Hiyama E, Piatyszek MA, Shay JW, Ishioka S, Yamakido M. (1995) Activation of telomerase in human lymphocytes and hematopoietic progenitor cells. *J. Immunol.* **155**: 3711–3715.

Horikawa I, Barrett JC. (2001) *cis*-Activation of the human telomerase gene (hTERT) by the hepatitis B virus genome. *J. Natl. Cancer Inst.* **93**: 1171–1173.

Horikawa I, Cable PL, Afshari C, Barrett JC. (1999) Cloning and characterization of the promoter region of human telomerase reverse transcriptase gene. *Cancer Res.* **59**: 826–830.

Horikawa I, Cable PL, Mazur SJ, Appella E, Afshari CA, Barrett JC. (2002) Downstream E-box-mediated regulation of the human telomerase reverse transcriptase (hTERT) gene transcription: evidence for an endogenous mechanism of transcriptional repression. *Mol. Biol. Cell.* **13**: 2585–2597.

Igarashi H, Sakaguchi N. (1997) Telomerase activity is induced in human peripheral B lymphocytes by the stimulation to antigen receptor. *Blood.* **89**: 1299–1307.

Kanaya T, Kyo S, Hamada K, Takakura M, Kitagawa Y, Harada H, Inoue M. (2000) Adenoviral expression of p53 represses telomerase activity through down-regulation of human telomerase reverse transcriptase transcription. *Clin. Cancer Res.* **6**: 1239–1247.

Kang SS, Kwon T, Kwon DY, Do SI. (1999) Akt protein kinase enhances human telomerase activity through phosphorylation of telomerase reverse transcriptase subunit. *J. Biol. Chem.* **274**: 13085–13090.

Katakura Y, Nakata E, Miura T, Shirahata S. (1999) Transforming growth factor beta triggers two independent-senescence programs in cancer cells. *Biochem. Biophys. Res. Commun.* **255**: 110–115.

Kataoka H, Tahara H, Watanabe T, Sugawara M, Ide T, Goto M, Furuichi Y, Sugimoto M. (1997) Immortalization of immunologically committed Epstein–Barr virus-transformed human B-lymphoblastoid cell lines accompanied by a strong telomerase activity. *Differentiation.* **62**: 203–211.

Kawauchi K, Ihjima K, Yamada O. (2005) IL-2 increases human telomerase reverse transcriptase activity transcriptionally and posttranslationally through phosphatidylinositol 3′-kinase/Akt, heat shock protein 90, and mammalian target of rapamycin in transformed NK cells. *J. Immunol.* **174**: 5261–5269.

Keith B, Simon MC. (2007) Hypoxia-inducible factors, stem cells, and cancer. *Cell.* **129**: 465–472.

Kilian A, Bowtell DD, Abud HE, Hime GR, Venter DJ, Keese PK, Duncan EL, Reddel RR, Jefferson RA. (1997) Isolation of a candidate human telomerase catalytic subunit gene, which reveals complex splicing patterns in different cell types. *Hum. Mol. Genet.* **6**: 2011–2019.

Kim NW, Piatyszek MA, Prowse KR, Harley CB, West MD, Ho PL, Coviello GM, Wright WE, Weinrich SL, Shay JW. (1994) Specific association of human telomerase activity with immortal cells and cancer. *Science.* **266**: 2011–2015.

Klingelhutz AJ, Foster SA, McDougall JK. (1996) Telomerase activation by the E6 gene product of human papillomavirus type 16. *Nature.* **380**: 79–82.

Klinz FJ, Schmidt A, Schinkothe T, Arnhold S, Desai B, Popken F, Brixius K, Schwinger R, Mehlhorn U, Staib P, Addicks K, Bloch W. (2005) Phospho-eNOS Ser-114 in human mesenchymal stem cells: constitutive phosphorylation, nuclear localization and upregulation during mitosis. *Eur. J. Cell. Biol.* **84**: 809–818.

Koshiji M, Kageyama Y, Pete EA, Horikawa I, Barrett JC, Huang LE. (2004) HIF-1alpha induces cell cycle arrest by functionally counteracting Myc. *EMBO J.* **23**: 1949–1956.

Kurz DJ, Hong Y, Trivier E, Huang HL, Decary S, Zang GH, Luscher TF, Erusalimsky JD. (2003) Fibroblast growth factor-2, but not vascular endothelial growth factor, upregulates telomerase activity in human endothelial cells. *Arterioscler. Thromb. Vasc. Biol.* **23**: 748–754.

Kusumoto M, Ogawa T, Mizumoto K, Ueno H, Niiyama H, Sato N, Nakamura M, Tanaka M. (1999) Adenovirus-mediated p53 gene transduction inhibits telomerase activity independent of its effects on cell cycle arrest and apoptosis in human pancreatic cancer cells. *Clin. Cancer Res.* **5**: 2140–2147.

Kyo S, Takakura M, Fujiwara T, Inoue M. (2008) Understanding and exploiting hTERT promoter regulation for diagnosis and treatment of human cancers. *Cancer Sci.* **99**: 1528–1538.

Kyo S, Takakura M, Kanaya T, Zhuo W, Fujimoto K, Nishio Y, Orimo A, Inoue M. (1999) Estrogen activates telomerase. *Cancer Res.* **59**: 5917–5921.

Kyo S, Takakura M, Taira T, Kanaya T, Itoh H, Yutsudo M, Ariga H, Inoue M. (2000) Sp1 cooperates with c-Myc to activate transcription of the human telomerase reverse transcriptase gene (hTERT). *Nucleic Acids Res.* **28**: 669–677.

Lacerte A, Korah J, Roy M, Yang XJ, Lemay S, Lebrun JJ. (2008) Transforming growth factor-beta inhibits telomerase through SMAD3 and E2F transcription factors. *Cell Signal.* **20**: 50–59.

Levine AJ, Oren M. (2009) The first 30 years of p53: growing ever more complex. *Nat. Rev. Cancer.* **9**: 749–758.

Li H, Liu JP. (2007) Mechanisms of action of TGF-beta in cancer: evidence for Smad3 as a repressor of the hTERT gene. *Ann. NY Acad. Sci.* **1114**: 56–68.

Li H, Xu D, Li J, Berndt MC, Liu JP. (2006a) Transforming growth factor beta suppresses human telomerase reverse transcriptase (hTERT) by Smad3 interactions with c-Myc and the hTERT gene. *J. Biol. Chem.* **281**: 25588–25600.

Li H, Xu D, Toh BH, Liu JP. (2006b) TGF-beta and cancer: is Smad3 a repressor of hTERT gene?. *Cell Res.* **16**: 169–173.

Lin SY, Elledge SJ. (2003) Multiple tumor suppressor pathways negatively regulate telomerase. *Cell.* **113**: 881–889.

Lindkvist A, Franzen A, Ren ZP, Heldin NE, Paulsson-Karlsson Y. (2005) Differential effects of TGF-beta1 on telomerase activity in thyroid carcinoma cell lines. *Biochem. Biophys. Res. Commun.* **338**: 1625–1633.

Lindkvist A, Ivarsson K, Jernberg-Wiklund H, Paulsson-Karlsson Y. (2006) Interferon-induced sensitization to apoptosis is associated with repressed transcriptional activity of the hTERT promoter in multiple myeloma. *Biochem. Biophys. Res. Commun.* **341**: 1141–1148.

Liu C, Fang X, Ge Z, Jalink M, Kyo S, Bjorkholm M, Gruber A, Sjoberg J, Xu D. (2007a) The telomerase reverse transcriptase (hTERT) gene is a direct target of the histone methyltransferase SMYD3. *Cancer Res.* **67**: 2626–2631.

Liu H, Luan F, Ju Y, Shen H, Gao L, Wang X, Liu S, Zhang L, Sun W, Ma C. (2007b) *In vitro* transfection of the hepatitis B virus PreS2 gene into the human hepatocarcinoma cell line HepG2 induces upregulation of human telomerase reverse transcriptase. *Biochem. Biophys. Res. Commun.* **355**: 379–384.

Liu L, Lai S, Andrews LG, Tollefsbol TO. (2004) Genetic and epigenetic modulation of telomerase activity in development and disease. *Gene.* **340**: 1–10.

Liu X, Dakic A, Zhang Y, Dai Y, Chen R, Schlegel R. (2009) HPV E6 protein interacts physically and functionally with the cellular telomerase complex. *Proc. Natl. Acad. Sci. USA.* **106**: 18780–18785.

Liu X, Yuan H, Fu B, Disbrow GL, Apolinario T, Tomaic V, Kelley ML, Baker CC, Huibregtse J, Schlegel R. (2005) The E6AP ubiquitin ligase is required for transactivation of the hTERT promoter by the human papillomavirus E6 oncoprotein. *J. Biol. Chem.* **280**: 10807–10816.

Luan F, Liu H, Gao L, Liu J, Sun Z, Ju Y, Hou N, Guo C, Liang X, Zhang L, Sun W, Ma C. (2009) Hepatitis B virus protein preS2 potentially promotes HCC development via its transcriptional activation of hTERT. *Gut.* **58**: 1528–1537.

Luke B, Lingner J. (2009) TERRA: telomeric repeat-containing RNA. *EMBO J.* **28**: 2503–2510.

Maida Y, Kyo S, Kanaya T, Wang Z, Yatabe N, Tanaka M, Nakamura M, Ohmichi M, Gotoh N, Murakami S, Inoue M. (2002) Direct activation of telomerase by EGF through Ets-mediated transactivation of TERT via MAP kinase signaling pathway. *Oncogene*. **21**: 4071–4079.

Marion RM, Blasco MA., Telomere rejuvenation during nuclear reprogramming. *Curr. Opin. Genet. Dev.* **20**: 190–196.

Meeker AK, Sommerfeld HJ, Coffey DS. (1996) Telomerase is activated in the prostate and seminal vesicles of the castrated rat. *Endocrinology*. **137**: 5743–5746.

Mei YP, Zhu XF, Zhou JM, Huang H, Deng R, Zeng YX. (2006) siRNA targeting LMP1-induced apoptosis in EBV-positive lymphoma cells is associated with inhibition of telomerase activity and expression. *Cancer Lett*. **232**: 189–198.

Minamino T, Mitsialis SA, Kourembanas S. (2001) Hypoxia extends the life span of vascular smooth muscle cells through telomerase activation. *Mol. Cell Biol.* **21**: 3336–3342.

Misiti S, Nanni S, Fontemaggi G, Cong YS, Wen J, Hirte HW, Piaggio G, Sacchi A, Pontecorvi A, Bacchetti S, Farsetti A. (2000) Induction of hTERT expression and telomerase activity by estrogens in human ovary epithelium cells. *Mol. Cell Biol.* **20**: 3764–3771.

Mitomo S, Maesawa C, Ogasawara S, Iwaya T, Shibazaki M, Yashima-Abo A, Kotani K, Oikawa H, Sakurai E, Izutsu N, Kato K, Komatsu H, Ikeda K, Wakabayashi G, Masuda T. (2008) Downregulation of miR-138 is associated with overexpression of human telomerase reverse transcriptase protein in human anaplastic thyroid carcinoma cell lines. *Cancer Sci.* **99**: 280–286.

Miura N, Sato R, Tsukamoto T, Shimizu M, Kabashima H, Takeda M, Takahashi S, Harada T, West JE, Drabkin H, Mejia JE, Shiota G, Murawaki Y, Virmani A, Gazdar AF, Oshimura M, Hasegawa J. (2009) A noncoding RNA gene on chromosome 10p15.3 may function upstream of hTERT. *BMC Mol Biol.* **10**: 5.

Moehren U, Papaioannou M, Reeb CA, Grasselli A, Nanni S, Asim M, Roell D, Prade I, Farsetti A, Baniahmad A. (2008) Wild-type but not mutant androgen receptor inhibits expression of the hTERT telomerase subunit: a novel role of AR mutation for prostate cancer development. *FASEB J.* **22**: 1258–1267.

Moyzis RK, Buckingham JM, Cram LS, Dani M, Deaven LL, Jones MD, Meyne J, Ratliff RL, WuJr. (1988) A highly conserved repetitive DNA sequence, (TTAGGG)*n*, present at the telomeres of human chromosomes. *Proc. Natl. Acad. Sci. USA*. **85**: 6622–6626.

Nair P, Jayaprakash PG, Nair MK, Pillai MR. (2000) Telomerase, p53 and human papillomavirus infection in the uterine cervix. *Acta Oncol.* **39**: 65–70.

Nakano K, Watney E, McDougall JK. (1998) Telomerase activity and expression of telomerase RNA component and telomerase catalytic subunit gene in cervical cancer. *Am. J. Pathol.* **153**: 857–864.

Nanni S, Benvenuti V, Grasselli A, Priolo C, Aiello A, Mattiussi S, Colussi C, Lirangi V, Illi B, D'Eletto M, Cianciulli AM, Gallucci M, De Carli P, Sentinelli S, Mottolese M, Carlini P, Strigari L, Finn S, Mueller E, Arcangeli G, Gaetano C, Capogrossi MC, Donnorso RP, Bacchetti S, Sacchi A, Pontecorvi A, Loda M, Farsetti A. (2009) Endothelial NOS, estrogen receptor beta, and HIFs cooperate in the activation of a prognostic transcriptional pattern in aggressive human prostate cancer. *J. Clin. Invest.* **119**: 1093–1108.

Nanni S, Narducci M, Della Pietra L, Moretti F, Grasselli A, De Carli P, Sacchi A, Pontecorvi A, Farsetti A. (2002) Signaling through estrogen receptors modulates telomerase activity in human prostate cancer. *J. Clin. Invest.* **110**: 219–227.

Nanni S, Priolo C, Grasselli A, D'Eletto M, Merola R, Moretti F, Gallucci M, De Carli P, Sentinelli S, Cianciulli AM, Mottolese M, Carlini P, Arcelli D, Helmer-Citterich M, Gaetano C, Loda M, Pontecorvi A, Bacchetti S, Sacchi A, Farsetti A. (2006) Epithelial-restricted gene profile of primary cultures from human prostate tumors: a molecular approach to predict clinical behavior of prostate cancer. *Mol. Cancer Res.* **4**: 79–92.

Narducci ML, Grasselli A, Biasucci LM, Farsetti A, Mule A, Liuzzo G, La Torre G, Niccoli G, Mongiardo R, Pontecorvi A, Crea F. (2007) High telomerase activity in neutrophils from unstable coronary plaques. *J. Am. Coll. Cardiol.* **50**: 2369–2374.

Nishi H, Nakada T, Kyo S, Inoue M, Shay JW, Isaka K. (2004) Hypoxia-inducible factor 1 mediates upregulation of telomerase (hTERT). *Mol. Cell Biol.* **24**: 6076–6083.

Oh S, Song Y, Yim J, Kim TK. (1999) The Wilms' tumor 1 tumor suppressor gene represses transcription of the human telomerase reverse transcriptase gene. *J. Biol. Chem.* **274**: 37473–37478.

Oh S, Song YH, Yim J, Kim TK. (2000) Identification of Mad as a repressor of the human telomerase (hTERT) gene. *Oncogene.* **19**: 1485–1490.

Paterlini-Brechot P, Saigo K, Murakami Y, Chami M, Gozuacik D, Mugnier C, Lagorce D, Brechot C. (2003) Hepatitis B virus-related insertional mutagenesis occurs frequently in human liver cancers and recurrently targets human telomerase gene. *Oncogene.* **22**: 3911–3916.

Pendino F, Dudognon C, Delhommeau F, Sahraoui T, Flexor M, Bennaceur-Griscelli A, Lanotte M, Segal-Bendirdjian E. (2003) Retinoic acid receptor alpha and retinoid-X receptor-specific agonists synergistically target telomerase expression and induce tumor cell death. *Oncogene.* **22**: 9142–9150.

Phipps SM, Love WK, White T, Andrews LG, Tollefsbol TO. (2009) Retinoid-induced histone deacetylation inhibits telomerase activity in estrogen receptor-negative breast cancer cells. *Anticancer Res.* **29**: 4959–4964.

Pouyssegur J, Dayan F, Mazure NM. (2006) Hypoxia signalling in cancer and approaches to enforce tumour regression. *Nature.* **441**: 437–443.

Qu ZL, Zou SQ, Cui NQ, Wu XZ, Qin MF, Kong D, Zhou ZL. (2005) Upregulation of human telomerase reverse transcriptase mRNA expression by *in vitro* transfection of hepatitis B virus X gene into human hepatocarcinoma and cholangiocarcinoma cells. *World J. Gastroenterol.* **11**: 5627–5632.

Rahman R, Latonen L, Wiman KG. (2005) hTERT antagonizes p53-induced apoptosis independently of telomerase activity. *Oncogene.* **24**: 1320–1327.

Renaud S, Loukinov D, Abdullaev Z, Guilleret I, Bosman FT, Lobanenkov V, Benhattar J. (2007) Dual role of DNA methylation inside and outside of CTCF-binding regions in the transcriptional regulation of the telomerase hTERT gene. *Nucleic Acids Res.* **35**: 1245–1256.

Rissanen TT, Vajanto I, Hiltunen MO, Rutanen J, Kettunen MI, Niemi M, Leppanen P, Turunen MP, Markkanen JE, Arve K, Alhava E, Kauppinen RA, Yla-Herttuala S. (2002) Expression of vascular endothelial growth factor and vascular endothelial growth factor receptor-2 (KDR/Flk-1) in ischemic skeletal muscle and its regeneration. *Am. J. Pathol.* **160**: 1393–1403.

Roos G, Nilsson P, Cajander S, Nielsen NH, Arnerlov C, Landberg G. (1998) Telomerase activity in relation to p53 status and clinico-pathological parameters in breast cancer. *Int. J. Cancer.* **79**: 343–348.

Rubin R, Baserga R. (1995) Insulin-like growth factor-I receptor. Its role in cell proliferation, apoptosis, and tumorigenicity. *Lab. Invest.* **73**: 311–331.

Seimiya H, Tanji M, Oh-hara T, Tomida A, Naasani I, Tsuruo T. (1999) Hypoxia up-regulates telomerase activity via mitogen-activated protein kinase signaling in human solid tumor cells. *Biochem. Biophys. Res. Commun.* **260**: 365–370.

Semenza GL. (2003) Targeting HIF-1 for cancer therapy. *Nat. Rev. Cancer.* **3**: 721–732.

Shats I, Milyavsky M, Tang X, Stambolsky P, Erez N, Brosh R, Kogan I, Braunstein I, Tzukerman M, Ginsberg D, Rotter V. (2004) p53-dependent down-regulation of telomerase is mediated by p21waf1. *J. Biol. Chem.* **279**: 50976–50985.

Shay JW, Bacchetti S. (1997) A survey of telomerase activity in human cancer. *Eur. J. Cancer.* **33**: 787–791.

Shin KH, Kang MK, Dicterow E, Park NH. (2003) Hypermethylation of the hTERT promoter inhibits the expression of telomerase activity in normal oral fibroblasts and senescent normal oral keratinocytes. *Br. J. Cancer.* **89**: 1473–1478.

Siegel PM, Massague J. (2003) Cytostatic and apoptotic actions of TGF-beta in homeostasis and cancer. *Nat. Rev. Cancer.* **3**: 807–821.

Soder AI, Going JJ, Kaye SB, Keith WN. (1998) Tumour specific regulation of telomerase RNA gene expression visualized by *in situ* hybridization. *Oncogene.* **16**: 979–983.

Soder AI, Hoare SF, Muir S, Going JJ, Parkinson EK, Keith WN. (1997) Amplification, increased dosage and in situ expression of the telomerase RNA gene in human cancer. *Oncogene.* **14**: 1013–1021.

Suzuki T, Kimura A, Nagai R, Horikoshi M. (2000) Regulation of interaction of the acetyltransferase region of p300 and the DNA-binding domain of Sp1 on and through DNA binding. *Genes Cells.* **5**: 29–41.

Sykorova E, Fajkus J. (2009) Structure–function relationships in telomerase genes. *Biol. Cell.* **101**: 375–392 371 p following 392.

Takahashi A, Higashino F, Aoyagi M, Yoshida K, Itoh M, Kyo S, Ohno T, Taira T, Ariga H, Nakajima K, Hatta M, Kobayashi M, Sano H, Kohgo T, Shindoh M. (2003) EWS/ETS fusions activate telomerase in Ewing's tumors. *Cancer Res.* **63**: 8338–8344.

Takakura M, Kyo S, Kanaya T, Hirano H, Takeda J, Yutsudo M, Inoue M. (1999) Cloning of human telomerase catalytic subunit (hTERT) gene promoter and identification of proximal core promoter sequences essential for transcriptional activation in immortalized and cancer cells. *Cancer Res.* **59**: 551–557.

Takakura M, Kyo S, Sowa Y, Wang Z, Yatabe N, Maida Y, Tanaka M, Inoue M. (2001) Telomerase activation by histone deacetylase inhibitor in normal cells. *Nucleic Acids Res.* **29**: 3006–3011.

Terrin L, Dal Col J, Rampazzo E, Zancai P, Pedrotti M, Ammirabile G, Bergamin S, Rizzo S, Dolcetti R, De Rossi A. (2008) Latent membrane protein 1 of Epstein–Barr virus activates the hTERT promoter and enhances telomerase activity in B lymphocytes. *J. Virol.* **82**: 10175–10187.

Teske E, Naan EC, van Dijk EM, Van Garderen E, Schalken JA. (2002) Canine prostate carcinoma: epidemiological evidence of an increased risk in castrated dogs. *Mol. Cell. Endocrinol.* **197**: 251–255.

Tu W, Zhang DK, Cheung PT, Tsao SW, Lau YL. (1999) Effect of insulin-like growth factor 1 on PHA-stimulated cord blood mononuclear cell telomerase activity. *Br. J. Haematol.* **104**: 785–794.

Ulaner GA, Hu JF, Vu TH, Giudice LC, Hoffman AR. (1998) Telomerase activity in human development is regulated by human telomerase reverse transcriptase (hTERT) transcription and by alternate splicing of hTERT transcripts. *Cancer Res.* **58**: 4168–4172.

Ulaner GA, Hu JF, Vu TH, Oruganti H, Giudice LC, Hoffman AR. (2000) Regulation of telomerase by alternate splicing of human telomerase reverse transcriptase (hTERT) in normal and neoplastic ovary, endometrium and myometrium. *Int. J. Cancer.* **85**: 330–335.

Ullrich A, Schlessinger J. (1990) Signal transduction by receptors with tyrosine kinase activity. *Cell.* **61**: 203–212.

Vale PR, Isner JM, Rosenfield K. (2001) Therapeutic angiogenesis in critical limb and myocardial ischemia. *J. Interv. Cardiol.* **14**: 511–528.

van der Geer P, Hunter T, Lindberg RA. (1994) Receptor protein-tyrosine kinases and their signal transduction pathways. *Annu. Rev. Cell. Biol.* **10**: 251–337.

Wang J, Xie LY, Allan S, Beach D, Hannon GJ. (1998) Myc activates telomerase. *Genes Dev.* **12**: 1769–1774.

Wang Z, Kyo S, Takakura M, Tanaka M, Yatabe N, Maida Y, Fujiwara M, Hayakawa J, Ohmichi M, Koike K, Inoue M. (2000) Progesterone regulates human telomerase reverse transcriptase gene expression via activation of mitogen-activated protein kinase signaling pathway. *Cancer Res.* **60**: 5376–5381.

Wasylyk B, Wasylyk C, Flores P, Begue A, Leprince D, Stehelin D. (1990) The c-ets proto-oncogenes encode transcription factors that cooperate with c-Fos and c-Jun for transcriptional activation. *Nature.* **346**: 191–193.

Wetterau LA, Francis MJ, Ma L, Cohen P. (2003) Insulin-like growth factor I stimulates telomerase activity in prostate cancer cells. *J. Clin. Endocrinol. Metab.* **88**: 3354–3359.

Wick M, Zubov D, Hagen G. (1999) Genomic organization and promoter characterization of the gene encoding the human telomerase reverse transcriptase (hTERT). *Gene.* **232**: 97–106.

Wu KJ, Grandori C, Amacker M, Simon-Vermot N, Polack A, Lingner J, Dalla-Favera R. (1999) Direct activation of TERT transcription by c-MYC. *Nat. Genet.* **21**: 220–224.

Xiao X, Athanasiou M, Sidorov IA, Horikawa I, Cremona G, Blair D, Barret JC, Dimitrov DS. (2003) Role of Ets/Id proteins for telomerase regulation in human cancer cells. *Exp. Mol. Pathol.* **75**: 238–247.

Xiao X, Phogat SK, Sidorov IA, Yang J, Horikawa I, Prieto D, Adelesberger J, Lempicki R, Barrett JC, Dimitrov DS. (2002) Identification and characterization of rapidly dividing U937 clones with differential telomerase activity and gene expression profiles: role of c-Myc/Mad1 and Id/Ets proteins. *Leukemia.* **16**: 1877–1880.

Xu D, Dwyer J, Li H, Duan W, Liu JP. (2008) Ets2 maintains hTERT gene expression and breast cancer cell proliferation by interacting with c-Myc. *J. Biol. Chem.* **283**: 23567–23580.

Xu D, Popov N, Hou M, Wang Q, Bjorkholm M, Gruber A, Menkel AR, Henriksson M. (2001) Switch from Myc/Max to Mad1/Max binding and decrease in histone acetylation at the telomerase reverse transcriptase promoter during differentiation of HL60 cells. *Proc. Natl. Acad. Sci. USA.* **98**: 3826–3831.

Xu D, Wang Q, Gruber A, Bjorkholm M, Chen Z, Zaid A, Selivanova G, Peterson C, Wiman KG, Pisa P. (2000) Downregulation of telomerase reverse transcriptase mRNA expression by wild type p53 in human tumor cells. *Oncogene.* **19**: 5123–5133.

Yang H, Kyo S, Takatura M, Sun L. (2001) Autocrine transforming growth factor beta suppresses telomerase activity and transcription of human telomerase reverse transcriptase in human cancer cells. *Cell. Growth Differ.* **12**: 119–127.

Yang J, Deng X, Deng L, Gu H, Fan W, Cao Y. (2004) Telomerase activation by Epstein–Barr virus latent membrane protein 1 is associated with c-Myc expression in human nasopharyngeal epithelial cells. *J. Exp. Clin. Cancer Res.* **23**: 495–506.

Yatabe N, Kyo S, Maida Y, Nishi H, Nakamura M, Kanaya T, Tanaka M, Isaka K, Ogawa S, Inoue M. (2004) HIF-1-mediated activation of telomerase in cervical cancer cells. *Oncogene.* **23**: 3708–3715.

Yi X, White DM, Aisner DL, Baur JA, Wright WE, Shay JW. (2000) An alternate splicing variant of the human telomerase catalytic subunit inhibits telomerase activity. *Neoplasia.* **2**: 433–440.

Zaccagnini G, Gaetano C, Della Pietra L, Nanni S, Grasselli A, Mangoni A, Benvenuto R, Fabrizi M, Truffa S, Germani A, Moretti F, Pontecorvi A, Sacchi A, Bacchetti S, Capogrossi MC, Farsetti A. (2005) Telomerase mediates vascular endothelial growth factor-dependent responsiveness in a rat model of hind limb ischemia. *J. Biol. Chem.* **280**: 14790–14798.

Zachary I. (2003) VEGF signalling: integration and multi-tasking in endothelial cell biology. *Biochem. Soc. Trans.* **31**: 1171–1177.

Zeng L, Xu Q. (2008) eNOS–ERalpha complex goes to telomerase. *Circ Res.* **103**: 10–12.

Zhang A, Zheng C, Lindvall C, Hou M, Ekedahl J, Lewensohn R, Yan Z, Yang X, Henriksson M, Blennow E, Nordenskjold M, Zetterberg A, Bjorkholm M, Gruber A, Xu D. (2000) Frequent amplification of the telomerase reverse transcriptase gene in human tumors. *Cancer Res.* **60**: 6230–6235.

Zhang H, Gao P, Fukuda R, Kumar G, Krishnamachary B, Zeller KI, Dang CV, Semenza GL. (2007) HIF-1 inhibits mitochondrial biogenesis and cellular respiration in VHL-deficient renal cell carcinoma by repression of C-MYC activity. *Cancer Cell.* **11**: 407–420.

Zhao J, Bilsland A, Hoare SF, Keith WN. (2003) Involvement of NF-Y and Sp1 binding sequences in basal transcription of the human telomerase RNA gene. *FEBS Lett.* **536**: 111–119.

Zhao JQ, Hoare SF, McFarlane R, Muir S, Parkinson EK, Black DM, Keith WN. (1998) Cloning and characterization of human and mouse telomerase RNA gene promoter sequences. *Oncogene.* **16**: 1345–1350.

Zinn RL, Pruitt K, Eguchi S, Baylin SB, Herman JG. (2007) hTERT is expressed in cancer cell lines despite promoter DNA methylation by preservation of unmethylated DNA and active chromatin around the transcription start site. *Cancer Res.* **67**: 194–201.

zur Hausen H. (2002) Papillomaviruses and cancer: from basic studies to clinical application. *Nat. Rev. Cancer.* **2**: 342–350.

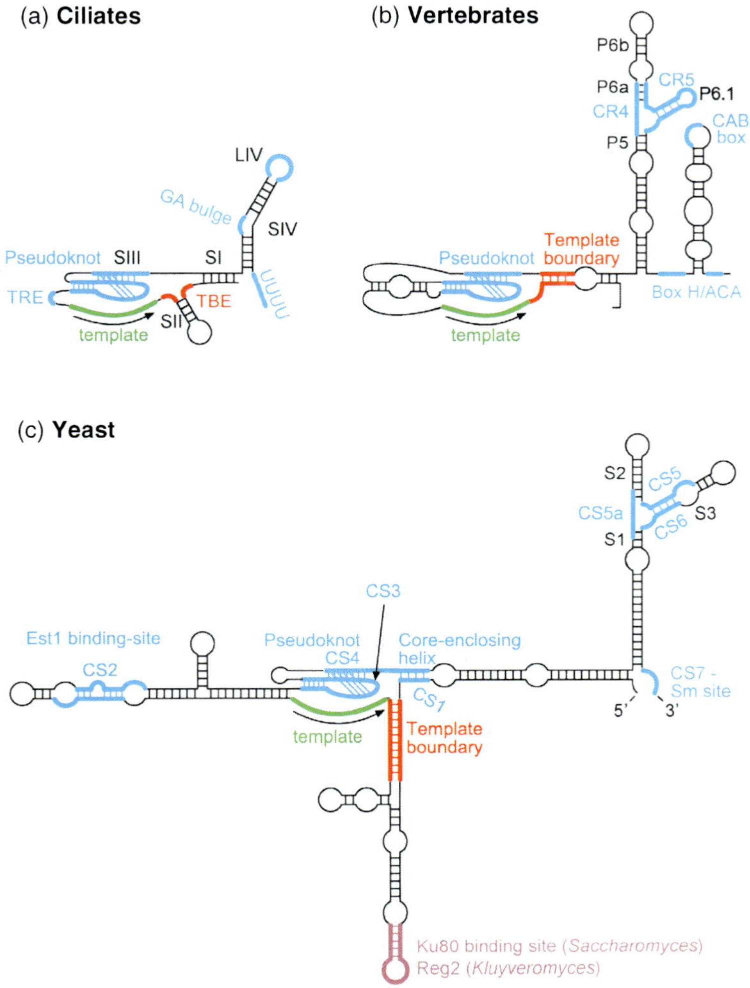

FIGURE 2.1 Common secondary structure models for ciliates, vertebrates, and budding yeast TERs. Indicated are the conserved regions/sequences (CR or CS), pairings/stems (P or S), loops (L), template recognition element (TRE), and template boundary element (TBE).

Telomerases: Chemistry, Biology, and Clinical Applications, First Edition.
Edited by Neal F. Lue and Chantal Autexier.

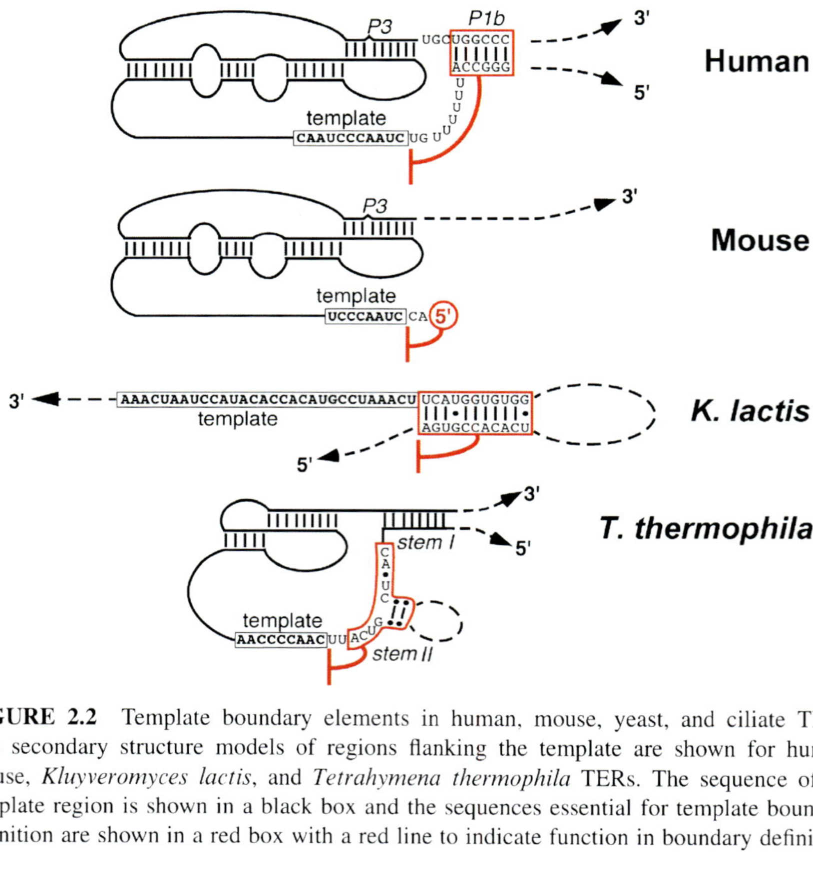

FIGURE 2.2 Template boundary elements in human, mouse, yeast, and ciliate TERs. The secondary structure models of regions flanking the template are shown for human, mouse, *Kluyveromyces lactis*, and *Tetrahymena thermophila* TERs. The sequence of the template region is shown in a black box and the sequences essential for template boundary definition are shown in a red box with a red line to indicate function in boundary definition.

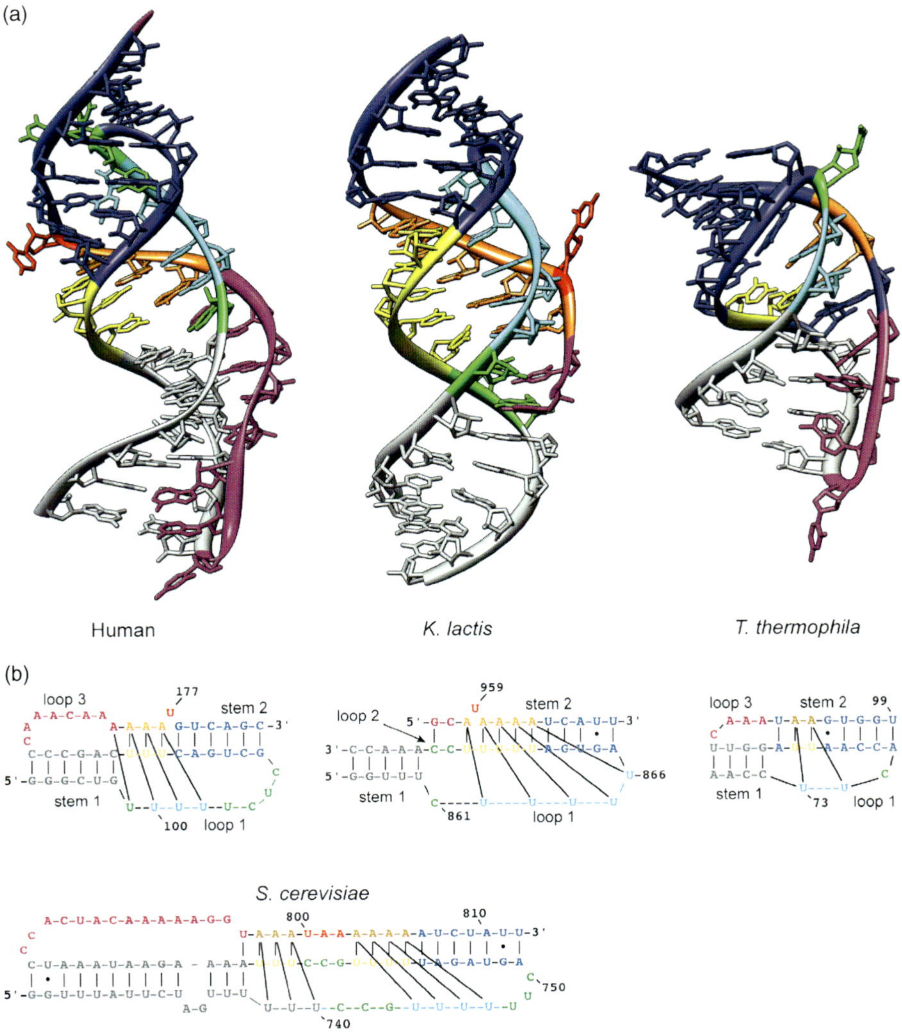

FIGURE 2.3 The pseudoknot structures of human, *K. lactis*, and *Tetrahymena thermophila* telomerase RNAs. (a) Ribbon representations of the three-dimensional solution structure of the human (Kim et al., 2008) pseudoknot, and the computer models of the *K. lactis* (Shefer et al., 2007) and *T. thermophila* (Ulyanov et al., 2007) pseudoknot, illustrated using the computer program *Chimera* (Couch et al., 2006). Stem 1 is shown in gray, residues of stem 2 not participating in base triples are shown in blue. Residues of stem 2 that are part of the triplex, are shown in orange (purines) and yellow (pyrimidines). Bulged-out U residues are shown in red. Residues of loop 1 that are part of the triplex, are shown in cyan. The rest of loop 1, as well as loop 2 if present, are shown in green. Loop 3 is shown in magenta. (b) A schematic representation of base pairing in the pseudoknot, including also the predicted scheme for the *S. cerevisiae* pseudoknot (Gunisova et al., 2009; Qiao and Cech, 2008). Vertical lines represent Watson–Crick interactions; tilted lines, Hoogsteen hydrogen bonds; and "•," a G:U wobble pair. Note that in the *K. lactis* pseudoknot, the region of the junction between stem 1 and 2 is illustrated as unpaired, since the interactions among these nucleotides are unknown.

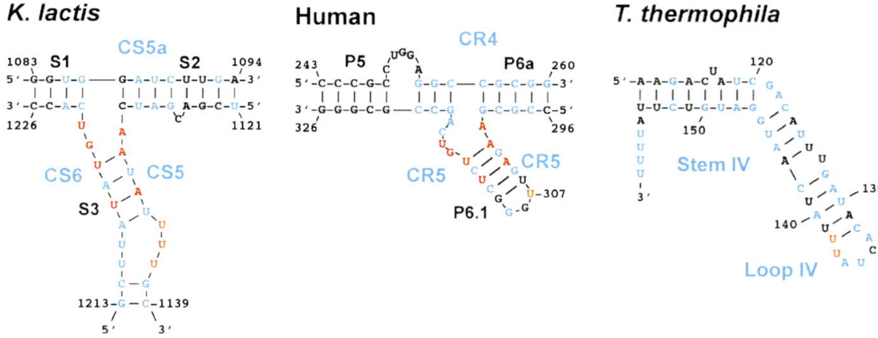

FIGURE 2.4 Conservation of the assembly/activation stem-loop elements in budding yeast, vertebrates, and *Tetrahymena* species. (*See text for full caption*).

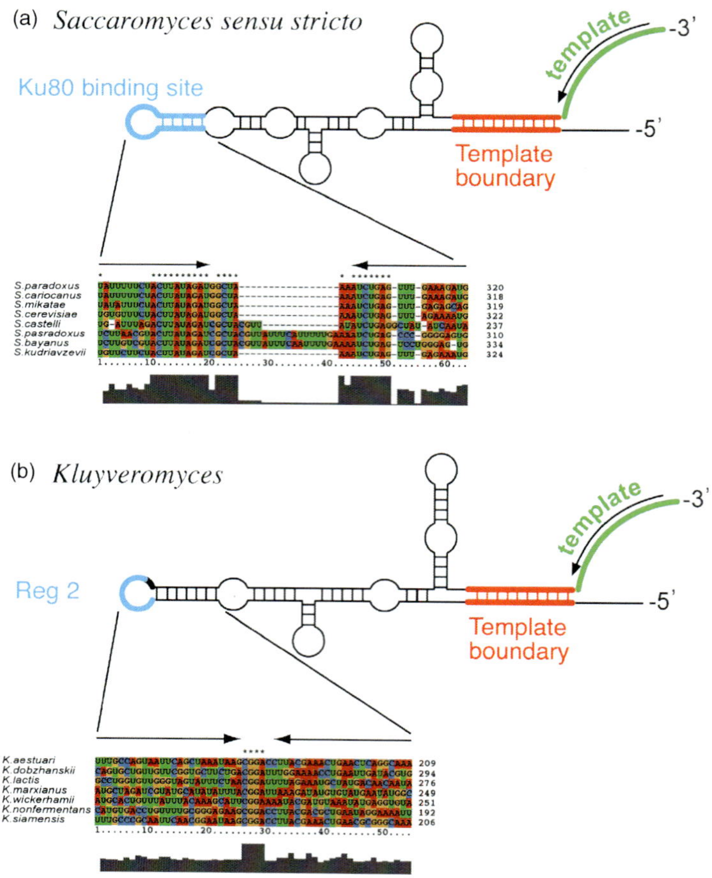

FIGURE 2.5 The 5′ arm of *Saccharomyces* and *Kluyveromyces* telomerase RNAs. Schemes illustrate the 5′ arm, template, and template boundary of *Saccharomyces sensu stricto* (a) and *Kluyveromyces* (b) TERs. Blue lines indicate sequence conservation. Sequence alignments showing the conservation of the Ku80-binding site (48 nt stem-loop) in *Saccharomyces* and a CGGA sequence motif in the *Kluyveromyces* Reg 2 element were made by the computer program *ClustalX* (Chenna et al., 2003).

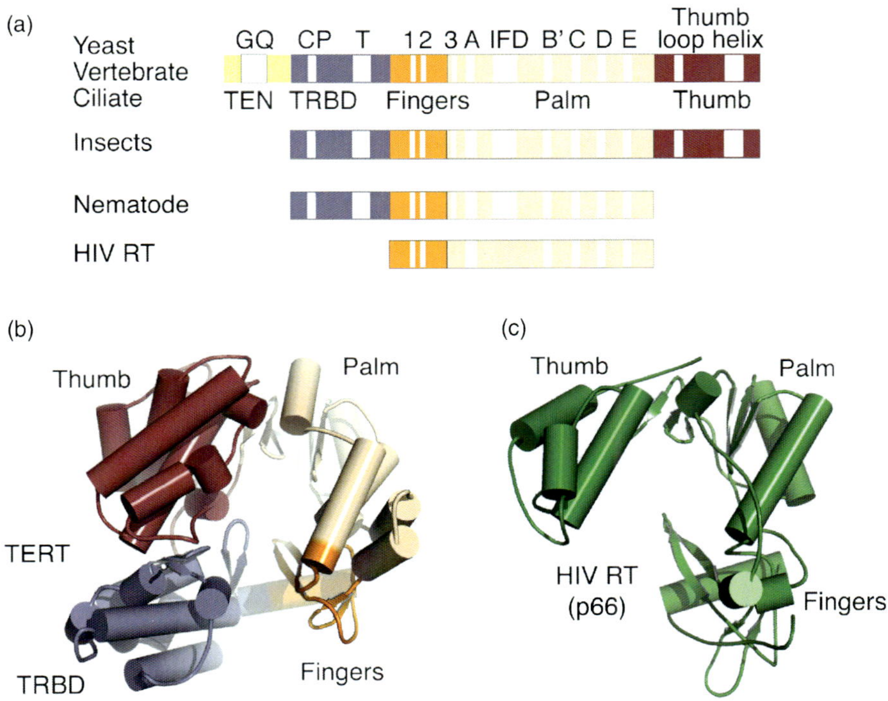

FIGURE 3.1 (a) The domain organizations of TERTs from different species are illustrated. (b) The structure of *Tc*TERT (PDB ID: 3DU6): the TRBD, fingers, palm, and thumb domains are colored in blue, orange, wheat, and red, respectively. (c) The structure of the p66 subunit of HIV-1 RT without the nuclease domain (PDB ID: 1RTD) is shown.

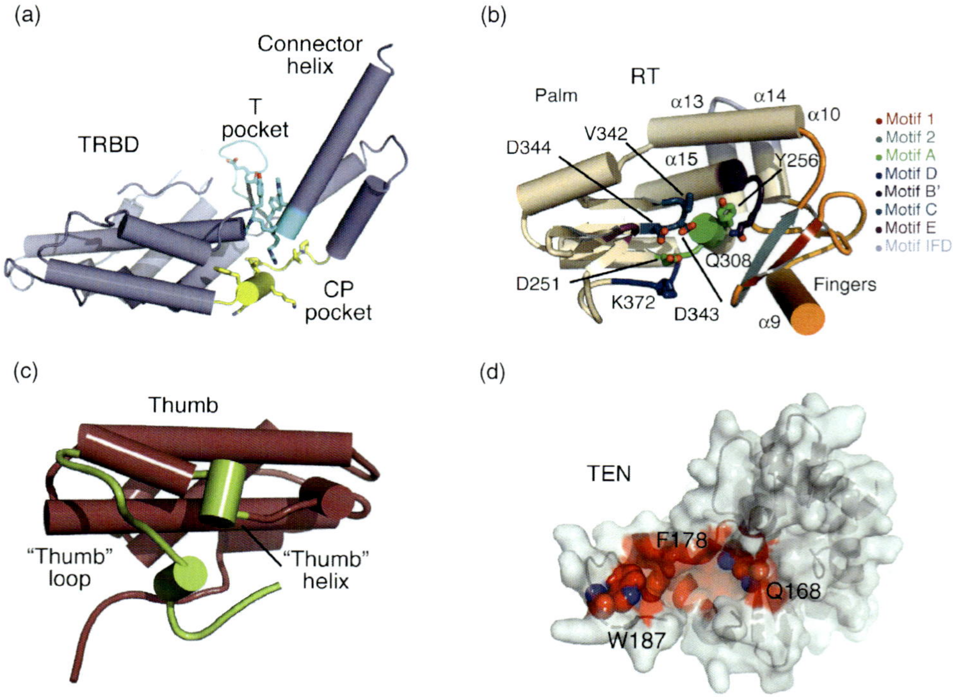

FIGURE 3.2 (a) The RNA-binding domain of *Tetrahymena thermophila* TERT with motif T in cyan and CP in yellow: conserved residues that comprise these motifs are shown in the stick representation. (b) The fingers (orange) and palm (wheat) subdomains of *Tc*TERT: conserved motifs implicated in nucleotide and nucleic acid binding and catalysis are displayed in the designated colors. (c) The thumb domain of *Tc*TERT with the two DNA-binding structural elements (the thumb loop and helix) highlighted in green. (d) The TEN domain of *T. thermophila* TERT is displayed in a surface representation; the putative DNA-binding groove and the residues implicated in DNA binding (Q168, F178, and W187) are accented.

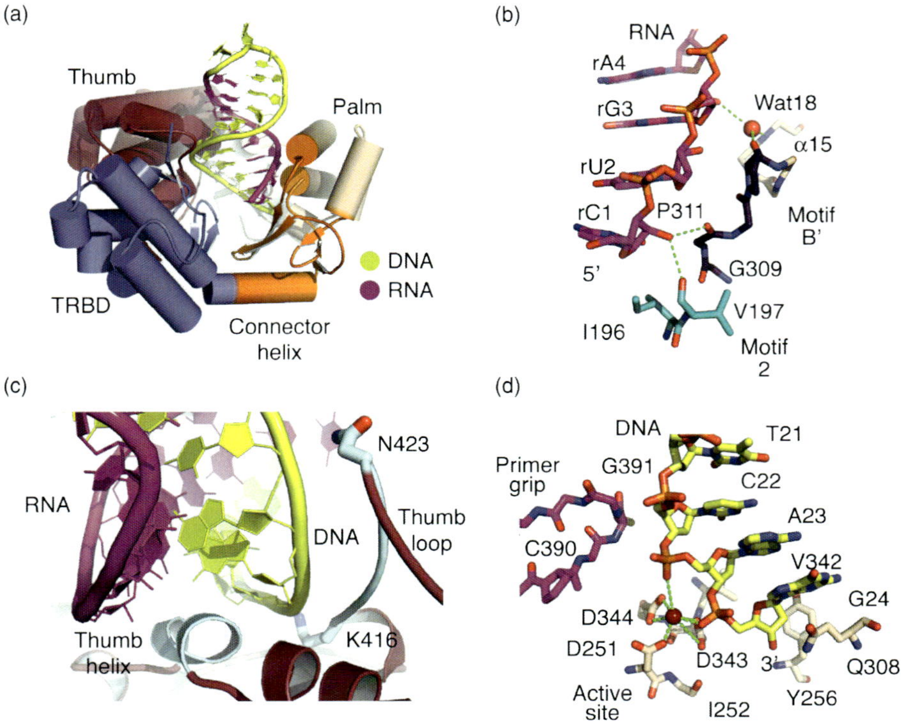

FIGURE 3.3 (a) A complex between *Tc*TERT and an RNA–DNA hairpin (PDB ID: 3KYL): the domain orientation and color scheme are similar to those shown in Figure 3.1B. (b) A close view of the contacts between the RNA template and motifs 2 and B′ of *Tc*TERT. (c) A close view of the contacts between the RNA–DNA hybrid and the thumb helix (light blue) and thumb loop (light blue) in the complex. (d) The primer grip region (motif E) is juxtaposed to the 3′-end of the DNA primer at the active site of the enzyme.

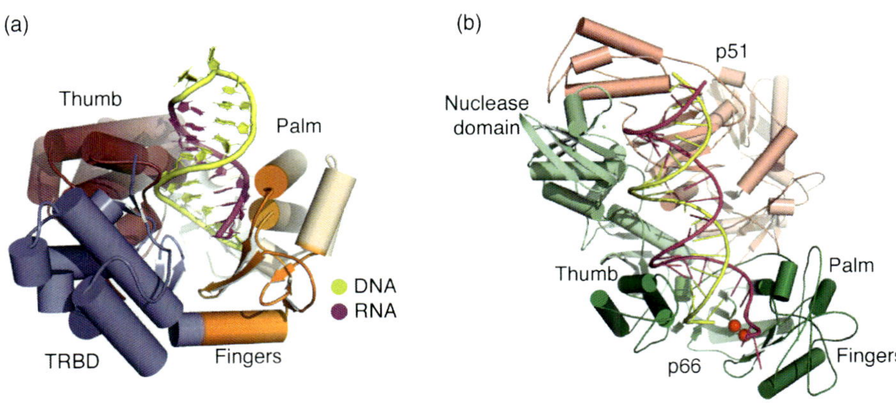

FIGURE 3.4 (a) Same as Figure 3.3a. (b) The structure of the HIV-1 RT bound to an RNA–DNA heteroduplex (PDB ID: 1RTD).

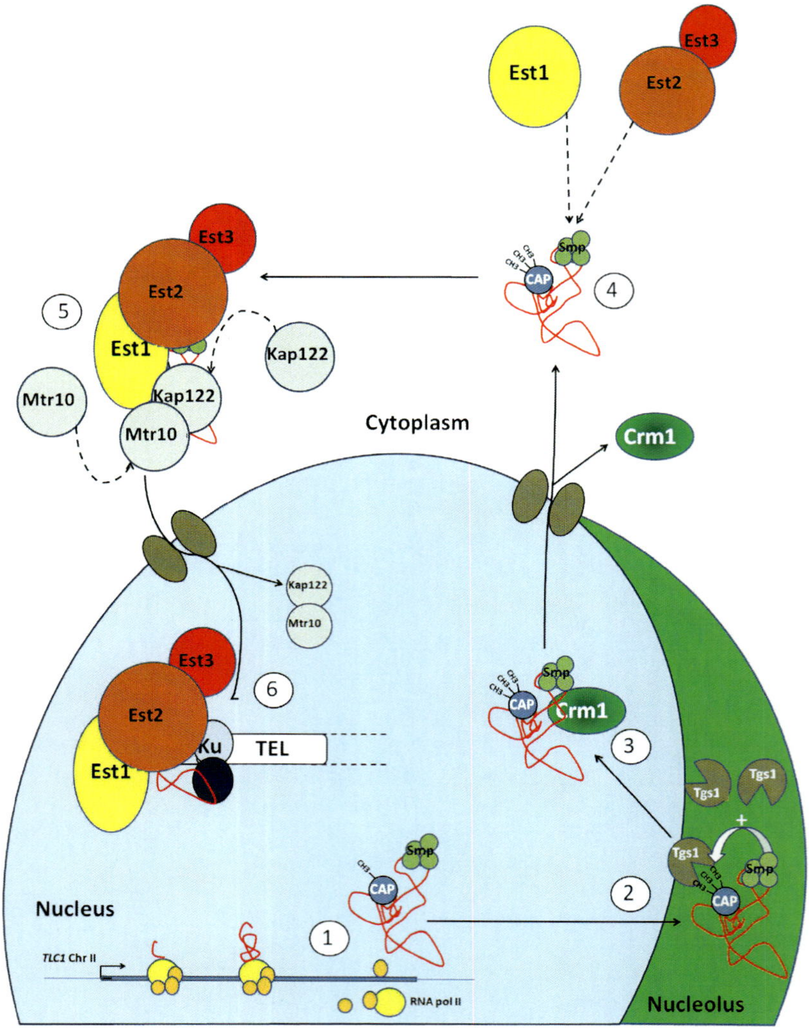

FIGURE 4.2 An integrated model of telomerase biogenesis in yeast. The *Saccharomyces cerevisiae* telomerase RNA *TLC1* is transcribed by the RNA polymerase II machinery (1) and targeted to the nucleolus where its 5′mono-methylguanosine cap is hypermethylated by Tgs1 (2). Following its 5′cap hypermethylation, the *TLC1* RNA is exported in the cytoplasm via the Crm1p-dependent pathway (3). In the cytoplasm, the *TLC1* RNA recruits the proteic components of the telomerase complex (4), assembles into a mature telomerase particle (5), and is imported back in the nucleus via a Mtr10/Kap122 pathway (5). Once in the nucleus, it can be recruited at the telomeres via the interaction between the *TLC1* RNA and the yKu heterodimer (6). (The telomerase holoenzyme at the telomere depicted in this figure would correspond to the one in S phase. As Est1 is actively degraded or not depending on the phase of the cell cycle, so the constitution of the telomerase recruited at the telomeres will vary accordingly). Taken from Gallardo and Chartrand (2008). ©Landes Bioscience.

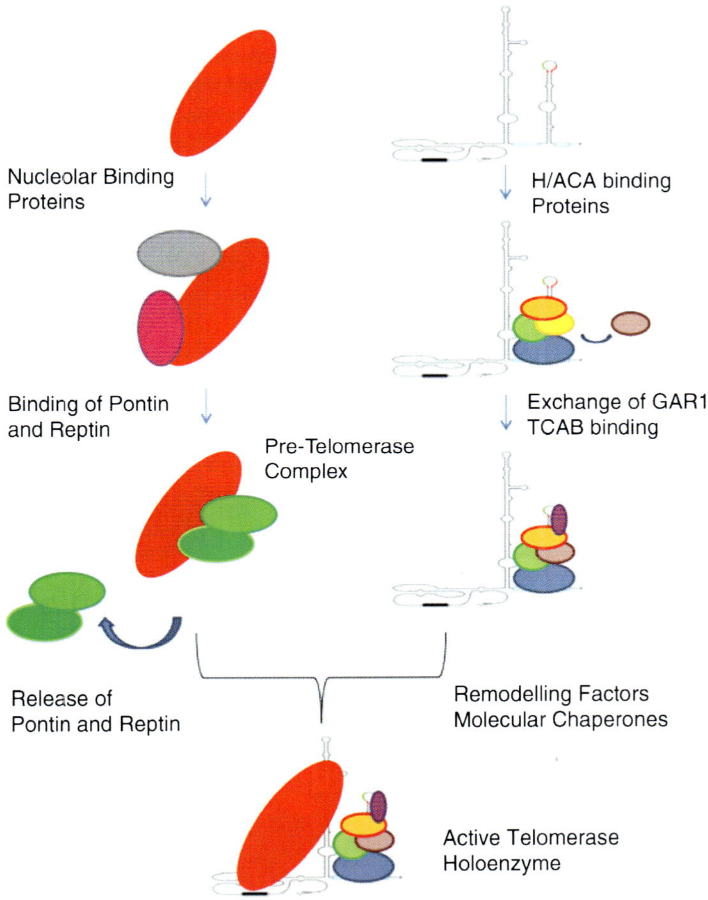

FIGURE 4.4 Assembly of the human telomerase complex. After the transcription and processing of the human telomerase RNA, the H/ACA proteins (dyskerin—blue, Nop10—green, NHP2—orange, and NAF1—yellow) bind to the 3′ end of the telomerase RNA. Subsequently, NAF1 is exchanged for GAR1 (burgundy), and TCAB1 (purple) binds to the hTR. After TERT (red) is localized to the nucleolus, mediated in part by interactions with 14-3-3 (gray) and nucleolin (pink) it is assembled with the ATPases pontin and reptin (shown in green), to form a pretelomerase complex. During S-phase, pontin and reptin are released from hTERT and the complex is remodeled or assembled with the help of additional factors (such as the molecular chaperones Nat10, GNL3L, heat shock proteins, and SMN) with hTR to form an active telomerase holoenzyme.

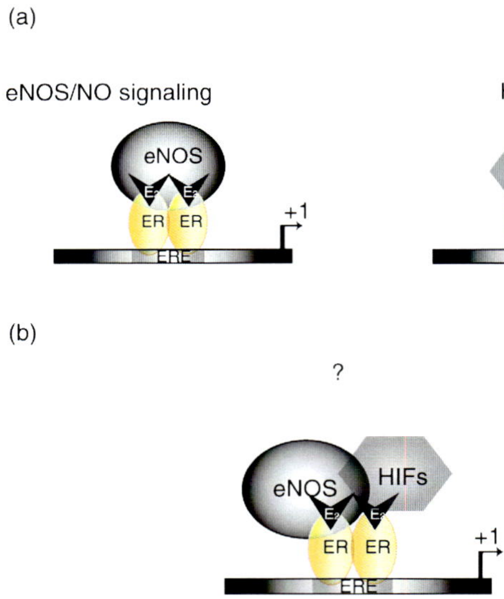

FIGURE 5.1 Cartoons illustrating the functional cooperation between the ERs, eNOS and HIFs pathways in the regulation of hTERT. (a) In primary cultures of human endothelium and in prostate cancer cell lines, ligand-activated ER and eNOS form a combinatorial complex on the estrogen response element (ERE) within the hTERT gene promoter (left panel). In prostate cancer cell lines with a constitutive hypoxic phenotype, ER/HIF-1α or ER/HIF-2α complexes are recruited upon estrogen treatment onto the hTERT–ERE (right panel). All these events lead to increased hTERT gene transcription and telomerase activity. (b) Speculative model of formation of a ER/eNOS/HIF trimeric complex. Since eNOS, ERs, and HIFs play a key role in prostate cancer progression, it is conceivable that they may cooperate in the tumor micro-enviroment by coregulating their transcriptional targets. We propose that in the presence of estrogen and of reduced O2 availability (hypoxia), these factors may form a trimeric complex recruited by the ERE. This event may induce a local chromatin remodeling significantly affecting the transcription of target genes.

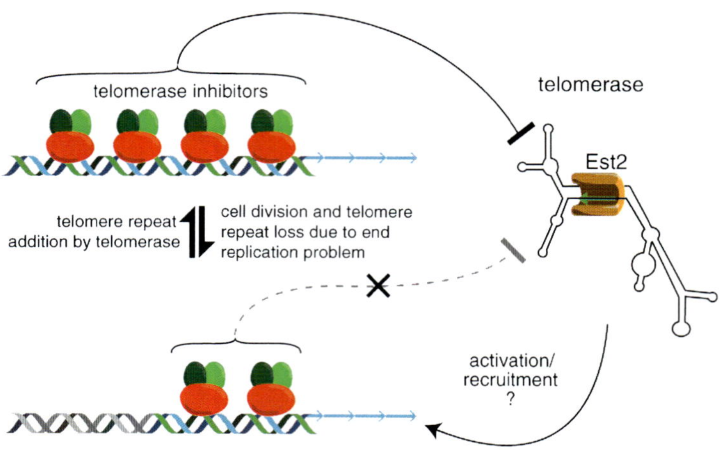

FIGURE 6.1 Schematic representation of the negative feedback "protein counting" model for telomere length regulation. Details described in the text.

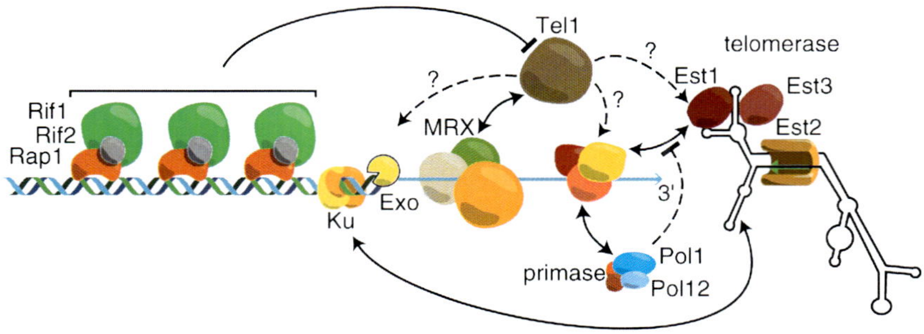

FIGURE 6.2 Proteins and interactions implicated in telomere length regulation in the budding yeast *S. cerevisiae*. Protein–protein interactions are indicated by double-headed arrows. See text for details.

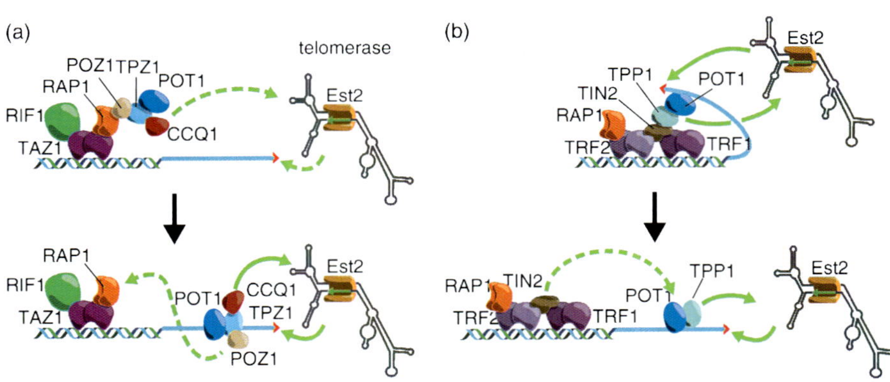

FIGURE 6.3 Schematic representations of models for telomerase activation at telomeres in the fission yeast *S. pombe* (a) and in human cells (b). See text for details.

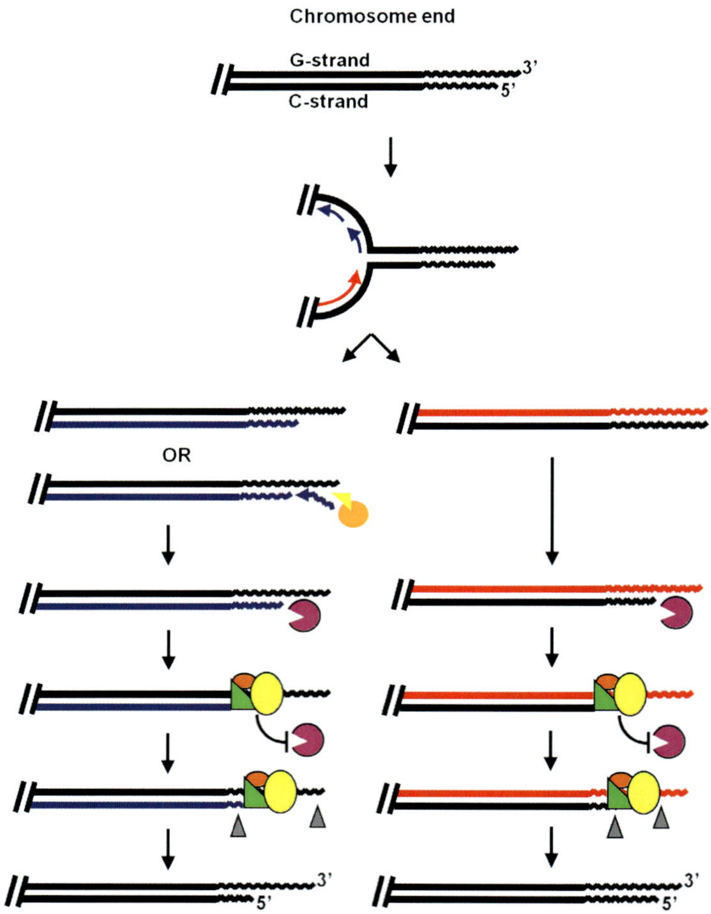

FIGURE 7.1 Telomere replication and the generation of a proper chromosome end structure. At telomeres, the G-strand always serves as a template for lagging-strand synthesis while the C-strand templates the telomere leading strand. Telomeric overhangs, the G-tails, serve as a substrate for telomerase annealing and are formed through different mechanisms on the leading and the lagging telomere sister chromatids. On the lagging strand, they may result simply from incomplete replication at the chromosomal ends or form the removal or the outmost RNA primer by the combined activity of specialized enzymes such as helicases, flap endonucleases or RNases (orange sphere and yellow triangle). The initial 3′-overhang may be further extended due to the activity of an exonuclease. On the leading-strand end, which is predicted to be blunt ended after replication, resection of the C-strand by exonucleolytic activities (magenta sphere) will generate the G-tail. Once overhangs of sufficient length are generated, the binding of ss telomeric DNA binding proteins (the assembly of yellow, orange ovals and green triangle) will obstruct more excessive nucleolytic degradation by blocking access to telomeric ends. Concomitantly, the G-tail bound proteins may modulate the cleavage sites for C- and G-strand specific endonucleases (small grey triangles) and dictate the composition of terminal nt on telomeric DNA and the different length of G-tails on leading versus lagging strand.

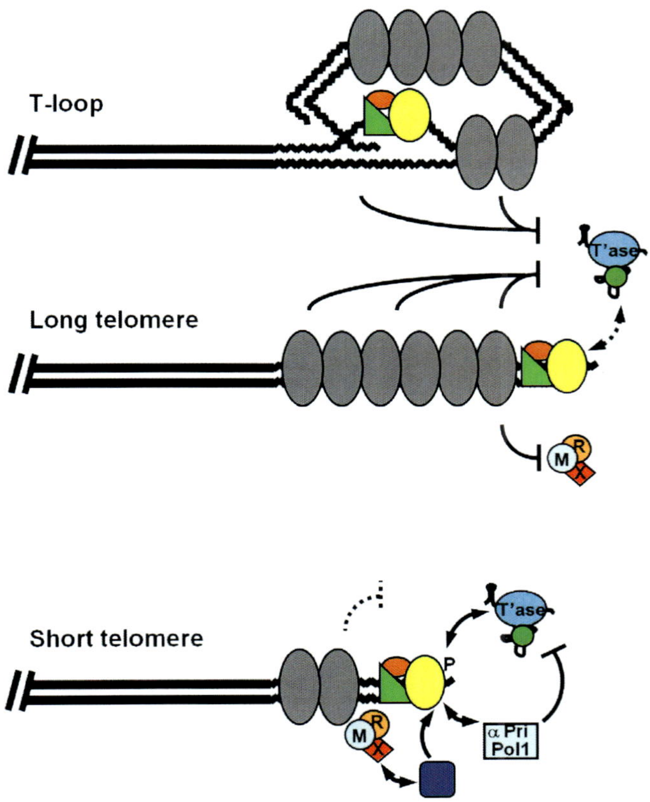

FIGURE 7.2 Regulation of telomerase by telomere-associated proteins. The current view is that the telomeric proteins (grey ovals) bound to the ds telomere repeats (duplex zig-zag line) negatively regulate telomerase. In several experimental systems, the activity of telomerase is reversely correlated with the number of ds telomeric repeats and, respectively, the number of telomeric proteins bound in *cis*, thereby establishing a negative feedback loop or a counting mechanism. In higher eukaryotes, additional negative regulation may be achieved by organization of sufficiently long telomeres into t-loops (illustrated on top). The G-tail-binding proteins (the assembly of yellow, orange ovals and green triangle) appear to facilitate telomerase access and positively regulate its activity at telomeres. Based on data primarily from budding yeast, it has been proposed that at long telomeres, the increased numbers of dsDNA-bound protein molecules inhibit telomerase access and the recruitment of factors promoting the activity of telomerase, such as Mre11 complex. Short telomeres, on the other hand, are permissive for Mre11 recruitment, which in turn recruits checkpoint kinases, like Tel1/ATM (blue square). Together they signal the presence of a short telomere by preparing telomere structure and modifying telomere proteins (P; phosphorylation) to facilitate the recruitment and extension of telomeric DNA by telomerase. The telomerase-mediated extension is tightly coordinated with the conventional replication machinery, which limits the addition of new telomeric repeats by telomerase.

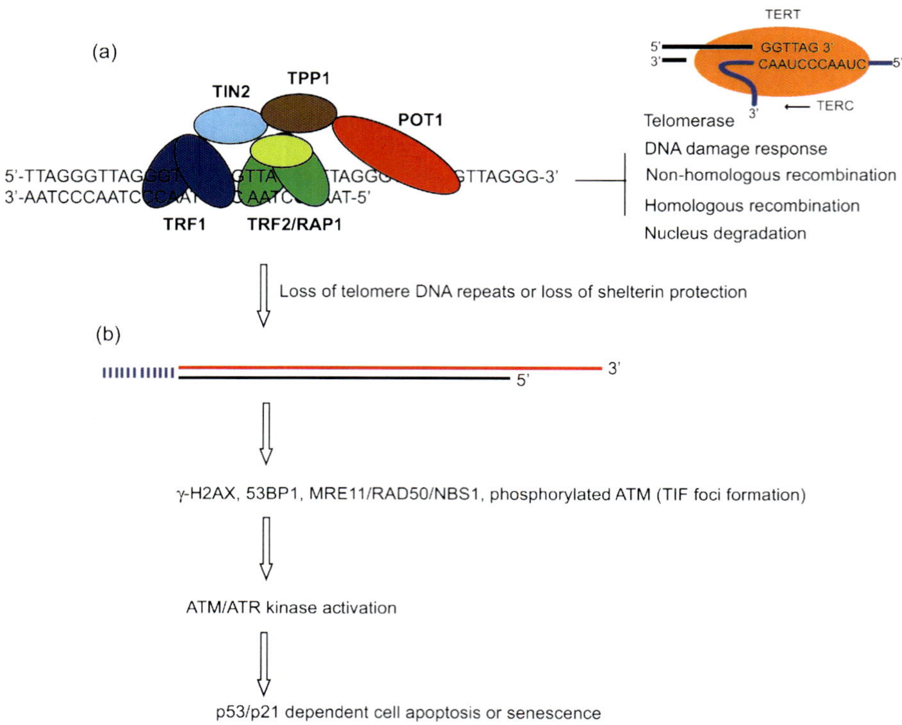

FIGURE 9.1 Telomere maintenance by telomerase and shelterin, and the consequences of telomere dysfunction. (a) Telomere DNA, telomerase, and shelterin. Telomeres cap the chromosome ends and protect against NHEJ, HR, DNA damage signaling, and nucleolytic degradation. The access of telomerase to the telomere is limited by telomere-bound POT1 and TRF1. (b) Dysfunctional telomeres arise via loss of telomere DNA repeats or loss of protection of shelterin, resulting in the induction of DNA damage foci at telomeres (TIF) and activation of ATM–ATR kinase pathways. These signaling cascades in turn can lead to p53/p21 dependent cell apoptosis, cell cycle arrest, and cellular senescence.

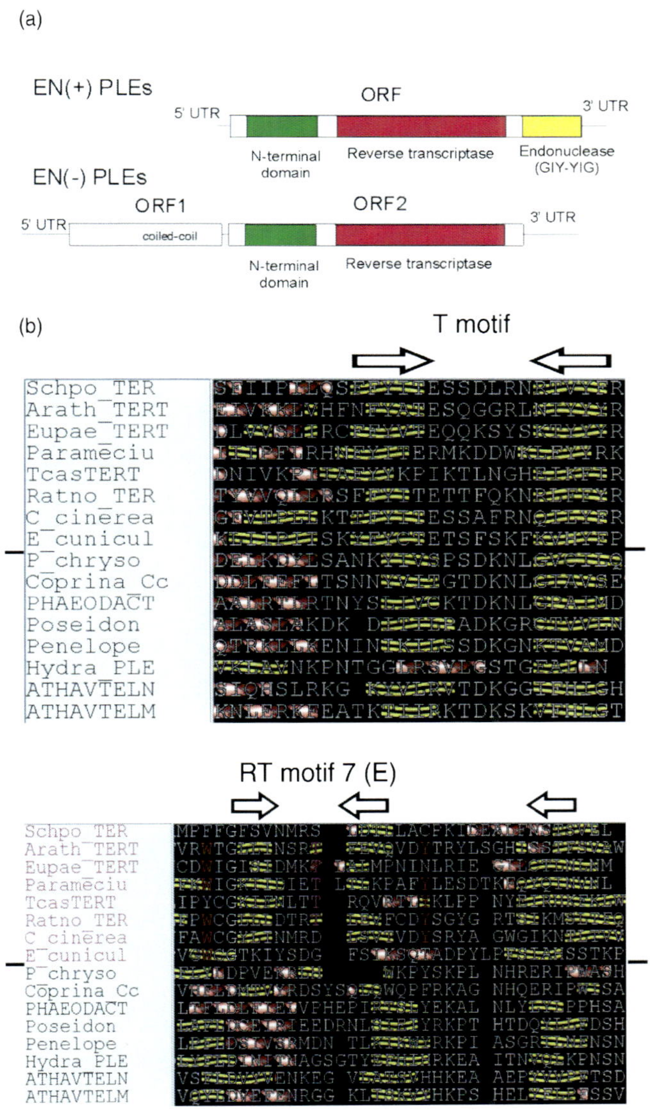

FIGURE 11.6 Structural organization of *Penelope*-like elements (a) and their similarity to telomerases (b). The EN(−) (endonuclease-deficient) PLEs exhibit specificity for telomeres in diverse eukaryotes. In panel (b), secondary structure predictions for representative TERT (top 8) and PLE (bottom 8) sequences are compared in selected portions of the RT amino acid alignment, showing the N-terminal T-motif region and the C-terminal motif 7 of the core RT domain. Arrows designate characteristic beta-hairpins in the secondary structure. Sequences were viewed with the aid of a structure-based sequence alignment program (STRAP) (http://www.bioinformatics.org/strap/).

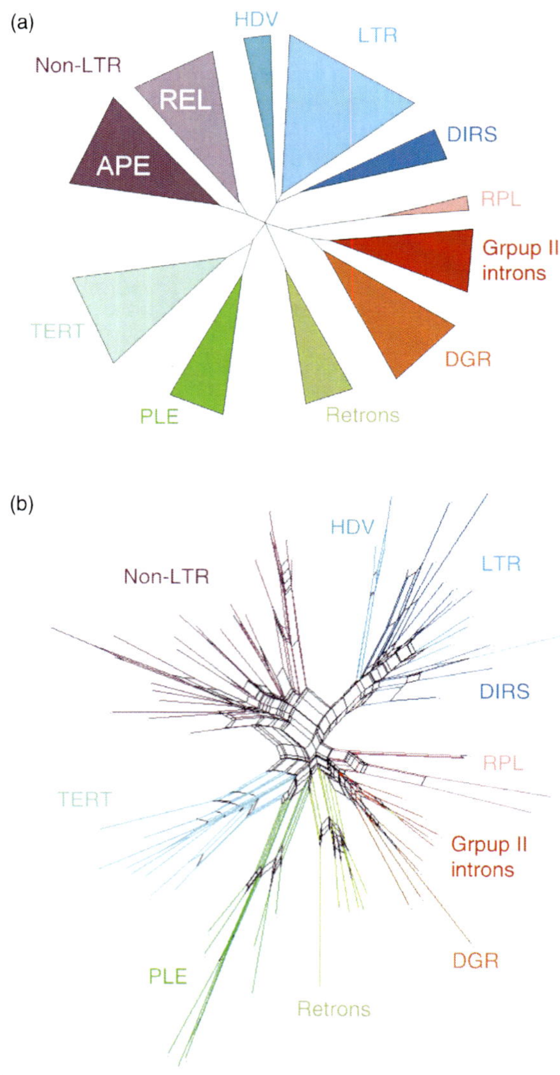

FIGURE 11.7 Phylogenetic relationships between different RT classes. (a) Diagrammatic representation of an unrooted phylogram showing each RT class as a triangle with the size approximately reflecting the diversity within the group, as in Eickbush and Malik (2002). The LTR group includes retroviruses, LTR-copia, LTR-gypsy, and caulimoviruses. RPL, mitochondrial retroplasmids, and RTL elements; HDV, hepadnaviruses; for other abbreviations, see text. (b) Phylogenetic network showing relative positions of each RT group and visualizing conflicting signals and areas of reticulate events in the overall tree-like phylogeny (maximum likelihood distance for an alignment of *ca* 600 RT amino acids) (SplitsTree4.1; Huson and Bryant 2006).

6

TELOMERASE REGULATION AND TELOMERE-LENGTH HOMEOSTASIS

JOACHIM LINGNER AND DAVID SHORE

6.1 INTRODUCTION

Although telomerase activity *in vitro* requires only a catalytic protein subunit (TERT) and an integral RNA template molecule (TER), *in vivo* this core complex is associated with numerous other proteins and undergoes an elaborate process of biogenesis before arriving at its site of action, a chromosome end. Telomerase structure and biogenesis are discussed in detail elsewhere in this volume (Chapters 2, 3 and 4). Here we focus primarily on mechanisms that regulate telomerase action such that cells maintain a fixed average length of telomeric repeat sequence sufficient for the reliable assembly of a stable protective ("capped") structure at telomeres.

Before entering into a more detailed discussion of molecules and mechanisms, it is useful to consider the problem of telomerase regulation from a more general perspective. As discussed elsewhere in this volume, the telomerase enzyme has emerged during evolution as a nearly universal solution to the problem of incomplete DNA-end replication by conventional DNA polymerases. However, the mechanism by which telomerase replenishes eroded DNA at chromosome ends, using a self-contained template RNA, poses another problem, since the enzyme itself would appear to be incapable of knowing precisely how much DNA to add to each individual chromosome end in order to maintain a (roughly) constant telomeric DNA length. This problem was recognized shortly after the discovery of telomere DNA structure and the telomerase enzyme by Murray and Szostak (Murray et al., 1988), who proposed that "... the constant average length of yeast telomeres implies a negative

Telomerases: Chemistry, Biology, and Clinical Applications, First Edition.
Edited by Neal F. Lue and Chantal Autexier.

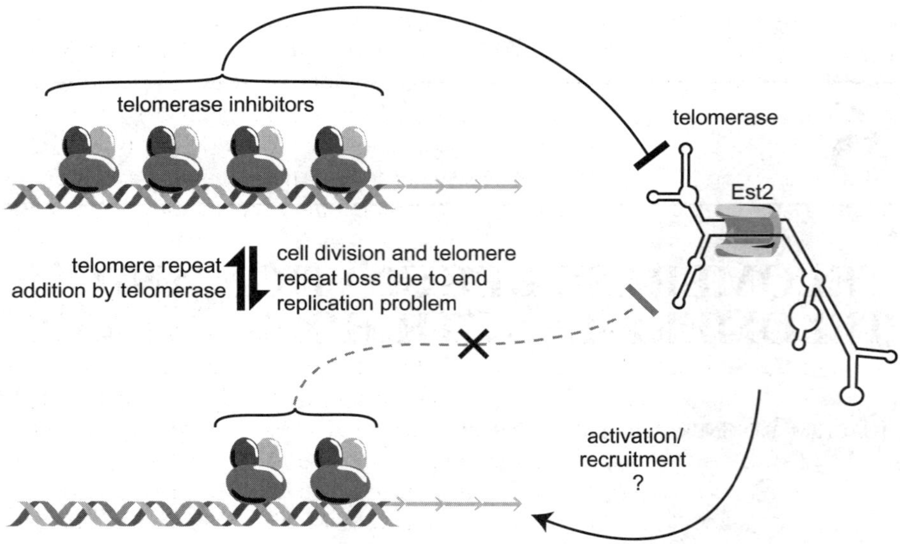

FIGURE 6.1 Schematic representation of the negative feedback "protein counting" model for telomere length regulation. Details described in the text. (See the color version of this figure in Color Plates section.)

feedback mechanism which senses the length of telomeric DNA and reduces the extent of non-template-directed DNA synthesis when the telomeric DNA exceeds a certain length." Nearly a decade later, experiments in yeast and human cells provided evidence for such a feedback mechanism, controlled, at least in part, by the proteins that bind to the duplex telomere repeat sequences themselves, Rap1 and TRF1, respectively (Marcand et al., 1997; van Steensel and de Lange, 1997). The proposal that emerged from these experiments, sometimes referred to as the "protein counting model", is that a metric for telomere DNA repeat length is generated by proteins that bind to the repeats themselves, such that repeat tract elongation, and thus increased protein binding, will tend to switch the telomere into a "off" state, or a state less likely to be acted upon by telomerase. This general scenario is illustrated in Figure 6.1. Subsequent experiments in yeast, using a site-specific recombination system to shorten a single marked telomere, showed that the average rate of telomere elongation, per cell cycle, is inversely proportional to telomere length, suggesting a rheostat-like mechanism as opposed to a strict, length-dependent on–off switch (Marcand et al., 1999). How such a negative feedback system might work was not at all clear at the time, and testable models of molecular mechanisms have emerged only recently, mostly in the budding yeast system.

An important advance in understanding the dynamics of telomerase action as a function of telomere repeat length came from studies in yeast where telomerase addition could be read out at the single-cell level through PCR-based sequence analysis of individual elongation events (Teixeira et al., 2004). This work, which took advantage of the irregular TG_{1-3} repeats characteristic of *Saccharomyces cerevisiae* telomeres, showed that telomerase does not act at every telomere during a single cell cycle, and that the probability of telomerase action at an individual telomere is inversely proportional to its TG_{1-3} tract length. Significantly, the length-dependent probability of telomerase action increases in cells mutated for either *RIF1* or *RIF2*, both of which encode negative regulators of telomere length (Hardy et al., 1992; Wotton and Shore, 1997). This and subsequent studies measuring elongation events during a single cell cycle found no evidence that telomere length affects telomerase processivity, except at extremely short telomeres (Arneric and Lingner, 2007; Chang et al., 2007).

Analysis of extension events in cells expressing two different RNA template subunits provided further insight into the dynamics of telomerase action *in vivo* by showing that a single telomere can be acted upon by two different telomerase molecules during a single cell cycle (Chang et al., 2007). This finding strongly argues that telomerase enzyme rapidly dissociates and reassociates with an individual telomere, implying a dynamic rather than static interaction of telomerase with telomere ends that could in principle be subject to regulation. These discoveries provide an important framework for evaluating other studies, to be discussed below, aimed more directly at addressing the molecular mechanisms of telomerase regulation at telomeres and how this regulation could be modulated by telomere repeat tract length.

This chapter is divided into two sections, the first of which discusses studies using the budding yeast (primarily *S. cerevisiae*) model system, but also touches upon recent work in the fission yeast (*Schizosaccharomyces pombe*), in which the structure and organization of telomere-specific proteins involved in telomerase regulation more closely resembles the situation in mammalian cells. The second part of this chapter focuses mostly on studies of mammalian cells in culture, but also contrasts these to studies in the mouse, as well as bringing together findings from human disease studies that impinge upon telomerase assembly and function. The work described in this section has relied heavily on bioimaging tools that allow one to localize telomerase components in cells, and has revealed an important role of telomerase biogenesis and trafficking in the regulation of telomere length. Studies in yeast, described in detail in Chapter 4, are beginning to reveal the general significance of telomerase maturation and localization to telomere maintenance. It is important to keep in mind that studies in both human and yeast cells point to the striking conclusion that overall telomerase levels in the cell are limiting, and have an important influence on telomere length homeostasis (Cristofari and Lingner, 2006b; Mozdy and Cech, 2006; Mozdy et al., 2008).

6.2 TELOMERASE REGULATION AT TELOMERES IN YEAST

6.2.1 Telomerase Holoenzyme in *S. cerevisiae*

The first gene-encoding a protein required for telomerase action, *EST1* (ever shorter telomeres 1), was identified in a pioneering genetic screen carried out in *S. cerevisiae* (Lundblad and Szostak, 1989), that later led to the identification of three other genes

(*EST2-4*) required for telomere maintenance in this organism (Lendvay et al., 1996). This and a subsequent study (Lingner et al., 1997b) showed that *EST2* encodes the catalytic protein core of telomerase (TERT), while the *EST4* mutation turned out to be a specific loss-of-function mutation in Cdc13, a telomere-specific single-stranded DNA-binding protein (Lendvay et al., 1996). Cdc13 functions in both telomerase recruitment and telomere end protection (described in more detail below and in Chapter 7), but does not appear to be a stable component of a telomerase holoenzyme. In contrast, both Est1 and Est3 proteins have been shown to physically associate with telomerase enzyme, based upon coimmunoprecipitation from cell extracts, though neither protein is required for telomerase activity *in vitro*, as is the case also for Cdc13 (Est4) (Lingner et al., 1997a). Finally, the second essential component of yeast telomerase, the template RNA encoded by the *TLC1* gene, was identified in a screen for genes which when overexpressed disrupt telomeric gene silencing (Singer and Gottschling, 1994). Est1 has been shown to physically associate with telomerase, at least in part, through an interaction with a conserved bulge-stem structure in the telomerase RNA, TLC1 (Seto et al., 2002). In contrast, Est3 protein instead appears to associate with telomerase through a direct interaction with the catalytic Est2 subunit (Hughes et al., 2000; Lee et al., 2010), through a putative OB-fold domain of the protein (Lee et al., 2008) that is conserved in the distantly related budding yeast (and human pathogen) *Candida albicans* (Yu et al., 2008).

6.2.2 Two Cell-Cycle Dependent Pathways Promote Association of Telomerase with Telomeres

The first hints of mechanisms governing the regulation of telomerase action came from pioneering studies by Lundblad and colleagues (Chandra et al., 2001; Evans and Lundblad, 1999; Pennock et al., 2001) that focused on the Cdc13 (Est4) protein. Through both genetic suppressor analysis and ingenious use of hybrid proteins, these authors developed several lines of evidence supporting the idea that an interaction between telomere-bound Cdc13 and the Est1 protein plays an instrumental role in promoting either the recruitment of telomerase holoenzyme to telomeres or its activation there (or both). These and other studies also suggested that the role of Cdc13 is complex, since its partner proteins Stn1 and Ten1 (as part of the Cdc13–Stn1–Ten1, or CST, complex) seem to exert a repressive effect on telomerase action, possibly through direct competition with Est1 for Cdc13 binding (Qi and Zakian, 2000).

The relatively recent advent of highly sensitive chromatin immunoprecipitation (ChIP) assays for telomerase components, as well as other proteins involved in telomere replication or capping, has allowed researchers to begin to address mechanistic questions and test predictions from genetic analyses. Initial studies revealed a curious pattern of cell-cycle dependent telomerase association (Taggart et al., 2002) in which the catalytic subunit (Est2), but not the accessory protein Est1, was found to be telomere associated in cells blocked in G1 phase. This G1 binding of Est2 decreased as cells released from the block and entered S phase, but returned to high levels, this time together with Est1, as cells passed through late S phase, the time at which telomeres are replicated and elongated by telomerase (Marcand et al., 2000). Subsequent studies showed that the G1 telomere association of Est2 requires a specific interaction between a stem-loop structure in TLC1 RNA and the telomere-bound Yku70/80 heterodimer (Fisher et al., 2004; Stellwagen et al., 2003). Why this interaction would appear to be abrogated, or weakened as cells exit G1 is still unclear.

At the same time, ChIP studies have provided new insights regarding the Cdc13–Est1 pathway uncovered by Lundblad and colleagues through genetic and hybrid protein studies. In a model system examining telomere formation at a double-strand break (DSB) flanked by TG_{1-3} repeats (Diede and Gottschling, 1999), or in which telomeric proteins are tethered to a DSB through a heterologous DNA-binding domain, point mutations in either Cdc13 or Est1 that confer an "ever shorter telomere" phenotype do indeed affect telomerase (Est2) recruitment to DNA ends (Bianchi et al., 2004). More recently, a comprehensive ChIP analysis of Est2 association at native telomeres throughout the cell cycle has provided a more refined view of the Cdc13–Est and Yku–TLC1 pathways (Chan et al., 2008). This study showed that normal Est1 association with telomeres in late S/G2 phase requires that it be a part of telomerase holoenzyme, since its binding is reduced by senescing mutations in TLC1 that abrogate its telomerase association (Seto et al., 2002), or abolished completely by deletion of *EST2*. This finding implies that the Cdc13–Est1 interaction, though apparently necessary for telomere maintenance, is not by itself sufficient in the absence of an additional interaction between Est1 and telomerase. Interestingly, mutations that lead to a short-telomere phenotype, either by interfering with the Yku–TLC1 interaction or removing Tel1 protein, cause a reduction in Est2 telomere association in late S/G2 that is indistinguishable from that observed in senescing (*est*) mutations (such as *cdc13-2*). Therefore, Est2 telomere association as measured by ChIP is not able to distinguish between some mutations that shorten telomere length, but maintain homeostasis, from those that lead to progressive telomere shortening and senescence, again pointing to regulation of telomerase at some step following its recruitment to telomeres. It is worth noting here that an additional single-stranded DNA-binding protein, the ubiquitous RPA complex, has also been proposed to play a role in telomerase recruitment, by acting together with Cdc13 to promote Est1 loading onto telomeres during S phase (Schramke et al., 2004).

Interestingly, the late S-phase association of Est1 with telomeres and telomerase (Est2/TLC1) correlates with an increase in Est1 protein abundance (Osterhage et al., 2006). Indeed, Est1 is normally targeted for proteosomal degradation preferentially during G1 phase. Overexpression of Est1 protein in G1 phase can drive its association with Est2, suggesting that its normal levels in G1 are insufficient to promote telomerase holoenzyme assembly. Despite the assembly of active telomerase in G1 cells overexpressing Est1, the enzyme is still incapable of elongating telomeres during this phase of the cell cycle (Osterhage et al., 2006). This striking finding again calls into question the significance of the G1 association of telomerase (Est2) in wild-type cells, and suggests instead that it might have no role in telomere elongation, but instead serve a protective role (Vega et al., 2007).

Apart from an increase in protein abundance, additional factors may promote Est1 association with telomerase (and telomeres) during late S/G2 phases of the cell cycle.

The Cdc13–Est1 interaction, discussed above, is promoted by cyclin-dependent CDK (Cdc28) phosphorylation of Cdc13 in S/G2 phase at a single residue within the Cdc13 "recruitment" domain. This phosphorylation event may help Est1 to overcome competitive (and inhibitory) binding to Cdc13 by Stn1 (Li et al., 2009).

6.2.3 Mechanisms Promoting Association of Telomerase at Short Telomeres

To address the question of why short telomeres are preferred substrates for telomerase action (Teixeira et al., 2004), two different groups adapted the telomere-shortening method developed earlier by Marcand et al. (1999) to quantify the association of telomerase and related proteins at short versus normal length telomeres. These ChIP studies both led to the conclusion that telomerase (both Est1 and Est2) association is increased at short telomeres whereas that of Cdc13 is not (Bianchi and Shore, 2007b; Sabourin et al., 2007). This conclusion, namely that telomere tract length regulates recruitment of telomerase to telomere ends, rests upon certain assumptions regarding the interpretation of ChIP assays, and in any event does not exclude the possibility of regulation at subsequent steps that might, for example, convert telomere-bound telomerase from an inactive to an active form (see below). Regarding the first point, an increased ChIP signal for telomerase (both Est1 and Est2) at short versus long telomeres might simply reflect the fact that the enzyme is engaged in nucleotide addition more frequently at the shorter ends, a state in which the enzyme might more efficiently cross-link to DNA. In this scenario, the shorter ends would promote the conversion of bound telomerase to an active state more efficiently than long ends, without affecting a "pre-equilibrium" binding reaction. However, the fact that the Cdc13–Est1 pathway also augments the ChIP signal for a catalytically inactive telomerase enzyme, at least at a telomeric DSB, argues that recruitment, or at least some step prior to nucleotide addition, is indeed regulated by telomere tract length (Bianchi et al., 2004).

Significantly, the yeast ATM kinase Tel1, required for normal telomere length homeostasis, is also more strongly associated with shortened telomeres (Bianchi and Shore, 2007b; Hector et al., 2007; Sabourin et al., 2007). Furthermore, the preferential elongation of short telomeres depends upon the presence of Tel1 (Arneric and Lingner, 2007). Because Tel1 (and Mre11) are required for normal telomerase association at telomeres (Goudsouzian et al., 2006), and because Tel1 phosphorylates Cdc13 *in vitro* at two sites within its recruitment domain that are required for telomere-length maintenance (Tseng et al., 2006), the ChIP data support a model in which increased Tel1 binding at short telomeres would promote increased telomerase recruitment through phosphorylation of Cdc13. However, more recent work argues strongly that Cdc13 is not actually phosphorylated by Tel1 *in vivo* (Gao et al., 2010). Thus, the putative target(s) of Tel1 relevant to increased association of telomerase at short telomeres remains to be identified.

How, then, is Tel1 association increased at short relative to long telomere tracts? A recent study (McGee et al., 2010) showed that Mre11, part of the MRX complex responsible for Tel1 recruitment to DNA ends (Nakada et al., 2003), is also present in higher amounts at short telomeres. Indeed, long TG_{1-3} tracts adjacent to a DSB

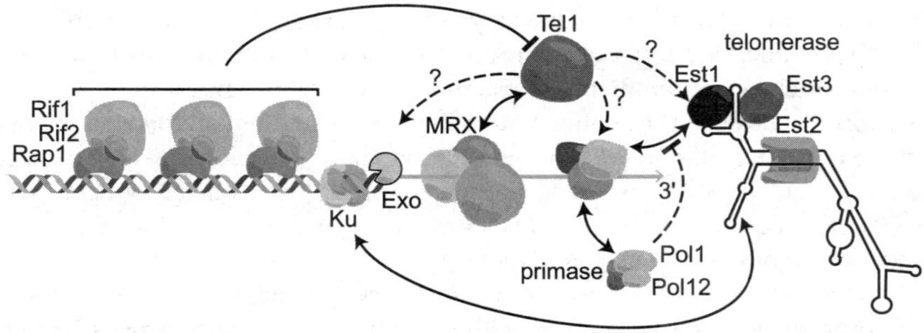

FIGURE 6.2 Proteins and interactions implicated in telomere length regulation in the budding yeast *S. cerevisiae*. Protein–protein interactions are indicated by double-headed arrows. See text for details. (See the color version of this figure in Color Plates section.)

strongly inhibit Mre11 binding at those ends (Negrini et al., 2007). The relative exclusion of Mre11 and Tel1 at long telomeres is correlated with increased binding of the negative regulator Rif2, and consistent with a direct role for this protein in regulating their binding. Consistent with this conclusion, Tel1 binds equally well to short and normal length telomeres in *rif2* mutant cells (McGee et al., 2010). *In vitro* binding experiments using GST hybrids in pull-down experiments from yeast extracts suggest that Rif2 might block Tel1 binding by competing for a binding site on the Xrs2 protein (Hirano et al., 2009), perhaps also affecting, at the same time, the stable association of the whole MRX complex at the telomere. Our current understanding of regulatory circuits acting *in cis* to control telomerase action at individual telomeres in *S. cerevisiae* is outlined in Figure 6.2.

Quite unexpectedly, telomere tract length also affects the cell-cycle dynamics of telomere replication (Bianchi and Shore, 2007a). Shortening a single telomere in yeast leads to its earlier replication due to activation of subtelomeric DNA replication origins. Earlier telomere replication correlates with earlier binding of telomerase holoenzyme and increased telomere elongation. This telomere-replication origin feedback effect may provide cells with a failsafe mechanism to rapidly elongate telomeres that have undergone near-catastrophic telomere repeat loss. The mechanism underlying this effect is at present unknown, but it is interesting to note that DSBs can also locally trigger firing of dormant replication origins (Doksani et al., 2009), suggesting that other pathways of DNA end repair might also be promoted by early replication.

6.2.4 Activation of Telomerase at Telomere Ends

As pointed out above, the conclusion that telomere tract length regulates recruitment of telomerase to telomere ends does not exclude the possibility of regulation at subsequent steps. In fact, a growing body of evidence points towards mechanisms that regulate the activity of telomere-bound telomerase enzyme. Early genetic (Evans and Lundblad, 2002) and hybrid protein (Evans and Lundblad, 1999) analysis first indicated an involvement of Est1 not only in telomerase recruitment, but also in a step (or steps) required for enzyme activity. For example, in cells where Est2 is directly tethered to telomeres through expression of a Cdc13–Est2 fusion protein (Cdc13 binds directly to the 3′ GT telomere overhang DNA) the essential function of Est1 is bypassed, yet expression of Est1 is still required for telomere overextension. Similarly, certain point mutations in Est1 abrogate this stimulatory activity and display a telomere length defect that is not bypassed by direct tethering of the mutant protein through fusion to Cdc13.

The first biochemical evidence for a role of Est1 protein in telomerase activation came from *in vitro* studies of the enzyme from *C. albicans*. This work (Singh and Lue, 2003), which compared the activity of partially purified telomerase from wild-type or *est1*-Δ strains on different telomeric primers, indicated that Est1 can promote telomerase synthesis through certain natural barriers within the long (23 bp) invariant telomere repeat of this organism. More recent biochemical analysis of the effect of the *S. cerevisiae* Est1 in a similar *in vitro* telomerase activity assay has provided further insights into the complex role of this holoenzyme component (DeZwaan and Freeman, 2009). This study also showed that *in vitro* Est1 stimulates telomerase extension activity, most likely through a direct protein–protein interaction with Est2, which earlier ChIP experiments suggested is independent of its interaction with TLC1 (Bianchi et al., 2004). Both the cdc13-2 and est1-60 mutant proteins, previously implicated in a salt–bridge interaction required for telomerase recruitment, were found to be defective in telomerase activation in this *in vitro* system (DeZwaan and Freeman, 2009). This work highlights the multifunctional nature of Est1 and suggests a more nuanced view of telomerase recruitment regulation. Dezwaan and Freeman (2010b) argue that strong telomerase association with telomeric DNA is (at least) a two-step process, the first of which involves an interaction between telomere-bound Yku70/80 and the TLC1 RNA, which brings a preassembled holoenzyme (containing Est1, Est3, and the TLC RNA) to the telomere. According to this model, Cdc13–Est1 interactions stabilize the complex, which is then converted to an active complex through Cdc13–Est1 interactions that involve the salt linkage between amino acids 252 in Cdc13 and 444 in Est1. Although it might prove possible to detect and measure these proposed intermediates *in vitro*, it is harder to imagine how this or other models can be tested *in vivo* by current approaches, due to the inability to know the extent to which quantitative changes in a ChIP signal are due to association effects or conformational changes within a complex.

In vitro telomerase assays have also recently revealed a direct role for the Cdc13 protein in telomerase activation (DeZwaan et al., 2009), independent of its DNA-binding activity and involving an N-terminal domain. This activation function of Cdc13 is completely inhibited by addition of the Stn1/Ten1 proteins (DeZwaan et al., 2009), which are proposed to form, together with Cdc13, an RPA-like heterotrimeric, telomere-specific single-stranded binding protein complex (Gao et al., 2007). Both Stn1 and Ten1, like Cdc13, perform an essential telomere capping function, and, consistent with the *in vitro* telomerase assay data, behave genetically like negative regulators of telomerase action (Grandin et al., 1997, 2001). Interestingly, addition of the yeast Hsp90-like protein Hsp82 converts this Cdc13–Stn1–Ten1 "capped" structure to a telomerase active state, apparently through promoting Cdc13 dissociation from the DNA (DeZwaan et al., 2009). Both Hsp82 and Hsc82 have been implicated previously *in vivo* in telomere function: their overexpression suppresses capping defects of *cdc13-1* and *stn1-1* mutants, and, in otherwise wild-type cells, leads to telomere shortening (Grandin and Charbonneau, 2001). The involvement of molecular chaperones in telomere function is an attractive notion, given the general observation that a large number of different and possibly competing proteins and protein complexes all appear to associate with telomeres during a brief window of time in late S/G2 phase (Chan et al., 2008; Puglisi et al., 2008; Schramke et al., 2004). The coordinated assembly and disassembly of different complexes in a highly dynamic fashion may be necessary to rapidly orchestrate the processes of conventional DNA replication, telomerase addition and DNA end capping (discussed more fully in DeZwaan and Freeman, 2010a).

6.2.5 Telomerase Regulation in the Fission Yeast *S. pombe*

Studies using the fission yeast *S. pombe* have begun to reveal details of telomerase recruitment in this organism that display significant differences to those in the budding yeast (Miyoshi et al., 2008). This in large part follows directly from the fact that telomere repeat-associated proteins in these two evolutionarily diverged yeasts are themselves quite different. Although both Rap1 and Rif1 are conserved telomere-associated proteins in *S. pombe*, Rap1 in this organism (as in mammals) does not bind directly to the telomere repeat DNA, but instead associates with the telomere through an interaction with the DNA-binding protein Taz1, a homodimeric Myb (SANT) domain-containing protein (Kanoh and Ishikawa, 2001; Park et al., 2002). Taz1 is also responsible for recruitment of *S. pombe* Rif1 protein (Kanoh and Ishikawa, 2001).

S. pombe Rap1 in turn interacts with the Poz1 protein, which itself interacts with a protein called Tpz1, a putative ortholog of the mammalian telomeric protein TPP1 (Miyoshi et al., 2008; see below). Tpz1 interacts with both Pot1 (Baumann and Cech, 2001; homologue of the mammalian single-strand telomere repeat binding protein POT1) and a protein called Ccq1, which until now appears to be unique to the fission yeast. Genetic and biochemical (both coimmunoprecipitation and ChIP) studies are consistent with a model in which the Poz1/Tpz1/Pot1/Ccq1 complex of proteins exists in two different states at the telomere, one refractory to telomerase recruitment and the other of which promotes telomerase association (see Fig. 6.3a). In the repressive state, Poz1 is proposed to interact with Rap1 at the duplex portion of the telomere repeats and thus somehow preventing Ccq1 from interacting with telomerase. In the active or "open" state, Poz1 and its partners are dissociated from Rap1 and bind to the single-stranded terminal portion of the telomere, through Pot1 and Tpz1, thus allowing Ccq1 to recruit (and perhaps activate) telomerase at the chromosome end (Miyoshi et al., 2008; reviewed in Bianchi and Shore, 2008). This scenario bears some resemblance to the current picture of telomerase regulation in mammalian cells through the partially related "shelterin" complex (see below and Fig. 6.3b). How telomere tract length influences the probability of transition to the

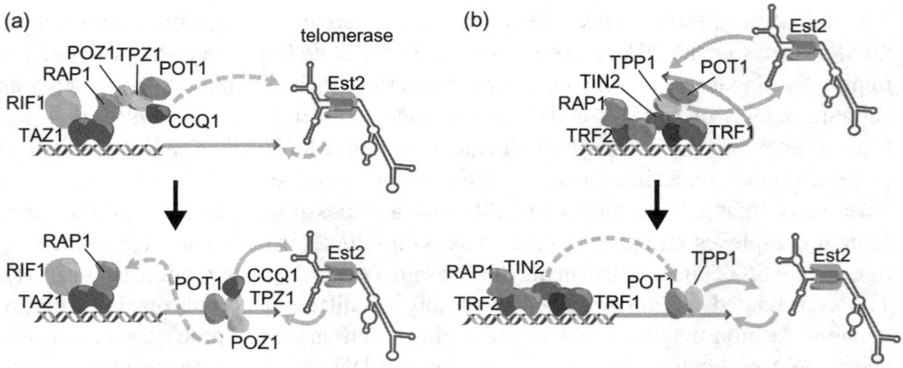

FIGURE 6.3 Schematic representations of models for telomerase activation at telomeres in the fission yeast *S. pombe* (a) and in human cells (b). See text for details. (See the color version of this figure in Color Plates section.)

proposed active state is still unknown. Interestingly, *S. pombe* Rif1, like its budding yeast homologue, is implicated in telomere length regulation (its deletion causes telomere over-elongation, similar to that observed in *S. cerevisiae*). However, Rif1 in *S. pombe* appears to act at least to some extent independently of Rap1 (Miller et al., 2005).

6.2.6 Mechanisms that Prevent Inappropriate Telomerase Action

Just as the cell has evolved elaborate systems to ensure that telomerase can maintain a stably capped DNA-repeat structure at chromosome ends, it has become equally clear that mechanisms must exist to prevent telomerase from acting inappropriately at DNA ends (accidental DSBs) that might arise elsewhere in the chromosome.

Early insights into such mechanisms came from studies of the DNA sequence requirements for (rare) spontaneous formation of telomeres at sites near an induced break in yeast (Kramer and Haber, 1993), as well as the identification of a helicase protein (Pif1) that inhibits *de novo* telomere formation (Schulz and Zakian, 1994). Significantly, the Pif1 helicase affects native telomeres too, with a slight over-elongation observed in its absence (Schulz and Zakian, 1994). Pif1 appears to be able to directly affect telomerase association at telomeres (Boule et al., 2005), perhaps acting *in vivo* through an interaction with the telomerase "finger" domain that weakens the association of telomerase with Est1 (Eugster et al., 2006). The finding that the frequency of spontaneous "gross chromosomal rearrangements" is greatly increased in *pif1* mutant cells (Myung et al., 2001), with nearly all of the events in these cells due to *de novo* telomere formation, dramatically highlighted the importance of Pif1 in global genome stability.

More recent work provides new insights into the regulation of Pif1 at DSBs by showing that a DNA-damage checkpoint kinase cascade, consisting of Mec1 (ATR in mammals), together with two downstream kinases, Rad53 and Dun1, is responsible for phosphorylating Pif1 following DNA damage (Makovets and Blackburn, 2009). Phosphorylation of a short (heptamer) peptide within a C-terminal region not conserved in other helicases, causes a specific decrease in the ability of Pif1 to inhibit telomere formation at DSBs (but no effect on telomere length). Phosphorylation of this region does not inhibit Pif1 binding to a DSB and at present the molecular cause of the defect is unknown.

A parallel study has recently revealed another mechanism by which Mec1 acts to prevent telomere formation at DSBs (Zhang and Durocher, 2010), by linking Mec1 to a phosphoregulatory loop, together with the Pph3 phosphatase and its Rrd1 regulatory subunit. In this regulatory system Mec1 phosphorylates residue 308 in Cdc13, which, through some unknown mechanism, blocks its accumulation at DSBs. This inhibition appears to increase the specificity of telomerase action towards ends that contain longer stretches of telomere-like TG-repeat sequences. The action of Mec1 on both Cdc13 and Pif1 may act cooperatively to restrict telomerase action to telomeres, though this has yet to be directly tested.

One striking and unexpected recent finding related to the above-mentioned studies is that telomerase enzyme binds avidly to DSBs in yeast, at least as measured by the ChIP assay (Oza et al., 2009; V. Ribaud, C. Ribeyre, and D.S., unpublished results), despite the fact that such ends are rarely, if ever, elongated by telomerase to produce functional telomeres in wild-type cells. How telomerase is recruited to DSBs is at present still unclear, but this result serves to reinforce the notion that telomerase regulation at steps following recruitment plays a very important role in promoting genome stability.

6.3 POST-TRANSLATIONAL REGULATION OF MAMMALIAN TELOMERASE

6.3.1 Assembly of Mammalian Telomerase

The human telomerase reverse transcriptase (hTERT) polypeptide together with the telomerase RNA moiety (hTERC) provides the catalytic core of telomerase, and together the two are sufficient for reconstitution of telomerase activity when assembled in rabbit reticulocyte lysates (Weinrich et al., 1997). Upon extensive purification of catalytically active human telomerase from HEK293 cells, mass spectrometric sequencing identified, together with TERT, the dyskerin protein (Cohen et al., 2007). In addition to TERT, TERC, and dyskerin, the telomerase holoenzyme contains a large number of other proteins that are involved in telomerase assembly, stability, trafficking, and possibly activation at chromosome ends. Dyskerin, NOP10, NHP2, and GAR1 are associated with telomerase RNA as well as other nuclear RNAs that share H/ACA motifs (Mitchell et al., 1999; Pogacic et al., 2000). The importance of these proteins for telomerase function is underlined by the fact that mutations in dyskerin, NHP2, and NOP10, in addition to hTERT and hTR give rise to the telomere disorder dyskeratosis congenita (DC) (see Lansdorp, 2009 for a review). DC patients have abnormally short telomeres and suffer from skin pigmentation abnormalities, nail dystrophy, and abnormalities of the oral mucosa. The most common fatal complications are bone marrow failure, pulmonary fibrosis, and cancer (Lansdorp, 2009).

The AAA + ATPases pontin and reptin are associated with several chromatin-remodeling complexes and have functions in transcriptional regulation. Pontin and reptin were also identified during telomerase purification as interacting factors of dyskerin (Venteicher et al., 2008). They have been implicated in a telomerase assembly step that precedes a mature and fully active TERT/TERC/dyskerin-containing complex. Pontin or reptin depletion impairs telomerase activity and reduces TERC abundance.

TCAB1 (telomerase Cajal body (CB) protein 1; Venteicher et al., 2009), which is also referred to as WDR79 (WD repeat domain 79; Tycowski et al., 2009), binds to human H/ACA small CB-specific RNPs (scaRNPs) in a manner that depends on the CAB box, a tetranucleotide RNA element that facilitates CB localization of scaRNPs. Telomerase RNA contains a CAB box (Richard et al., 2003) and TCAB1 is found to be associated with the bulk of human telomerase activity and telomerase RNA in cellular extracts (Venteicher et al., 2009). Importantly, depletion of TCAB1 by RNA interference prevents localization of hTR to Cajal bodies and to telomeres. TCAB1 depletion also leads to continuous telomere shortening. However, telomerase catalytic activity is unaffected by TCAB1-depletion. This indicates that TCAB1 executes its essential functions for telomere extension through controlling telomerase trafficking (see below). In addition to the telomerase proteins discussed here, a large number of proteins have been found associated with telomerase. Their mechanisms of action are not yet well understood and the reader is referred to other reviews (e.g., Cristofari and Lingner, 2006a) and to Chapter 4.

6.3.2 Regulation of Human Telomerase by Subnuclear Trafficking

Cytological studies demonstrated S-phase specific assembly of human telomerase with telomeres (Jady et al., 2006; Tomlinson et al., 2006). During interphase, hTR was detected by fluorescence in situ hybridization (FISH) in Cajal bodies. In early in S-phase, Cajal bodies containing hTR are present at the periphery of nucleoli. During S-phase, hTERT (detected by immunofluorescence) and hTR both localize to foci adjacent to Cajal bodies. Some Cajal bodies, hTERT, and hTR are also found in association with telomeres during S-phase. Live imaging indicates that the association between Cajal bodies and telomeres lasts about 30 min. (Jady et al., 2006). Human telomerase RNA accumulates in telomerase-positive cancer cells in CBs in a CAB-box and TCAB1-dependent manner (Cristofari et al., 2007; Venteicher et al., 2009). CB localization of hTR is also dependent on hTERT (Tomlinson et al., 2008). Despite this close association of hTR and hTERT with Cajal bodies, hTR accumulation in CBs is not needed for assembly of the catalytically active core of telomerase (Cristofari et al., 2007). However, mutations in the CAB box or depletion of TCAB1 both interfere with telomere association of telomerase during S phase of the cell cycle and telomere extension (Cristofari et al., 2007; Venteicher et al., 2009). This therefore indicates that CB localization of telomerase is not needed for assembly of the catalytic core of telomerase, but is required to render telomerase competent for telomere association and extension *in vivo*. The nature of this putative modification and activation step is not known.

Intriguingly, telomerase RNA is not enriched in Cajal bodies in mouse cells at steady state, but instead it accumulates in separate nuclear foci of unknown identity (Tomlinson et al., 2010). On the other hand, ectopically expressed human telomerase RNA also accumulates in Cajal bodies in the mouse. Thus, distinct trafficking pathways may precede telomere association in these mammals. It will be interesting to determine if TCAB1 and the CAB-box are also important for telomerase-mediated telomere extension in the mouse, as is the case in humans.

6.3.3 Recruitment of Mammalian Telomerase to Telomeres by Shelterin

As discussed above, FISH-experiments revealed that hTR colocalizes with human telomeres in S phase. Human telomeres are assembled with the six shelterin components (Palm and de Lange, 2008): TRF1, TRF2, RAP1, and TIN2 associate with the double-stranded region of the telomere. POT1 binds the short, single-stranded DNA overhang of the telomere but also interacts with the double-stranded region through protein–protein interaction with TPP1. TPP1 associates with both regions through interactions with TIN2 and POT1, and is thus viewed as central to telomerase recruitment to its site of action at the 3′ overhang, as described below.

The requirement of the telomeric shelterin protein components for telomerase association with telomeres has been assayed by ChIP of hTERT and analysis of hTR by FISH in HeLa cells in which telomerase was overexpressed (Abreu et al., 2010). Depletion of shelterin proteins TPP1 and TIN2 (but not POT1) prevented association of telomerase with telomeres. These findings indicate that the majority of telomerase is recruited to telomeres by TPP1 that is bound to telomeres via TIN2 in humans, and likely other vertebrates. Thus, the bulk of telomerase is tethered to the double-stranded region of the telomere through TPP1–TIN2–TRF1–TRF2 interactions. Recruitment to the telomeric 3′ overhang via POT1 was not detected in this study, possibly due to the lower number of telomeric TPP1–POT1 versus TPP1–TIN2–TRF2 complexes. The single-stranded 3′ overhang of telomeric DNA to which POT1 directly binds is generally much less extensive (typically 0.1–0.3 kb) than the double-stranded tract (typically 2–20 kb). Thus the fraction of TPP1 that may be present at the 3′ overhang in a POT1-dependent manner (see Fig. 6.3b) would be expected to be small relative to that bound to the double-stranded part of the telomere via TIN2. TPP1 had also been suspected to play a role in the recruitment of telomerase to telomeres based on its physical interaction with telomerase (Xin et al., 2007). Xin et al. demonstrated that TAP-tagged hTERT and GST-tagged TPP1 copurify when fractionated from cellular extracts derived from cells coexpressing the tagged proteins. In addition, both GST-tagged TPP1 and the oligonucleotide oligosaccharide (OB)-fold of TPP1 pull down *in vitro* translated HA-tagged TERT and telomerase activity, indicating that the TPP1 OB-fold is important for association of TPP1 with telomerase. Interestingly, the TPP1 OB-fold was also required for telomerase-association with telomeres *in vivo* (Abreu et al., 2010).

The significance of telomerase association with telomeres via TPP1–TIN2–TRF2 is supported by the recent identification of TIN2 mutations in patients suffering from the short telomere disease DC (Savage et al., 2008). However, it is uncertain whether inefficient telomerase recruitment contributes to the pathogenesis of DC in these patients. Interestingly, several studies also unraveled negative roles of the shelterin components in telomere length control. For example, depletion of TRF1 leads to telomere elongation and overexpression of TRF1 causes telomere shortening in human telomerase-positive cells without affecting *in vitro*-assayed telomerase activity, suggesting that reinforcement of the shelterin complex inhibits telomerase function (Smogorzewska and de Lange, 2004). Similarly, depletion of TPP1 by RNAi or disruption of the TPP1–POT1 interaction (which are both accompanied by loss of the POT1 signal at telomeres) also results in telomere lengthening (Liu et al., 2004; Ye et al., 2004). At the same time however, in addition to the recruitment roles of TPP1 and TIN2, several other findings support positive roles of shelterin in telomere length regulation. In particular, TPP1 together with POT1 has been shown to improve telomerase activity and processivity *in vitro* (Wang et al., 2007) by slowing primer dissociation and aiding telomerase translocation (Latrick and Cech, 2010). Dissection of the apparently opposing roles of the shelterin complex components on telomerase function awaits further investigation.

TPP1 deletion in mouse embryonic fibroblasts has also been reported to reduce mTERT binding to chromatin and to telomeres (Tejera et al., 2010). However, in this study the mTERT–chromatin association occurred even in MEFs that carried a homozygous deletion of the telomerase RNA gene. This finding seems to contradict the studies in human cells, which demonstrated a requirement of telomerase RNA expression for hTERT recruitment and the requirement of TCAB1, which mediates its effects on telomerase by binding to the CAB-box of hTR (Cristofari et al., 2007; Venteicher et al., 2009). The discrepancies between these studies might reflect differences in telomere biology between mouse and humans, which were alluded to above, or they may be due to experimental differences.

6.3.4 Telomere Elongation During the Cell Cycle

Cell-cycle dependent telomere elongation by telomerase has been measured using sophisticated molecular biological techniques in human cancer cells (Zhao et al., 2009). This analysis demonstrated that telomerase extends 70% (HeLa cervical carcinoma cells) to 100% (H1299 nonsmall cell lung carcinoma cells) of lagging-strand telomeres in every cell cycle by approximately 100 nucleotides, after semi-conservative DNA replication. Fill-in synthesis occurs in a temporally delayed manner later in S phase. In this study, telomerase action at leading strand telomeres could not be directly measured, but circumstantial evidence supported the notion that most leading strand telomeres were also extended in every cell cycle by telomerase in these cancer cell lines. This discovery that in human cancer cells most telomeres are extended by telomerase in every cell cycle contrasts the finding in *S. cerevisiae*

(Teixeira et al., 2004; see above). However, the study by Wright and colleagues does not provide a satisfactory mechanism for the maintenance of telomere length homeostasis in cancer cells. If the lagging-strand telomeres were extended in every cell cycle, they would be expected to grow continuously unless they would be again resected at other points during the cell cycle, by as yet undefined mechanisms.

In a separate study, Baird and colleagues (Britt-Compton et al., 2009) assessed telomerase-mediated telomere elongation in primary MRC5 human fibroblast cell upon ectopic expression of hTERT. Single telomere length analysis (STELA) revealed that under these nonequilibrium conditions, telomerase preferentially extended the shortest telomeres. Similarly, restoration of telomerase in mice that inherited short telomeres from telomerase-negative parents led to preferential elongation of the shortest telomeres in the analyzed splenocytes (Hemann et al., 2001). Overall, differences in telomerase abundance and telomere length distribution in the different studies may have influenced the experimental outcome. Telomerase regulatory mechanisms at chromosome ends might also differ between normal and cancer cells. However, the Baird and the Greider studies both support the notion that in humans and mice telomerase does recognize short telomeres and trigger their preferential elongation.

6.3.5 TERRA as a Putative Regulator of Telomerase

Telomeric repeat containing RNA (TERRA) is a large noncoding RNA that is transcribed at mammalian, yeast, and plant telomeres by RNA polymerase II (Azzalin et al., 2007; Luke et al., 2008; Schoeftner and Blasco, 2008; Vrbsky et al., 2010). Cell fractionation and FISH demonstrated that TERRA is present in the nucleoplasm as well as being partially associated with telomeric chromatin (Azzalin et al., 2007; Luke et al., 2008; Schoeftner and Blasco, 2008). TERRA transcription starts in the subtelomeric region and proceeds towards chromosome ends. Therefore, TERRA molecules contain near their 3′ ends telomeric repeat sequences in the form of RNA (5′-UUAGGG-3′ in mammals) that are complementary to the template region of the telomerase RNA moiety. This suggests that TERRA might bind to telomerase, modulating its activity. Indeed, a fraction of endogenous TERRA is bound to human telomerase in cell extracts (Redon et al., 2010). Furthermore, using *in vitro* reconstituted telomerase and synthetic TERRA molecules it was demonstrated that the 5′-UUAGGG-3′ repeats of TERRA base-pair with the RNA template of the telomerase RNA moiety. In addition TERRA contacts the telomerase reverse transcriptase (TERT) protein subunit independently of hTR (Redon et al., 2010). The TERRA–hTERT interaction site remains to be characterized. Overall, the affinity of telomerase for TERRA is remarkable as it exceeds the affinity for telomeric DNA. *In vitro* studies further demonstrate that TERRA is not used as a telomerase substrate. Instead, TERRA inhibits telomerase (Redon et al., 2010; Schoeftner and Blasco, 2008) acting as a potent competitive inhibitor for telomeric DNA in addition to exerting an uncompetitive mode of inhibition (Redon et al., 2010). The observed mixed inhibition by TERRA indicates that TERRA can bind to telomerase even while bound to the telomere substrate. This may be explainable through a direct hTERT–TERRA interaction, which may be distinct from the previously reported direct TERT–DNA contacts that are elicited through the so-called anchor site.

Remarkably, human TERRA is displaced and/or degraded at telomeres by factors involved in RNA decay, which physically interact with the telomeric chromatin (Azzalin et al., 2007). Among these factors, EST1A/SMG6 was also identified through its sequence similarity with the *S. cerevisiae* telomerase recruiter Est1 and human EST1A/SMG6, as yeast Est1 physically interacts with telomerase (Redon et al., 2007; Reichenbach et al., 2003; Snow et al., 2003). Thus, human EST1/SMG6 may elicit its effects on telomerase through TERRA, rather than playing a role in telomerase recruitment.

Genetic experiments in *S. cerevisiae* provide further support for the notion that TERRA regulates telomerase *in vivo*. In the *rat1-1* mutant background in which the function of the 5′–3 exonuclease Rat1 is reduced, TERRA is upregulated and telomeres are shorter than in wild-type cells due to impairment of telomerase-mediated telomere elongation (Luke et al., 2008). Further support for the role of TERRA in inhibiting telomerase *in vivo* stems from an observation that forced telomere transcription (through the use of the strong Gal-promoter) leads to telomere shortening of the transcribed telomere *in cis* (Sandell et al., 1994).

It is currently unknown under which physiological circumstances TERRA may contribute to telomerase regulation *in vivo*. Notably, TERRA levels are cell-cycle regulated and are the lowest in S phase (Porro et al., 2010), at the time when telomerase is able to extend chromosome ends. Thus, it is conceivable that TERRA contributes to the cell-cycle regulation of telomerase.

6.4 CONCLUDING REMARKS

Telomerase is a multicomponent enzyme, containing both protein and RNA subunits, which undergoes a remarkably complex process of biogenesis that is still incompletely understood (described in more detail elsewhere in this volume). Since telomerase levels appear to be limiting for telomere elongation in many different cellular contexts, regulation of telomerase biogenesis, or defects caused by mutation, can have important implications for telomere maintenance. As highlighted in this chapter, even after assembly of a minimally functional enzyme, a variety of mechanisms are in place to control telomerase association and catalytic activity at individual chromosome ends. Together, these mechanisms help to assure that chromosome ends are kept in a capped state through the maintenance by telomerase of an appropriate length buffer of telomeric repeat sequences. Related mechanisms, equally important for the preservation of genome stability, ensure that telomerase is prevented from acting inappropriately at DNA ends generated by accidental chromosome breaks. Genetic and biochemical studies over the past 25 years, together with more recent ChIP-based insights into protein–DNA interactions occurring in living cells, as well as imaging studies of telomere and telomerase components, have revealed a multitude of new players and molecular mechanisms involved in this complex regulatory network, and are likely to continue to do so in the coming years.

In addition, we anticipate that biochemical approaches using increasingly purified systems, together with structural studies, will begin to play a leading role in uncovering molecular details behind the complex and dynamic events that underlie the regulation of telomerase activity at telomere ends.

REFERENCES

Abreu E, Aritonovska E, Reichenbach P, Cristofari G, Culp B, Terns RM, Lingner J, Terns MP. (2010) TIN2-tethered TPP1 recruits human telomerase to telomeres *in vivo*. *Mol. Cell Biol.* **30**: 2971–2982.

Arneric M, Lingner J. (2007) Tel1 kinase and subtelomere-bound Tbf1 mediate preferential elongation of short telomeres by telomerase in yeast. *EMBO Rep.* **8**: 1080–1085.

Azzalin CM, Reichenbach P, Khoriauli L, Giulotto E, Lingner J. (2007) Telomeric repeat containing RNA and RNA surveillance factors at mammalian chromosome ends. *Science.* **318**: 798–801.

Baumann P, Cech TR. (2001) Pot1, the putative telomere end-binding protein in fission yeast and humans. *Science.* **292**: 1171–1175.

Bianchi A, Negrini S, Shore D. (2004) Delivery of yeast telomerase to a DNA break depends on the recruitment functions of Cdc13 and Est1. *Mol. Cell.* **16**: 139–146.

Bianchi A, Shore D. (2007a) Early replication of short telomeres in budding yeast. *Cell.* **128**: 1051–1062.

Bianchi A, Shore D. (2007b) Increased association of telomerase with short telomeres in yeast. *Genes Dev.* **21**: 1726–1730.

Bianchi A, Shore D. (2008) How telomerase reaches its end: mechanism of telomerase regulation by the telomeric complex. *Mol. Cell.* **31**: 153–165.

Boule JB, Vega LR, Zakian VA. (2005) The yeast Pif1p helicase removes telomerase from telomeric DNA. *Nature.* **438**: 57–61.

Britt-Compton B, Capper R, Rowson J, Baird DM. (2009) Short telomeres are preferentially elongated by telomerase in human cells. *FEBS Lett.* **583**: 3076–3080.

Chan A, Boule JB, Zakian VA. (2008) Two pathways recruit telomerase to *Saccharomyces cerevisiae* telomeres. *PLoS Genet.* **4**: e1000236.

Chandra A, Hughes TR, Nugent CI, Lundblad V. (2001) Cdc13 both positively and negatively regulates telomere replication. *Genes Dev.* **15**: 404–414.

Chang M, Arneric M, Lingner J. (2007) Telomerase repeat addition processivity is increased at critically short telomeres in a Tel1-dependent manner in *Saccharomyces cerevisiae*. *Genes Dev.* **21**: 2485–2494.

Cohen SB, Graham ME, Lovrecz GO, Bache N, Robinson PJ, Reddel RR. (2007) Protein composition of catalytically active human telomerase from immortal cells. *Science.* **315**: 1850–1853.

Cristofari G, Adolf E, Reichenbach P, Sikora K, Terns RM, Terns MP, Lingner J. (2007) Human telomerase RNA accumulation in Cajal bodies facilitates telomerase recruitment to telomeres and telomere elongation. *Mol. Cell.* **27**: 882–889.

Cristofari G, Lingner J. (2006a) The telomerase ribonucleoprotein particle. In: *Telomeres*, 2nd edition, CSH Laboratory Press, Cold Spring Harbor, N.Y. (USA). Titia de Lange, Vicki Lundblad, Elizabeth Blackburn (editors): 21–47.

Cristofari G, Lingner J. (2006b) Telomere length homeostasis requires that telomerase levels are limiting. *EMBO J.* **25**: 565–574.

DeZwaan DC, Freeman BC. (2009) The conserved Est1 protein stimulates telomerase DNA extension activity. *Proc. Natl. Acad. Sci. USA.* **106**: 17337–17342.

DeZwaan DC, Freeman BC. (2010a) HSP90 manages the ends. *Trends Biochem. Sci.* **35**: 384–391.

DeZwaan DC, Freeman BC. (2010b) Is there a telomere-bound 'EST' telomerase holoenzyme? *Cell Cycle* 9.

DeZwaan DC, Toogun OA, Echtenkamp FJ, Freeman BC. (2009) The Hsp82 molecular chaperone promotes a switch between unextendable and extendable telomere states. *Nat. Struct. Mol. Biol.* **16**: 711–716.

Diede SJ, Gottschling DE. (1999) Telomerase-mediated telomere addition *in vivo* requires DNA primase and DNA polymerases alpha and delta. *Cell.* **99**: 723–733.

Doksani Y, Bermejo R, Fiorani S, Haber JE, Foiani M. (2009) Replicon dynamics, dormant origin firing, and terminal fork integrity after double-strand break formation. *Cell.* **137**: 247–258.

Eugster A, Lanzuolo C, Bonneton M, Luciano P, Pollice A, Pulitzer JF, Stegberg E, Berthiau AS, Forstemann K, Corda Y, Lingner J, Geli V, Gilson E. (2006) The finger subdomain of yeast telomerase cooperates with Pif1p to limit telomere elongation. *Nat. Struct. Mol. Biol.* **13**: 734–739.

Evans SK, Lundblad V. (1999) Est1 and Cdc13 as comediators of telomerase access. *Science.* **286**: 117–120.

Evans SK, Lundblad V. (2002) The Est1 subunit of *Saccharomyces cerevisiae* telomerase makes multiple contributions to telomere length maintenance. *Genetics.* **162**: 1101–1115.

Fisher TS, Taggart AK, Zakian VA. (2004) Cell cycle-dependent regulation of yeast telomerase by Ku. *Nat. Struct. Mol. Biol.* **11**: 1198–1205.

Gao H, Cervantes RB, Mandell EK, Otero JH, Lundblad V. (2007) RPA-like proteins mediate yeast telomere function. *Nat. Struct. Mol. Biol.* **14**: 208–214.

Gao H, Toro TB, Paschini M, Braunstein-Ballew B, Cervantes RB, Lundblad V. (2010) Telomerase recruitment in *Saccharomyces cerevisiae* is not dependent on Tel1-mediated phosphorylation of Cdc13. *Genetics.* **186**: 1147–1159.

Goudsouzian LK, Tuzon CT, Zakian VA. (2006) S. cerevisiae Tel1p and Mre11p are required for normal levels of Est1p and Est2p telomere association. *Mol. Cell.* **24**: 603–610.

Grandin N, Charbonneau M. (2001) Hsp90 levels affect telomere length in yeast. *Mol. Genet. Genomics.* **265**: 126–134.

Grandin N, Damon C, Charbonneau M. (2001) Ten1 functions in telomere end protection and length regulation in association with Stn1 and Cdc13. *EMBO J.* **20**: 1173–1183.

Grandin N, Reed SI, Charbonneau M. (1997) Stn1, a new *Saccharomyces cerevisiae* protein, is implicated in telomere size regulation in association with Cdc13. *Genes Dev.* **11**: 512–527.

Hardy CF, Sussel L, Shore D. (1992) A RAP1-interacting protein involved in transcriptional silencing and telomere length regulation. *Genes Dev.* **6**: 801–814.

Hector RE, Shtofman RL, Ray A, Chen BR, Nyun T, Berkner KL, Runge KW. (2007) Tel1p preferentially associates with short telomeres to stimulate their elongation. *Mol. Cell.* **27**: 851–858.

Hemann MT, Strong MA, Hao LY, Greider CW. (2001) The shortest telomere, not average telomere length, is critical for cell viability and chromosome stability. *Cell.* **107**: 67–77.

Hirano Y, Fukunaga K, Sugimoto K. (2009) Rif1 and rif2 inhibit localization of tel1 to DNA ends. *Mol. Cell.* **33**: 312–322.

Hughes TR, Evans SK, Weilbaecher RG, Lundblad V. (2000) The est3 protein is a subunit of yeast telomerase. *Curr. Biol.* **10**: 809–812.

Jady BE, Richard P, Bertrand E, Kiss T. (2006) Cell cycle-dependent recruitment of telomerase RNA and Cajal bodies to human telomeres. *Mol. Biol. Cell.* **17**: 944–954.

Kanoh J, Ishikawa F. (2001) spRap1 and spRif1, recruited to telomeres by Taz1, are essential for telomere function in fission yeast. *Curr. Biol.* **11**: 1624–1630.

Kramer KM, Haber JE. (1993) New telomeres in yeast are initiated with a highly selected subset of TG1-3 repeats. *Genes Dev.* **7**: 2345–2356.

Lansdorp PM. (2009) Telomeres and disease. *EMBO J.* **28**: 2532–2540.

Latrick CM, Cech TR. (2010) POT1–TPP1 enhances telomerase processivity by slowing primer dissociation and aiding translocation. *EMBO J.* **29**: 924–933.

Lee J, Mandell EK, Rao T, Wuttke DS, Lundblad V. (2010) Investigating the role of the Est3 protein in yeast telomere replication. *Nucl. Acids Res.* **38**: 2279–2290.

Lee J, Mandell EK, Tucey TM, Morris DK, Lundblad V. (2008) The Est3 protein associates with yeast telomerase through an OB-fold domain. *Nat. Struct. Mol. Biol.* **15**: 990–997.

Lendvay TS, Morris DK, Sah J, Balasubramanian B, Lundblad V. (1996) Senescence mutants of *Saccharomyces cerevisiae* with a defect in telomere replication identify three additional EST genes. *Genetics.* **144**: 1399–1412.

Li S, Makovets S, Matsuguchi T, Blethrow JD, Shokat KM, Blackburn EH. (2009) Cdk1-dependent phosphorylation of Cdc13 coordinates telomere elongation during cell-cycle progression. *Cell.* **136**: 50–61.

Lingner J, Cech TR, Hughes TR, Lundblad V. (1997a) Three ever shorter telomere (EST) genes are dispensable for in vitro yeast telomerase activity. *Proc. Natl. Acad. Sci. USA.* **94**: 11190–11195.

Lingner J, Hughes TR, Shevchenko A, Mann M, Lundblad V, Cech TR. (1997b) Reverse transcriptase motifs in the catalytic subunit of telomerase. *Science.* **276**: 561–567.

Liu D, Safari A, O'Connor MS, Chan DW, Laegeler A, Qin J, Songyang Z. (2004) PTOP interacts with POT1 and regulates its localization to telomeres. *Nat. Cell. Biol.* **6**: 673–680.

Luke B, Panza A, Redon S, Iglesias N, Li Z, Lingner J. (2008) The Rat1p 5′ to 3′ exonuclease degrades telomeric repeat-containing RNA and promotes telomere elongation in *Saccharomyces cerevisiae*. *Mol. Cell.* **32**: 465–477.

Lundblad V, Szostak JW. (1989) A mutant with a defect in telomere elongation leads to senescence in yeast. *Cell.* **57**: 633–643.

Makovets S, Blackburn EH. (2009) DNA damage signalling prevents deleterious telomere addition at DNA breaks. *Nat. Cell. Biol.* **11**: 1383–1386.

Marcand S, Brevet V, Gilson E. (1999) Progressive *cis*-inhibition of telomerase upon telomere elongation. *EMBO J.* **18**: 3509–3519.

Marcand S, Brevet V, Mann C, Gilson E. (2000) Cell cycle restriction of telomere elongation. *Curr. Biol.* **10**: 487–490.

Marcand S, Gilson E, Shore D. (1997) A protein-counting mechanism for telomere length regulation in yeast. *Science.* **275**: 986–990.

McGee JS, Phillips JA, Chan A, Sabourin M, Paeschke K, Zakian VA. (2010) Reduced Rif2 and lack of Mec1 target short telomeres for elongation rather than double-strand break repair. *Nat. Struct. Mol. Biol.*

Miller KM, Ferreira MG, Cooper JP. (2005) Taz1, Rap1 and Rif1 act both interdependently and independently to maintain telomeres. *EMBO J.* **24**: 3128–3135.

MitchellJr, Wood E, Collins K. (1999) A telomerase component is defective in the human disease dyskeratosis congenita. *Nature*. **402**: 551–555.

Miyoshi T, Kanoh J, Saito M, Ishikawa F. (2008) Fission yeast Pot1–Tpp1 protects telomeres and regulates telomere length. *Science*. **320**: 1341–1344.

Mozdy AD, Cech TR. (2006) Low abundance of telomerase in yeast: implications for telomerase haploinsufficiency. *RNA*. **12**: 1721–1737.

Mozdy AD, Podell ER, Cech TR. (2008) Multiple yeast genes, including Paf1 complex genes, affect telomere length via telomerase RNA abundance. *Mol. Cell. Biol.* **28**: 4152–4161.

Murray AW, Claus TE, Szostak JW. (1988) Characterization of two telomeric DNA processing reactions in *Saccharomyces cerevisiae. Mol. Cell. Biol.* **8**: 4642–4650.

Myung K, Chen C, Kolodner RD. (2001) Multiple pathways cooperate in the suppression of genome instability in *Saccharomyces cerevisiae. Nature*. **411**: 1073–1076.

Nakada D, Matsumoto K, Sugimoto K. (2003) ATM-related Tel1 associates with double-strand breaks through an Xrs2-dependent mechanism. *Genes Dev*. **17**: 1957–1962.

Negrini S, Ribaud V, Bianchi A, Shore D. (2007) DNA breaks are masked by multiple Rap1 binding in yeast: implications for telomere capping and telomerase regulation. *Genes Dev*. **21**: 292–302.

Osterhage JL, Talley JM, Friedman KL. (2006) Proteasome-dependent degradation of Est1p regulates the cell cycle-restricted assembly of telomerase in *Saccharomyces cerevisiae. Nat. Struct. Mol. Biol.* **13**: 720–728.

Oza P, Jaspersen SL, Miele A, Dekker J, Peterson CL. (2009) Mechanisms that regulate localization of a DNA double-strand break to the nuclear periphery. *Genes Dev*. **23**: 912–927.

Palm W, de Lange T. (2008) How shelterin protects mammalian telomeres. *Annu. Rev. Genet.* **42**: 301–334.

Park MJ, Jang YK, Choi ES, Kim HS, Park SD. (2002) Fission yeast Rap1 homolog is a telomere-specific silencing factor and interacts with Taz1p. *Mol. Cells.* **13**: 327–333.

Pennock E, Buckley K, Lundblad V. (2001) Cdc13 delivers separate complexes to the telomere for end protection and replication. *Cell*. **104**: 387–396.

Pogacic V, Dragon F, Filipowicz W. (2000) Human H/ACA small nucleolar RNPs and telomerase share evolutionarily conserved proteins NHP2 and NOP10. *Mol. Cell Biol.* **20**: 9028–9040.

Porro A, Feuerhahn S, Reichenbach P, Lingner J. (2010) Molecular dissection of telomeric repeat-containing RNA biogenesis unveils the presence of distinct and multiple regulatory pathways. *Mol. Cell Biol.* **30**: 4808–4817.

Puglisi A, Bianchi A, Lemmens L, Damay P, Shore D. (2008) Distinct roles for yeast Stn1 in telomere capping and telomerase inhibition. *EMBO J.* **27**: 2328–2339.

Qi H, Zakian VA. (2000) The *Saccharomyces* telomere-binding protein Cdc13p interacts with both the catalytic subunit of DNA polymerase alpha and the telomerase-associated est1 protein. *Genes Dev*. **14**: 1777–1788.

Redon S, Reichenbach P, Lingner J. (2007) Protein RNA and protein protein interactions mediate association of human EST1A/SMG6 with telomerase. *Nucl. Acids Res.* **35**: 7011–7022.

Redon S, Reichenbach P, Lingner J. (2010) The non-coding RNA TERRA is a natural ligand and direct inhibitor of human telomerase. *Nucl. Acids Res.* **38**: 5797–5806.

Reichenbach P, Hoss M, Azzalin CM, Nabholz M, Bucher P, Lingner J. (2003) A human homolog of yeast est1 associates with telomerase and uncaps chromosome ends when overexpressed. *Curr. Biol.* **13**: 568–574.

Richard P, Darzacq X, Bertrand E, Jady BE, Verheggen C, Kiss T. (2003) A common sequence motif determines the Cajal body-specific localization of box H/ACA scaRNAs. *EMBO J.* **22**: 4283–4293.

Sabourin M, Tuzon CT, Zakian VA. (2007) Telomerase and Tel1p preferentially associate with short telomeres in *S. cerevisiae*. *Mol Cell.* **27**: 550–561.

Sandell LL, Gottschling DE, Zakian VA. (1994) Transcription of a yeast telomere alleviates telomere position effect without affecting chromosome stability. *Proc. Natl. Acad. Sci. USA.* **91**: 12061–12065.

Savage SA, Giri N, Baerlocher GM, Orr N, Lansdorp PM, Alter BP. (2008) TINF2, a component of the shelterin telomere protection complex, is mutated in dyskeratosis congenita. *Am. J. Hum. Genet.* **82**: 501–509.

Schoeftner S, Blasco MA. (2008) Developmentally regulated transcription of mammalian telomeres by DNA-dependent RNA polymerase II. *Nat. Cell. Biol.* **10**: 228–236.

Schramke V, Luciano P, Brevet V, Guillot S, Corda Y, Longhese MP, Gilson E, Geli V. (2004) RPA regulates telomerase action by providing Est1p access to chromosome ends. *Nat. Genet.* **36**: 46–54.

Schulz VP, Zakian VA. (1994) The *Saccharomyces* PIF1 DNA helicase inhibits telomere elongation and *de novo* telomere formation. *Cell.* **76**: 145–155.

Seto AG, Livengood AJ, Tzfati Y, Blackburn EH, Cech TR. (2002) A bulged stem tethers Est1p to telomerase RNA in budding yeast. *Genes Dev.* **16**: 2800–2812.

Singer MS, Gottschling DE. (1994) *TLC1*: template RNA component of *Saccharomyces cerevisiae* telomerase. *Science.* **266**: 404–409.

Singh SM, Lue NF. (2003) Ever shorter telomere 1 (EST1)-dependent reverse transcription by *Candida* telomerase *in vitro*: evidence in support of an activating function. *Proc. Natl. Acad. Sci. USA.* **100**: 5718–5723.

Smogorzewska A, de Lange T. (2004) Regulation of telomerase by telomeric proteins. *Annu. Rev. Biochem.* **73**: 177–208.

Snow BE, Erdmann N, Cruickshank J, Goldman H, Gill RM, Robinson MO, Harrington L. (2003) Functional conservation of the telomerase protein est1p in humans. *Curr. Biol.* **13**: 698–704.

Stellwagen AE, Haimberger ZW, Veatch JR, Gottschling DE. (2003) Ku interacts with telomerase RNA to promote telomere addition at native and broken chromosome ends. *Genes Dev.* **17**: 2384–2395.

Taggart AK, Teng SC, Zakian VA. (2002) Est1p as a cell cycle-regulated activator of telomere-bound telomerase. *Science.* **297**: 1023–1026.

Teixeira MT, Arneric M, Sperisen P, Lingner J. (2004) Telomere length homeostasis is achieved via a switch between telomerase- extendible and -nonextendible states. *Cell.* **117**: 323–335.

Tejera AM, Stagno d'Alcontres M, Thanasoula M, Marion RM, Martinez P, Liao C, Flores JM, Tarsounas M, Blasco MA. (2010) TPP1 is required for TERT recruitment, telomere elongation during nuclear reprogramming, and normal skin development in mice. *Dev. Cell.* **18**: 775–789.

Tomlinson RL, Abreu EB, Ziegler T, Ly H, Counter CM, Terns RM, Terns MP. (2008) Telomerase reverse transcriptase is required for the localization of telomerase RNA to Cajal bodies and telomeres in human cancer cells. *Mol. Biol. Cell.* **19**: 3793–3800.

Tomlinson RL, Li J, Culp BR, Terns RM, Terns MP. (2010) A Cajal body-independent pathway for telomerase trafficking in mice. *Exp. Cell. Res.* **316**: 2797–2809.

Tomlinson RL, Ziegler TD, Supakorndej T, Terns RM, Terns MP. (2006) Cell cycle-regulated trafficking of human telomerase to telomeres. *Mol. Biol. Cell.* **17**: 955–965.

Tseng SF, Lin JJ, Teng SC. (2006) The telomerase-recruitment domain of the telomere binding protein Cdc13 is regulated by Mec1p/Tel1p-dependent phosphorylation. *Nucl. Acids Res.* **34**: 6327–6336.

Tycowski KT, Shu MD, Kukoyi A, Steitz JA. (2009) A conserved WD40 protein binds the Cajal body localization signal of scaRNP particles. *Mol. Cell.* **34**: 47–57.

van Steensel B, de Lange T. (1997) Control of telomere length by the human telomeric protein TRF1. *Nature.* **385**: 740–743.

Vega et al., 2007 Vega LR, Phillips JA, Thornton BR, Benanti JA, Onigbanjo MT, Toczyski DP, Zakian VA. (2007) Sensitivity of yeast strains with long G-tails to levels of telomere-bound telomerase. *PLoS Genet.* **3**: e105.

Venteicher et al., 2009 Venteicher AS, Abreu EB, Meng Z, McCann KE, Terns RM, Veenstra TD, Terns MP, Artandi SE. (2009) A human telomerase holoenzyme protein required for Cajal body localization and telomere synthesis. *Science.* **323**: 644–648.

Venteicher et al., 2008 Venteicher AS, Meng Z, Mason PJ, Veenstra TD, Artandi SE. (2008) Identification of ATPases pontin and reptin as telomerase components essential for holoenzyme assembly. *Cell.* **132**: 945–957.

Vrbsky et al., 2010 Vrbsky J, Akimcheva S, Watson JM, Turner TL, Daxinger L, Vyskot B, Aufsatz W, Riha K. (2010) siRNA-mediated methylation of *Arabidopsis* telomeres. *PLoS Genet.* **6**: e1000986.

Wang F, Podell ER, Zaug AJ, Yang Y, Baciu P, Cech TR, Lei M. (2007) The POT1–TPP1 telomere complex is a telomerase processivity factor. *Nature.* **445**: 506–510.

Weinrich SL, Pruzan R, Ma L, Ouellette M, Tesmer VM, Holt SE, Bodnar AG, Lichtsteiner S, Kim NW, Trager JB, Taylor RD, Carlos R, Andrews WH, Wright WE, Shay JW, Harley CB, Morin GB. (1997) Reconstitution of human telomerase with the template RNA component hTR and the catalytic protein subunit hTRT. *Nat. Genet.* **17**: 498–502.

Wotton D, Shore D. (1997) A novel Rap1p-interacting factor, Rif2p, cooperates with Rif1p to regulate telomere length in *Saccharomyces cerevisiae*. *Genes Dev.* **11**: 748–760.

Xin H, Liu D, Wan M, Safari A, Kim H, Sun W, O'Connor MS, Songyang Z. (2007) TPP1 is a homologue of ciliate TEBP-beta and interacts with POT1 to recruit telomerase. *Nature.* **445**: 559–562.

Ye JZ, Hockemeyer D, Krutchinsky AN, Loayza D, Hooper SM, Chait BT, de Lange T. (2004) POT1-interacting protein PIP1: a telomere length regulator that recruits POT1 to the TIN2/TRF1 complex. *Genes Dev.* **18**: 1649–1654.

Yu EY, Wang F, Lei M, Lue NF. (2008) A proposed OB-fold with a protein-interaction surface in *Candida albicans* telomerase protein Est3. *Nat. Struct. Mol. Biol.* **15**: 985–989.

Zhang W, Durocher D. (2010) *De novo* telomere formation is suppressed by the Mec1-dependent inhibition of Cdc13 accumulation at DNA breaks. *Genes Dev.* **24**: 502–515.

Zhao Y, Sfeir AJ, Zou Y, Buseman CM, Chow TT, Shay JW, Wright WE. (2009) Telomere extension occurs at most chromosome ends and is uncoupled from fill-in in human cancer cells. *Cell.* **138**: 463–475.

7

TELOMERE STRUCTURE IN TELOMERASE REGULATION

MOMCHIL D. VODENICHAROV AND RAYMUND J. WELLINGER

7.1 INTRODUCTION

7.1.1 Telomere Structure and Function

The ends of linear eukaryotic chromosomes, called telomeres, are defined by a unique chromatin region, consisting of the underlying telomere characteristic DNA and its associated proteins. At the earliest dawn of telomere research, it was recognized that these specialized nucleoprotein complexes fulfill an essential function for genome stability. Classic genetic experiments by Muller and McClintock showed that native chromosome ends possess a protective ability allowing their distinction from accidental chromosome breaks and avoidance of consecutive joining/breakage reactions (McClintock, 1938, 1941; Muller, 1938). Later, and after the discovery of the semiconservative nature of DNA replication, it was predicted that incomplete replication of DNA ends would result in strand-specific losses of information with successive cycles of cell division (Olovnikov, 1971, 1973; Watson, 1972). One solution to this problem became apparent after the discovery of the repeated nature of terminal DNA sequences and the activity of the telomerase enzyme, which elongates chromosomal ends by synthesizing new telomeric repeats (Blackburn et al., 2006). Ever since, telomere structure and telomere elongation have been closely interrelated. Today, there is mounting evidence that, in addition to a tightly regulated expression, biogenesis and intracellular trafficking of telomerase enzyme components, the telomere complex and telomere structure also affect its activity *in vivo*. Here we provide a brief overview of

Telomerases: Chemistry, Biology, and Clinical Applications, First Edition.
Edited by Neal F. Lue and Chantal Autexier.

telomere organization in various experimental organisms, describe how the telomere structure and telomere-bound proteins modulate telomerase activity at telomeres, and discuss the dynamic interplay between the cell-cycle progression, conventional DNA replication, and telomerase action at telomeres.

7.1.2 Telomeric DNA

Telomeric DNA structures are fairly well conserved among eukaryotes. In virtually all organisms examined so far, chromosomal ends comprise small tandem arrays of direct DNA repeats. In most of the species studied to date, the strand running 5′ to 3′ from the centromere towards the telomere is guanine-rich (G-strand) and extends beyond its complementary C-rich strand to form a single-stranded (ss) overhang, also referred to as G-tail (LeBel and Wellinger, 2005). One notable exception of this rule is the nematode *Caenorhabditis elegans*, in which half of the telomeres appear to bear C-rich overhangs (Raices et al., 2008). The telomeric repeat unit, G-tail lengths, and the overall telomere lengths vary between species and may also considerably differ between individuals of the same species. In vertebrates, for instance, telomeric DNA is composed of $5'$-$[T_2AG_3]_n$-$3'$ repeats that can span between 2 and 100 kilobase pairs (kbp) and end with a G-tail of up to 500 nucleotides (nt) in length. Plant telomeres are built of a slightly different $5'$-$[T_3AG_3]_n$-$3'$ repeat unit and can also reach 50 kbp and more in length with an overhang of 20–30 nt. In some but not all fungi, telomeric repeats can be much more extensive and complex (Lue, 2010). In perhaps the best-characterized system, that of the budding yeast *Saccharomyces cerevisiae*, telomeres bear approximately 300 bp of heterogeneous $5'$-$[TG_{1-3}]_n$-$3'$ double-stranded (ds) repeats and a $3'$-protrusion of 12–14 nt (Larrivee et al., 2004; LeBel and Wellinger, 2005). Fission yeast telomeric DNA also measures roughly 300 bp in length but consists of a more complex $5'$-$[T_{1-2}ACA_{0-1}C_{0-1}G_{1-6}]_n$-$3'$ repeat sequence (Sugawara and Szostak, 1986). Despite the differences in composition and length, the most widely used method for maintaining this repetitive DNA at telomeres is the activity of telomerase. Telomerase compensates for the inability of the conventional DNA-replication machinery to fully replicate linear DNA molecules, which otherwise would lead to telomere erosion with progressive cell divisions (see below).

Studies in a variety of systems that include humans, mice, plants, worms, and ciliated protozoans have demonstrated that G-tails can invade upstream homologous duplex telomeric tracts to form a large lasso-like secondary structure, known as telomeric loop (t-loop) (Cesare and Griffith, 2004; Cesare et al., 2003; Griffith et al., 1999; Munoz-Jordan et al., 2001; Murti and Prescott, 1999; Raices et al., 2008). In addition to the fact that this structure requires the presence of sufficiently long telomeric tracts, which may not be the case for all organisms, the question of whether t-loops are constitutively present on all chromosomal ends as well as their precise functions are still a matter of debate. Besides their proposed role in telomere protection, a putative function of the t-loops would be to regulate the access of telomerase to telomeres (reviewed in de Lange, 2004; Verdun and Karlseder, 2007).

7.1.3 Telomere Replication

Both double- and ss telomeric DNA have to be maintained within a defined, species-specific size range to ensure recruitment of telomere-binding factors and proper assembly of the protective cap. As mentioned above, the activities involved in semiconservative replication alone are unable to do so. For example, DNA synthesis relies on a short 8–12 nt RNA primer for replication of the lagging DNA strand. At telomeres, lagging-strand replication always occurs on the G-strand and the enzymatic removal of the outmost RNA primer is expected to leave at least primer-sized gaps in newly synthesized C-strands (Fig. 7.1). While the removal of RNA primers during such lagging-strand replication naturally creates at least some parts of the required G-tails, the replication of the leading strand, which always synthesizes the telomeric G-strand, is expected to produce blunt-ended termini. The lack of 3′-extensions at the leading-strand ends adds at least two important aspects to the end-replication problem mentioned above; these ends cannot recruit ss telomere DNA-binding proteins to ensure proper protection and they cannot serve as a substrate for extension by telomerase (Lingner et al., 1995). This imposes an obligatory enzymatic processing of leading strands that must involve resection of the C-strand of newly synthesized blunt-ended molecules (Fig. 7.1; Wellinger et al., 1996). Thus, in the absence of compensatory mechanisms, chromosome ends will incur losses of DNA sequences after each round of DNA duplication, a phenomenon known as "the end-replication problem" (for reviews see Chakhparonian and Wellinger, 2003; Hug and Lingner, 2006).

This gradual loss of terminal sequences is counteracted by telomerase (for reviews see Cech, 2004; Hug and Lingner, 2006; Osterhage and Friedman, 2009). As described in other chapters of this book, telomerase is a two-component ribonucleoprotein enzyme containing a conserved reverse transcriptase catalytic subunit and an associated RNA component (Autexier and Lue, 2006). Telomerase aligns to and uses the G-tail at the chromosome terminus as a primer to copy a short template sequence residing within telomerase RNA (Shippen-Lentz and Blackburn, 1990). Repeated alignment and extension of the G-tail by telomerase creates the typical arrays of direct telomeric repeats at chromosome ends. Hence, the ss G-tails are key elements in both telomere replication and telomere protection mechanisms (see below). Despite this central role G-tails play in modulating telomere homeostasis, the identity of cellular activities that impinge upon their formation for a long time has remained a matter of conjecture. Several recent studies have made inroads into this issue and are discussed below. Telomerase activity is generally responsible for telomere maintenance in eukaryotic cells but there are exceptions to this rule. For example, while most telomerase-lacking mutant yeast cells senesce and cease to divide after approximately 45–60 generations (Lundblad and Szostak, 1989), a small fraction of cells overcomes this proliferation block. In these postsenescence cells, called "survivors," homologous recombination (HR) is essential to repair and maintain telomeric DNA (Lundblad and Blackburn, 1993). Evidence for the existence of similar recombination-based mechanisms has also been found in some immortalized human cell lines that are able to maintain telomeres for many population doublings in the absence of telomerase (Bryan et al., 1995; Murnane et al., 1994). Furthermore, it has become clear that although the majority of

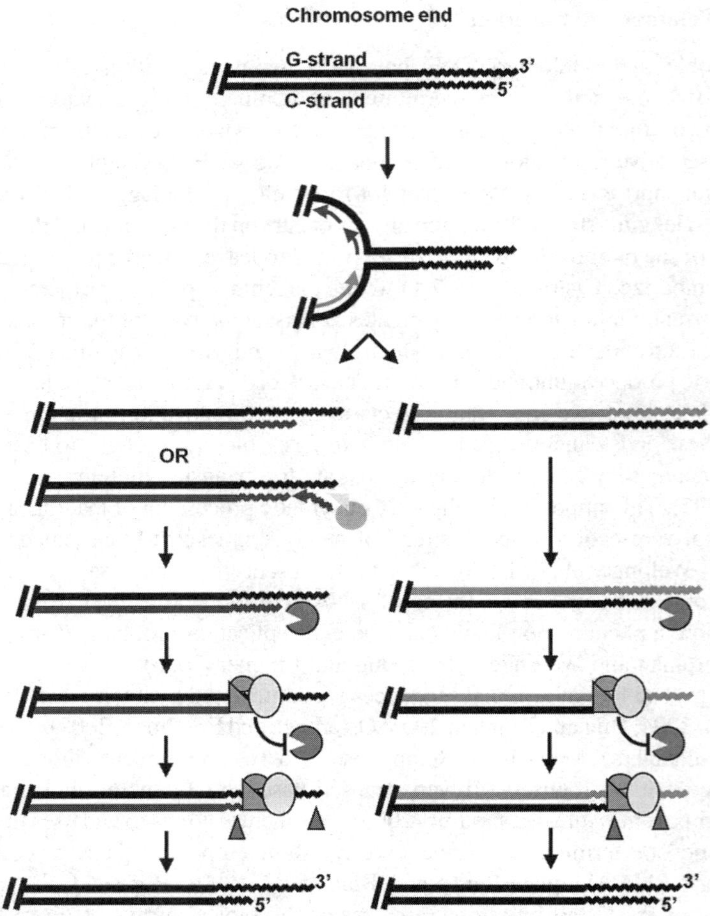

FIGURE 7.1 Telomere replication and the generation of a proper chromosome end structure. At telomeres, the G-strand always serves as a template for lagging-strand synthesis while the C-strand templates the telomere leading strand. Telomeric overhangs, the G-tails, serve as a substrate for telomerase annealing and are formed through different mechanisms on the leading and the lagging telomere sister chromatids. On the lagging strand, they may result simply from incomplete replication at the chromosomal ends or form the removal or the outmost RNA primer by the combined activity of specialized enzymes such as helicases, flap endonucleases or RNases (orange sphere and yellow triangle). The initial 3′-overhang may be further extended due to the activity of an exonuclease. On the leading-strand end, which is predicted to be blunt ended after replication, resection of the C-strand by exonucleolytic activities (magenta sphere) will generate the G-tail. Once overhangs of sufficient length are generated, the binding of ss telomeric DNA binding proteins (the assembly of yellow, orange ovals and green triangle) will obstruct more excessive nucleolytic degradation by blocking access to telomeric ends. Concomitantly, the G-tail bound proteins may modulate the cleavage sites for C- and G-strand specific endonucleases (small grey triangles) and dictate the composition of terminal nt on telomeric DNA and the different length of G-tails on leading versus lagging strand. (See the color version of this figure in Color Plates section.)

cancerous cells avoid telomere shortening through increasing telomerase activity, some tumor cells have the capacity to maintain their telomeres without telomerase by employing homology-directed repair mechanisms jointly referred to as alternative lengthening of telomeres (ALT) (reviewed in Cesare and Reddel, 2010).

7.2 TELOMERE-BOUND PROTEINS

Telomeric DNA serves as a scaffold for the binding of sequence-specific DNA-binding proteins that associate with the repeats in duplex form or with the G-tails, respectively (de Lange, 2005; LeBel and Wellinger, 2005; Shore, 2001; Vega et al., 2003). These direct DNA binders have been shown to recruit additional proteins via protein–protein interactions (O'Sullivan and Karlseder, 2010; Palm and de Lange, 2008). The resulting nucleoprotein complex conceals the chromosome end from being recognized as a damaged DNA end and therefore attenuates cellular responses, such as cell-cycle arrest and DNA repair that would normally ensue after this type of DNA damage has occurred in cells. The inhibition of DNA damage response (DDR) at chromosome ends is a remarkable and indeed essential feature of telomeres, which is often referred to as "chromosome capping" (Lydall, 2009; Wellinger, 2010). In addition to this protection function, telomere-specific proteins also orchestrate telomerase action at telomeres. Here, we will briefly introduce the major telomere-associated proteins in various organisms, with special emphasis on yeast and humans, and their role in controlling telomerase will be discussed in the following sections.

7.2.1 Telomeric Proteins in Budding Yeast

The major ds telomere repeat binding factor in budding yeast is Rap1, a multifunctional protein which regulates telomere length and represses telomere–telomere fusions but also functions in gene expression and silencing at other genomic loci (Marcand et al., 2008; Pardo and Marcand, 2005; Pina et al., 2003; Vodenicharov et al., 2010). Important for the diverse Rap1 functions is its conserved C-terminal (RCT) domain. Mutation analyses and recent crystallographic studies have revealed that this RCT domain in *S. cerevisiae* Rap1 (scRap1) serves as a platform for recruiting Rif proteins that mediate Rap1 function in telomere-length regulation as well as Sir proteins involved in its silencing functions (Feeser and Wolberger, 2008; Hardy et al., 1992a,b; Moretti et al., 1994; Moretti and Shore, 2001; Wotton and Shore, 1997). scRap1 binds directly to a consensus 13 bp telomeric repeat unit via its central MYB domain. Although the MYB domain is also conserved in fission yeast and human Rap1 homologs, these latter proteins lack direct DNA-binding activity and depend on protein–protein interactions for their localization to telomeres (Hanaoka et al., 2001; Li and de Lange, 2003; Li et al., 2000). Human Rap1 association with telomeres requires its RCT domain for an interaction with the telomere repeat binding protein TRF2 (Li and de Lange, 2003; Li et al., 2000). Similarly, the fission yeast Rap1 homolog is recruited to telomeres via an interaction with Taz1, the major ds telomere repeats binding protein in this organism (Kanoh and Ishikawa, 2001).

In budding yeast, the ss G-tail is bound by the heterotrimeric CST complex, composed of three essential proteins: Cdc13, Stn1, and Ten1. Cdc13 recognizes and avidly binds to an 11-nt long ss G-rich telomeric sequence (Hughes et al., 2000; Nugent et al., 1996), and while Stn1 and Ten1 may also be able to associate with G-tail DNA, they have a much lower affinity (Gao et al., 2007). The essential function of the complex consists in protection of telomeres from extensive C-strand degradation which will cause a robust cell-cycle checkpoint response (Garvik et al., 1995; Grandin et al., 1997, 2001; Lydall, 2003; Lydall and Weinert, 1995). Besides this function in telomere protection, Cdc13 participates in telomerase regulation at telomeres and Stn1 is involved in the coordination between the conventional DNA-replication machinery and telomerase (Lundblad, 2003; Puglisi et al., 2008).

7.2.2 The Mammalian Shelterin Complex

Mammalian telomeres are associated with and protected by a complex composed of six different proteins that has been dubbed "shelterin" (de Lange, 2005). Three of the complex members have the capacity to bind DNA directly: TRF1 and TRF2 bind to ds telomeric repeats, whereas the POT1 protein can associate with the ss G-tails. Two other members of the complex, TIN2 and TPP1, interact with TRF1–TRF2 and POT1, respectively, and stabilize their associations with DNA. In addition to its interactions with TRF1 and TRF2, TIN2 can also associate with TPP1, thereby bridging the shelterin components that bind to ds and ss telomeric DNA. As mentioned earlier, RAP1 is recruited to the complex via its interaction with TRF2. For a recent and more comprehensive overview of the shelterin complex the reader is referred to Palm and de Lange (2008).

7.2.3 A Shelterin-Like Complex in Fission Yeast

The components of the telomeric cap in the fission yeast *Schizosacharomyces pombe* appear closely related to those in mammalian cells. Telomeric repeats are bound by Taz1, which is orthologous to the mammalian TRF proteins. Similar to TRF2, Taz1 recruits Rap1 to telomeres (Chikashige and Hiraoka, 2001; Kanoh and Ishikawa, 2001). Furthermore, the G-tail binding protein Pot1 in fact was first identified in fission yeast (Baumann and Cech, 2001) and recently a shelterin-like complex was reported to associate with fission yeast telomeres through interactions with spPot1 and spRap1 (Miyoshi et al., 2008; Tomita and Cooper, 2008). In addition to Pot1, the complex includes Tpz1 (an ortholog of mammalian TPP1), Ccq1 (involved in establishment of telomere heterochromatin and telomere length regulator), and Poz1 (functionally similar to TIN2, connects G-tail bound Pot1 to duplex telomere repeat bound Taz1 and Rap1).

7.2.4 Telomeric Proteins in Other Organisms

Although protein composition and telomere complex architecture can vary considerably amongst different phyla, telomere-specific proteins associated with ds and ss portions of telomeric DNA have been identified in virtually all examined species. The chromosomal ends in ciliates are protected by the TEBPα/β-dimer complex (Gottschling and Zakian, 1986; Price and Cech, 1987) bound to the telomeric overhang via conserved ssDNA binding domain, called OB (oligonucleotide- or oligosaccharide binding) fold. The conserved nature of the OB-fold has aided the identification of orthologous proteins such as Cdc13 and Pot1 in yeasts and humans. Furthermore, similar ss telomeric DNA-binding proteins have been found also in mice, plants, and worms (Raices et al., 2008; Shakirov et al., 2005; Surovtseva et al., 2007). Although most mammalian species contain a single Pot1 gene, rodents posses two closely related Pot1 paralogs that have partially diverged in their function (Hockemeyer et al., 2006). Multiple Pot1 homologs have also been identified in plants; for instance, *Arabidopsis thaliana* genome encodes three Pot proteins—Pot1A, Pot1B, and Pot1C (Shakirov et al., 2005; Surovtseva et al., 2007). Curiously, in the case of the nematode *C. elegans*, ss G-tails have been found on both the G- and C-rich strands and two different OB-fold containing, strand-specific proteins, CeOB1 and CeOB2, are needed to maintain telomere homeostasis (Raices et al., 2008). In addition, although real functional counterparts of the human ds telomeric DNA-binding proteins TRF1 and TRF2 have not yet been found in plants or worms, several TRF-like proteins have surfaced from *in silico* searches in *Arabidopsis* (Karamysheva et al., 2004). These TRF-like proteins were identified by virtue of their sequence homology with the Myb/SAINT domain, a conserved DNA binding domain found in TRF1, TRF2, and Taz1. Telomeres of other species, such as the fruit fly *Drosophila melanogaster*, are organized and maintained in a much more unusual way. Remarkably, this species does not normally use telomerase to maintain their telomeres but rather employs a retrotransposition-based mechanism. Nonetheless, a telomere-specific complex, termed terminin, has been reported to associate with *Drosophila* telomeres, which includes the two founding members heterochromatin protein 1 (HP1), HP1-associated protein/Carravagio (HAOP/Cav) and the recently discovered new members Moigliani (Moi), Verrocchio (Ver), and HP1- and HAOP-interacting protein (HipHop) (Gao et al., 2010; Raffa et al., 2009, 2010; Rong, 2008). As opposed to other organisms, the assembly of terminin at *Drosophila* telomeres does not require specific telomeric repeat sequences and it is thought to conceal them from DNA damage surveillance and repair by purely epigenetic mechanisms.

The putative role in telomere length regulation (via telomerase modulation or not) for the majority of these species-specific factors is either unknown or ill-defined, thus remaining a challenge for future investigations. For this reason, we mainly focus here on studies relevant to telomerase regulation by telomeric proteins in yeast and human cells.

7.2.5 End Protection and Consequences of Telomere Dysfunction

Most of the features of chromosome uncapping are evolutionarily conserved since very similar consequences to telomere dysfunction have been documented in very distant organisms (Karlseder, 2009; Lydall, 2009; Wellinger, 2010). Human and mouse cells in which TRF2 is depleted rapidly arrest in G1 (Karlseder et al., 1999), telomeres attract DDR factors, such as activated ATM, γH2A-X, and 53BP1 (Takai et al., 2003) and suffer massive increases of telomere fusions (Celli and de Lange, 2005; Smogorzewska et al., 2002; van Steensel et al., 1998). Similarly, the Taz1 and Rap1 proteins are required to suppress fusions of telomeres in yeast cells (Ferreira and Cooper, 2001; Miller et al., 2005; Pardo and Marcand, 2005) and human RAP1 also contributes to the inhibition of telomere fusions (Bae and Baumann, 2007; Sarthy et al., 2009). Moreover, experimental interference with Pot1 protein function elicits a DNA damage checkpoint response and ATR- and Chk1-dependent cell-cycle arrest in human and chicken cells (Churikov and Price, 2008; Hockemeyer et al., 2005). In mouse cells, the abrogation of Pot1a leads to C-strand degradation, elevated telomere recombination, and cell-cycle arrest (Hockemeyer et al., 2006; Wu et al., 2006), responses that closely resemble those observed after inactivation of members of the CST complex in budding yeast. Based solely on unsuccessful attempts to identify homologs of the budding yeast CST proteins, for a long time it has been thought that in terms of telomeric protein composition, the budding yeast species have diverged significantly from other organisms such as fission yeast, plants, and mammals (Wellinger, 2009). However, the recent discovery of proteins with high functional similarities to Stn1 and Ten1 in fission yeast and humans, followed by the isolation of heterotimeric CST-like complexes from *Arabidopsis* plants and human cells has revealed that shelterin and CST-like activities may coexist and cooperate in telomere protection in many species (Miyake et al., 2009; Surovtseva et al., 2009).

Finally, it has become clear that the Ku proteins which normally are involved in the non-homologous end joining (NHEJ) pathway for DNA double-strand break (DSB) repair, unexpectedly also reside at telomeres in cells of various species, including trypanosomes, yeasts, plants, flies, and mammals (Fisher and Zakian, 2005). For example, budding yeast Yku is required for proper telomerase localization, regulates *de novo* telomere addition at DSB, protects telomeres from degradation and recombination, promotes silencing of telomere-proximal genes, and participates in anchoring telomeres to the nuclear periphery (reviewed in Fisher and Zakian, 2005). Deletion of either YKU70 or YKU80 genes in yeast leads to telomere shortening and ssDNA accumulation at telomeres (Gravel et al., 1998; Nugent et al., 1998; Polotnianka et al., 1998; Porter et al., 1996). Interestingly, a very recent report demonstrates that human Ku86 performs an essential protective role in human somatic cells that involved repression of dramatic telomere loss and DDRs (Wang et al., 2009).

7.3 TELOMERE ELONGATION AND DNA DAMAGE RESPONSES

Analyses of the cellular responses to experimentally induced telomere dysfunction or gradual telomere erosion in senescent cells have suggested that DNA damage surveillance and repair mechanisms are induced, similar to those triggered by DSB. In fact, even in a normal situation of telomere maintenance and in spite of the disastrous consequences that an induced DDR and DNA repair may cause at telomeres, many proteins that usually monitor and repair DNA damage are also implicated in proper telomere maintenance and regulation of telomerase activity at telomeres (reviewed in d'Adda di Fagagna et al., 2004; Longhese, 2008; Wellinger, 2010). An indication for the involvement of DDR proteins in telomere maintenance first came from genetic studies in yeast. One of the earliest events after a DSB formation is the activation of two proteins, Tel1 and Mec1 that are orthologs of the mammalian DNA-damage checkpoint kinases ATM and ATR, respectively. Interestingly, cells lacking one of these checkpoint kinases display telomere shortening in both budding and fission yeast (Craven et al., 2002; Lustig and Petes, 1986; Nakamura et al., 2002; Ritchie et al., 1999; Ritchie and Petes, 2000). Moreover, simultaneous inactivation of both kinases leads to catastrophic loss of telomeric DNA and rampant increase in genome instability (Chan and Blackburn, 2003; Craven et al., 2002; Mieczkowski et al., 2003; Naito et al., 1998; Nakamura et al., 2002). In addition, a wealth of studies have uncovered that proteins implicated in DSB repair, such as the Ku70-80 heterodimer, and the Mre11-complex (composed of Mre11, Rad50, and Xrs2 in budding yeast; Mre11, Rad32-Rad50, Nbs1 in fission yeast, plants, and mammals), a multitasking complex functioning in HR, NHEJ, and meiosis, are bound to telomeres and necessary for normal telomere function (Boulton and Jackson, 1996, 1998; Gravel et al., 1998; Laroche et al., 1998; Mishra and Shore, 1999; Nakamura et al., 2002; Nugent et al., 1998; Polotnianka et al., 1998; Porter et al., 1996; Ritchie and Petes, 2000). Parallel studies in nematodes, plants, and vertebrates have revealed that the involvement of these checkpoint and DNA-repair proteins in telomere maintenance is not limited to fungal species (Ahmed and Hodgkin, 2000; d'Adda di Fagagna et al., 2001; Hsu et al., 1999; Riha and Shippen, 2003; Riha et al., 2002; Vespa et al., 2005; Zhu et al., 2000). Furthermore, an exceedingly high number of other activities normally involved in DNA transactions such as DNA recombination, repair, or replication associate with telomere proteins, contribute to proper telomere structure formation, and may influence telomere maintenance by telomerase. These include nucleases such as Sae2/Ctp1/CtIP, Exo1, and Dna2 (Bonetti et al., 2009; Tomita et al., 2004), XPF/ERCC1 (Zhu et al., 2003), Appolo (Lenain et al., 2006; van Overbeek and de Lange, 2006), RecQ family helicases such as Sgs1/BLM and WRN (Bonetti et al., 2009; Crabbe et al., 2004; Kibe et al., 2007; Opresko et al., 2002, 2004; Rog et al., 2009), recombination factors such as Rad51D (Tarsounas et al., 2004), and replication protein A (RPA) (Mallory et al., 2003; Ono et al., 2003; Schramke et al., 2004).

Although these proteins can contribute to the regulation of telomere structure and telomerase in a plethora of different ways, the emerging picture points to two major mechanisms. First, they facilitate in some way telomerase access to telomeres, and second, they aid in the generation of G-tails which is necessary for telomerase loading. Undoubtedly, the Mre11-complex contributes to G-tail formation in various organisms. Budding yeast cells lacking any member of this complex have short telomeres, shortened G-tails, and the initial epitasis analysis has established that the complex acts in the telomerase pathway for telomere maintenance, presumably by creating G-tail substrate for telomerase annealing or helping telomerase recruitment to telomeres (Diede and Gottschling, 2001; Larrivee et al., 2004; Nugent et al., 1998; Tsukamoto et al., 2001). Consistently, the Mre11-complex localizes to telomeres in late S phase

(Takata et al., 2005), the time when G-tail length increases and telomeres are elongated in budding yeast (Marcand et al., 2000; Wellinger et al., 1996). Furthermore, the Mre11-complex is very important for the recruitment of Tel1 to short telomeres via an interaction with the Xrs2 C-terminus, presumably in the same manner as it does at a DSB (Goudsouzian et al., 2006; Sabourin et al., 2007; Takata et al., 2005; Tsukamoto et al., 2005). Mre11 indeed binds exclusively to leading-strand telomeric ends, which are more akin to unprocessed DSB, but not to lagging-strand telomeres (Faure et al., 2010). Similarly, human cells derived from patients carrying mutations in the NBS1 or ATM genes exhibit accelerated telomere shortening (Ranganathan et al., 2001; Vaziri et al., 1997) and the human ATM and Mre11-complex are recruited to telomeres in late S and G2 phases of the cell cycle (Verdun et al., 2005). In this case, the recognition of functional human telomeres by DNA damage signaling and repair machineries is also required to ensure adequate telomere protection perhaps via processing of chromosomal ends and restoration of a t-loop structure (Verdun et al., 2005; Verdun and Karlseder, 2006). As a further testimony to the evolutionarily conserved nature of these processes, fission yeast Tel1 and Rad3 (Mec1/ATR) promote telomere protection and telomerase recruitment (Moser et al., 2009) and ATM and ATR contribute to chromosome end protection and regulation of individual telomeric tract lengths in *A. thaliana* (Vespa et al., 2005, 2007). In essence, the data accumulated to date and derived from studies with various model systems reveal that the DDR machinery plays an indispensable role in the telomerase pathway for telomere maintenance.

7.4 SINGLE-STRANDED TELOMERE OVERHANGS AS TELOMERASE SUBSTRATES

7.4.1 Cell-Cycle Regulation of Telomere Overhangs

The central importance of the ss 3′ extensions on the telomeres, the G-tails, is emphasized by the fact that this structure is found in virtually all organisms studied today. As mentioned above, telomerase requires a free 3′ end as a substrate for extension, suggesting that mechanisms influencing the generation of these telomeric G-tails would also have an impact on telomerase activity on telomeres. Moreover, in organisms where the G-tail can invade internal telomeric tracts and initiate t-loop formation, the activities regulating the availability of G-tails would also be expected to regulate telomerase activity. As outlined above, due to the semiconservative nature of conventional DNA replication, after replication the two ends of a linear chromosome would predictably have different termini—one bearing a 3′ overhang left after RNA primer removal and the other that is blunt (Fig. 7.1). However, in ciliates, yeast, and human cells, eventually G-tails can be detected on both types of chromosomal ends, suggesting that this is an essential structure and its generation may be highly regulated (Chai et al., 2006a; Jacob et al., 2001; Makarov et al., 1997; Wellinger et al., 1996).

Early studies examining the molecular events related to G-tail generation have shown that budding yeast telomeres transiently acquire long (50–100 bases) G-tails during a short time window in late S, subsequent to the passage of the conventional replication fork through telomeres (Wellinger et al., 1993a,b). These extended G-tails in late S phase are detected on both ends of a linear minichromosome (Wellinger et al., 1996), and importantly, the arrival of replication fork is necessary for their formation in this narrow time window in the cell cycle (Dionne and Wellinger, 1998). Furthermore, since they can also be detected in telomerase negative cells, these transient G-tails are not due to telomerase activity (Dionne and Wellinger, 1996). These observations have lead to the idea that the blunt end left after completion of leading-strand replication is subjected to end resection, presumably by nuclease/helicase activities, to generate long G-tails in late S (Dionne and Wellinger, 1996; Wellinger et al., 1996). In mammalian cells, chromosomes bear relatively long G-tails throughout the cell cycle (Makarov et al., 1997; McElligott and Wellinger, 1997) and these constitutively long 3′ extensions apparently do not require conventional replication as in yeast, because they have also been detected in quiescent human cells (McElligott and Wellinger, 1997). Similar to the situation in budding yeast, the formation of G-tails in vertebrates does not rely on telomerase activity because again, G-tails are detectable in cells of telomerase knockout mice and telomerase-negative primary human cells (Chai et al., 2006a; Hemann and Greider, 1999), suggesting that nuclease-mediated end-processing may represent a well-conserved step in telomere maintenance (Fig. 7.1). Further studies in ciliate and human cells have lent direct support for the hypothetic involvement of end-processing activity in shaping the telomere G-tails. In *Tetrahymena thermophila* lacking telomerase, both the G-strand and the C-strand are processed accurately to give rise to G-tails of defined length and sequence at both chromosome ends (Jacob et al., 2003). Similarly, in human cells, both strands, but particularly the C-rich one, have a strong bias towards terminating with a precise sequence (Sfeir et al., 2005). Both studies also show that telomerase expression can alter the precision of the terminal cleavage events, providing evidence for continued association of telomerase with telomere ends even after repeat addition, which potentially interferes with processing activities. However, artificially changed telomeric sequences did not alter the accuracy of cleavage, suggesting that the nucleolytic activities involved do not possess sequence specificity (Jacob et al., 2003). Instead, the data is consistent with the notion that G-tail-binding proteins determine the exact patterns of terminal nucleotide processing, a view that has received support from experiments showing that knocking down the expression of shelterin component POT1 in human cancer cells randomizes the 5′-end nt of the G-tail (Fig. 7.1; Hockemeyer et al., 2005).

The above studies on telomere end-processing in human and ciliate cells in the absence of telomerase have also revealed two other intriguing features of this process. First, the processing events of the C- and the G-strands appear tightly coordinated (Fan and Price, 1997; Jacob et al., 2001) and second, leading- and lagging-strand telomeres eventually carry overhangs of different lengths (Fig. 7.1; Chai et al., 2006a; Sfeir et al., 2005). These findings underscore previous results that postulated differential G-tail formation on the two different types of chromosome ends (Bailey et al., 2001; Parenteau and Wellinger, 2002) and it is even possible that several distinct DNA-processing events or perhaps several different nucleolytic activities act at telomeres (Bonetti et al., 2009). Accordingly, budding yeast Mre11 protein appears to be differentially recruited at leading versus lagging-strand telomeric ends (Faure et al., 2010) and distinct multi-step mechanisms are responsible for generation of mature G-tails at leading and lagging daughter telomeres in human cells (Dai et al., 2010). Finally, C-strand synthesis by the conventional DNA-replication machinery and the G-strand synthesis by telomerase must be coupled in a certain way. Thus the formation of proper G-tails is a process tightly regulated during the cell cycle involving the activities of the replication machinery, telomeric proteins, and nucleases as discussed below in more details.

7.4.2 Mechanisms for Telomere Overhang Formation

For a long time, the identity of the nucleolytic activities responsible for G-tail generation has remained enigmatic. At least for lagging-strand telomeres, some processing activities may be part of the replication apparatus itself. Formally, a compromised Okazaki fragment synthesis or even just the RNA primer removal (in budding yeast for example) can account for generation of G-tails of sufficient length to serve for annealing of the telomerase enzyme. A reconstituted linear SV40 DNA-replication system has demonstrated that while the newly replicated leading strand is blunt ended, the lagging-strand replication is gradually halted in the terminal 250–500 nt, leaving long 3′-overhangs of sizes typically observed in human cells (Ohki et al., 2001). Consistently, mutations affecting the budding yeast FEN1 flap-endonuclease homolog Rad27, which normally functions in Okazaki fragment maturation, cause a significant G-tail length deregulation (Parenteau and Wellinger, 2002) and fission yeast Dna2, which has a similar function during lagging-strand replication as FEN1, is required for generation of cell-cycle-regulated ss extensions at telomeres (Tomita et al., 2004). Moreover, the inability of the conventional lagging-strand synthesis machinery to displace and remove the last RNA primer poses an additional complication at the very tip of the lagging-strand telomeres that may require separate activities, including RNaseH and the Pif1 helicase (Budd et al., 2006; Jeong et al., 2004; Qiu et al., 1999). Thus, although G-tails on the lagging-strand ends could result simply from incomplete DNA replication, additional processing events are very likely to occur.

At the leading-strand end, which is predicted to be blunt ended and closely resembling an accidental DSB, the Mre11-complex has emerged as a principle candidate for initiating end processing. The absence of the Mre11-complex greatly suppresses new telomere formation in a telomere healing assay that monitors the addition of new telomeric repeats onto a seed sequence next to a HO endonuclease-induced DSB *in vivo* (Diede and Gottschling, 2001). Because the HO nuclease generates only a 4 nt 3′ overhang, which may be suboptimal for annealing with the telomerase RNA template, additional C-strand resection orchestrated by the Mre11-complex presumably is required to generate a substrate for telomerase binding and promote repeat addition. However, chromatin-immunoprecipitation studies did not detect a reduction in Cdc13 accumulation at telomeres of Mre11-deficient yeast cells and, based on that observation, it was inferred that the Mre11-complex does not affect G-tail length but rather collaborates with Tel1 in telomerase recruitment (Tsukamoto et al., 2001). More precise measurements of G-tail lengths in *mre11*Δ cells have revealed that the Mre11-complex is in fact required for the generation of proper constitutive G-tails (Larrivee et al., 2004).

Indeed, it appears that the overall dynamics of the late S-phase-specific telomeric G-rich overhang elongation is not affected in calls lacking Mre11-complex components, but the average length of G-tails is reduced in these cells. A very recent report has suggested that the role of this complex is limited to telomeric ends produced by leading-strand replication because the absence of Mre11 abolishes telomerase binding only to the leading but not the lagging telomeric strand (Faure et al., 2010). In human cells, the Mre11-complex is also implicated in telomeric overhang formation (Chai et al., 2006b). Somewhat surprisingly, mutations in *MRE11* that diminish its *in vitro* nuclease activity are not defective in telomere length regulation and in the telomere healing assay (Frank et al., 2006), suggesting that the Mre11-complex is not the only factor responsible for G-overhang generation at least at budding yeast telomeres. Indeed, there is recent evidence for the involvement of multiple other activities, suggesting that telomere processing may have numerous parallels with the processing of DNA ends at a DSB (Bonetti et al., 2009, 2010; Vodenicharov et al., 2010). At such an accidental DSB, both the processing events and the choice of repair pathway differ depending on the cell-cycle stage in which a DSB arises. The decisive mechanism turns out to be DSB resection, which is regulated by cyclin-dependent kinases (CDKs) (Aylon et al., 2004; Frank et al., 2006; Ira et al., 2004; Jazayeri et al., 2006). In this respect, one confirmed target of CDKs appears to be the budding yeast protein Sae2 and its homologs in fission yeast and humans—Ctp1 and CtIP, respectively (Huertas et al., 2008; Huertas and Jackson, 2009; Limbo et al., 2007; Sartori et al., 2007). Intriguingly, Sae2 functions with the Mre11-complex in the initial trimming of the DSB ends to generate short 3′-overhangs of approximately 50–100 nt. This is followed by a secondary processing that exposes extensive 3′-ss tails and is redundantly executed by either the Sgs1 helicase in collaboration with the Dna2 nuclease or the 5′–3′ exonuclease Exo1 (Gravel et al., 2008; Mimitou and Symington, 2008; Nimonkar et al., 2008; Zhu et al., 2008). Significantly, the generation of ssDNA at uncapped telomeres also requires the activity of CDK1 and is limited to late S and G2 phases (Vodenicharov and Wellinger, 2006). Importantly, that CDK1 may control telomerase-mediated telomere elongation is consistent with the time window, in which budding yeast telomeres are replicated, the length of G-tails increases, and telomerase can elongate telomeric DNA (reviewed in Vodenicharov and Wellinger, 2007). Indeed, CDK1 inhibition completely blocks telomere healing at telomere seed sequences and G-tail formation at native telomeres (Frank et al., 2006; Vodenicharov and Wellinger, 2006). Consistently, the generation of telomeric G-tails has recently been shown to have similar requirements in terms of nucleases and CDK1-dependent Sae2 phosphorylation as those acting at DSB mentioned above (Bonetti et al., 2009), reinforcing the notion that telomere and DSB ends share common requirements for processing activities. In spite of these similarities, it is striking that while the simultaneous inactivation of Sae2 and Sgs1 completely abolishes resection, *de novo* telomere addition, and homologous-directed repair at DSB, the initially critically short telomeres in the double mutant are gradually elongated with consecutive generations to reach and stabilize at wild-type lengths (Bonetti et al., 2009). These results suggest that in yeast yet another activity, one that is redundant with the Sae2/Sgs1 pathway, may contribute to the creation of the ss overhangs, thus providing a substrate needed for extension by telomerase. Redundant processing activities may well be present in mammalian cells as well, as indicated by recent reports implicating the 5′–3′ exonuclease Apollo in generation of the 3′ ss overhangs at leading-strand telomeres in mouse embryonic fibroblasts (Lam et al., 2010; Wu et al., 2010).

7.4.3 Overhang Secondary Structures in Telomerase Regulation

An alternative way by which the G-rich overhangs may regulate telomerase is by inducing structural changes at telomeres that limit telomerase access to telomeric DNA ends. In higher eukaryotes such as humans, mice, chicken, and plants, the formation of the t-loop at chromosomal ends likely regulates the ability of telomerase to add new telomeric repeats (reviewed in Verdun and Karlseder, 2007). Consequently, several shelterin proteins, for example TRF2 and TIN2 that promote and/or stabilize t-loop formation probably also inhibit telomerase via enhancement of this high-order telomeric structure and sequestration of the ss overhang. Evidently, even if constitutively present, these structures will have to be opened to allow telomerase access, which could be compared to the switching between nonextendable and extendable states in budding yeast discussed below. Consistent with this view, evidence collected in human cells indicates that t-loops are unfolded during telomere replication and there is a concomitant recruitment of processing factors in late S and G2 phase at telomeres (Verdun et al., 2005; Verdun and Karlseder, 2006).

Another potential structural impediment for telomerase action at telomeres are so-called G-quadruplex structures (also called G-quartets) consisting of particular G–G base-pairs that could form between folded parts of the ss overhang (reviewed in Lipps and Rhodes, 2009). ss telomeric DNA-binding proteins such as POT1 in humans and TEBP1 in ciliates are known for their ability to disrupt G-quartet structures, providing a potential means for facilitating telomerase access to the end (Paeschke et al., 2005; Zaug et al., 2005).

In summary, the available data indicate postreplicative nucleolytic processing of chromosomal ends that is necessary for the generation of the proper substrate for telomerase, the G-tails, and, at the same time, for restoration of appropriate structures of the ends of eukaryotic chromosomes. Some data point towards the possibility that leading and lagging telomeric ends are prepared in a different fashion. Presently it is not clear whether the same processing activities act on both leading and lagging ends or whether both are processed by the same activities but with different efficiency or to a different extent, thus leading to the observed phenotypic differences between leading and lagging ends (Fig. 7.1). However from studies in yeast, it has become increasingly clear that multiple and redundant activities are responsible for these processing events, highlighting their importance for the maintenance of telomeres and chromosome stability.

7.5 TELOMERASE REGULATION BY TELOMERIC PROTEINS

7.5.1 Regulation By dsDNA Telomeric Repeat Binding Proteins

Currently, the best-understood example of telomerase regulation by telomeric proteins is that of the budding yeast telomerase. The most important findings are:

first, a negative feedback mechanism that senses the length of ds telomeric tracts restricts new telomeric repeat addition, thus maintaining a roughly constant average telomere length in cells. Second, the association of telomerase with telomeres is separable from its catalytic activation at the extreme chromosome terminus, an event that actually results in synthesis of new telomeric repeats. Third, telomeres must adopt a particular "open" state in order for telomerase to gain access and elongate them. Fourth, a signal, similar to the one evoked by broken DNA ends, emanates from the shortest telomeres, leading to their preferential elongation by telomerase. In this section, we provide a summary of major discoveries that have lead to the establishment of these principles, some of which are illustrated in Figure 7.2.

Pioneering studies have suggested that budding yeast Rap1 can act as a telomerase regulator based on the observed telomere-length changes and increased heterogeneity in cells expressing hypomorphic or truncated alleles of the Rap1 protein (Conrad et al., 1990; Lustig et al., 1990). More detailed mutational and genetic analyses have revealed that Rap1 is a negative regulator of telomerase and that this function is mediated by its C-terminal RCT-domain. Point mutations in this domain lead to moderately elongated and stable telomeres while a complete deletion will lead to very promiscuous telomere elongation and highly unstable telomeric tracts (Kyrion et al., 1992; Sussel and Shore, 1991). However, overexpression of the RCT-domain alone also caused telomere elongation, an observation that was interpreted as an indication that this protein fragment now titrates negative telomerase regulators that normally are recruited via the RCT to the telomeres away from telomeres (Conrad et al., 1990). Indeed, the identification of two RCT-interacting partners, Rif1 and Rif2, validated this prediction (Hardy et al., 1992b; Wotton and Shore, 1997). The studies showed that the two Rif proteins function via separate mechanisms, that they are recruited by Rap1 to telomeres, and the effect of their combined deletion recapitulates that of an RCT deletion. Shortly after the role of Rap1 as negative regulator was uncovered, a series of elegant experiments demonstrated that the number of repeats at an individual telomere was reduced proportionally to the number of RCT domains artificially targeted to this telomere but not other telomeres in the same cell (Marcand et al., 1997). Moreover, the involvement of a degradation activity in this process was excluded since the tract-length regulation was achieved solely by progressive *cis*-inhibition of telomerase upon telomere elongation rather than by achieving equilibrium between the rates of repeat loss and addition in the course of cell division (Marcand et al., 1999). Based on these experiments, the existence of a negative feedback counting mechanism for telomere-length regulation that can discriminate the precise number of Rap1 molecules bound to a chromosomal end was proposed (Fig. 7.2). Subsequently, similar experiments, in which Rif proteins instead of RCT-domains were tethered to telomeric ends substantiated that what appears to be counted are in fact the Rif-proteins, and not Rap1 (Levy and Blackburn, 2004). Interestingly, varying the number of telomere-bound Rap1 molecules does not affect telomere length in cells lacking the checkpoint protein Tel1 (Ray and Runge, 1999) and telomeres remain short after RCT deletion in a *tel1* background (Craven and Petes, 1999). These findings suggested that Rap1 and Rif proteins bound to the ds telomere repeats may modulate telomerase activity by inhibiting positive telomerase

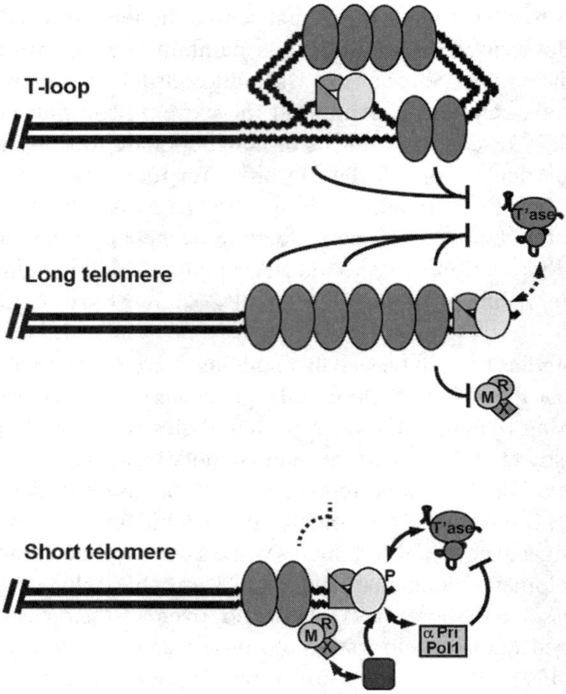

FIGURE 7.2 Regulation of telomerase by telomere-associated proteins. The current view is that the telomeric proteins (grey ovals) bound to the ds telomere repeats (duplex zig-zag line) negatively regulate telomerase. In several experimental systems, the activity of telomerase is reversely correlated with the number of ds telomeric repeats and, respectively, the number of telomeric proteins bound in *cis*, thereby establishing a negative feedback loop or a counting mechanism. In higher eukaryotes, additional negative regulation may be achieved by organization of sufficiently long telomeres into t-loops (illustrated on top). The G-tail-binding proteins (the assembly of yellow, orange ovals and green triangle) appear to facilitate telomerase access and positively regulate its activity at telomeres. Based on data primarily from budding yeast, it has been proposed that at long telomeres, the increased numbers of dsDNA-bound protein molecules inhibit telomerase access and the recruitment of factors promoting the activity of telomerase, such as Mre11 complex. Short telomeres, on the other hand, are permissive for Mre11 recruitment, which in turn recruits checkpoint kinases, like Tel1/ATM (blue square). Together they signal the presence of a short telomere by preparing telomere structure and modifying telomere proteins (P; phosphorylation) to facilitate the recruitment and extension of telomeric DNA by telomerase. The telomerase-mediated extension is tightly coordinated with the conventional replication machinery, which limits the addition of new telomeric repeats by telomerase. (See the color version of this figure in Color Plates section.)

regulators, such as the Mre11-complex and Tel1. Support for such a view came form experiments showing that deletion of Rif1 and Rif2 allows telomerase to work even in Tel1/Mec1-deficient cells, suggesting that these kinases promote telomerase by altering telomere structure and increasing accessibility, rather than by direct activation of its enzymatic activity (Chan et al., 2001). Indeed, the molecular mechanism behind these observations has just begun to be revealed. First, long telomeric repeat tracts attract less Mre11 and are poorly resected as compared to short tracts, an effect that apparently depends on Rap1 binding because the difference in Mre11 binding and resection between short and long tracts is lost if the long tract has been engineered to bear a mutation preventing Rap1 binding (Negrini et al., 2007). Second, a recent report shows that Rif proteins attenuate telomerase through inhibition of Tel1 localization to long telomeric DNA ends. Indeed, Rif2 competes with Tel1 for binding to the Xrs2 C-terminus and, after Tel1 delocalization, the Mre11-complex does not associate efficiently with Rap1-covered telomeric repeat tracts (Hirano et al., 2009). Together, these results point to a mechanism, in which the number of Rap1/Rif molecules bound to ds telomeric tracts may influence telomerase activity by either masking chromosomal ends from Tel1 recognition and Mre11-complex-dependent resection at long telomeres while allowing these events to take place at short telomeres (Fig. 7.2). Indeed, when the telomere-elongation events were scored at individual telomeres in a cell during a single cell cycle, not all telomeres but only a few amongst the shortest were extended by the telomerase (Teixeira et al., 2004). Deletion of either Rif1 or Rif2 in this setting increased the number of productive association events of telomerase with short telomeres per cell cycle but not the extent of elongation per telomeric end. Although telomerase can execute multiple rounds of repeat addition at one telomeric end, it becomes measurably processive only at telomeres with lengths below 125 bp and this threshold length is signaled by Tel1 (Arneric and Lingner, 2007; Chang et al., 2007). Collectively, these data imply that telomere tract length and associated proteins modulate telomerase by switching between an extendable and a nonextendable state at telomeres (Hug and Lingner, 2006).

The mechanism of telomerase inhibition by ds telomeric repeat binding factors appears to be conserved because there is evidence of a similar mechanism in fission yeast and human cells. Initial studies in human cells uncovered roles for TRF1 and TRF2 as telomerase inhibitors, since interference with their binding to telomeres resulted in elongation while their overexpression resulted in telomere shortening (Smogorzewska et al., 2000; van Steensel et al., 1998). Tethering experiments have provided direct evidence for the inhibitory effect of TRF1 and TRF2 on telomerase activity in mammalian cells (Ancelin et al., 2002). This negative *cis* regulation of human telomerase by TRF proteins is at least in part mediated by TIN2 and RAP1, both members of the shelterin complex, and probably involves structural changes at telomeres that modulate telomerase accessibility. TIN2 clearly behaves as a negative regulator of telomere elongation (Houghtaling et al., 2004; Kim et al., 1999), a function that could stem directly from its role in sheltering complex organization and stabilization via its independent interactions with TRF1, TRF2, and TPP1 (Houghtaling et al., 2004; Kim et al., 2004; Ye et al., 2004a,b). Since a functional shelterin complex relies on a precise level and stoichiometry of its components within telomeric chromatin (O'Connor et al., 2006; Takai et al., 2010; Ye et al., 2004a), it is possible that TIN2 inhibits telomerase by stabilizing the higher-order t-loop structure at mammalian telomeres. That this higher-order structure might be the major impediment for telomerase at mammalian telomeres is also consistent with a recent report showing that human telomerase acts immediately after telomeres are replicated and when, presumably, this structure is temporarily dissolved (Zhao et al., 2009). One striking difference reported by these authors as compared to the yeast model for shortest telomere elongation is that telomerase acts indiscriminately on most telomeres in each cell cycle (Zhao et al., 2009). Whether this difference reflects genuinely distinctive mechanisms operating in lower versus higher eukaryotes, or it reflects disequilibrium of the steady-state situation due to transformation process of the human cancer cell lines used in this study, remains an open question.

The precise role of human RAP1 is unclear but it has been suggested that its role as a negative regulator of telomerase has been conserved in higher eukaryotes (Li et al., 2000; O'Connor et al., 2004). However, a comparison of different Rap1 proteins also suggests that the domain functions in RAP1 have undergone extensive evolutionary changes, which could be due to the changes from DNA binder to interacting with telomere-bound proteins (Li and de Lange, 2003). The association of hRap1 with telomeres requires its RCT domain for its interaction with TRF2 (Li and de Lange, 2003; Li et al., 2000), while the acquisition of two MYB domains connected through a flexible linker might have provided a selective advantage for shifting Rap1 from indirect DNA binder form, present in most of the yeast species and mammals, to a direct DNA-binder in budding yeast (Lue, 2010). Contrary to the situation in budding yeast, mammalian Rif1 does not normally reside at telomeres and is recruited only to dysfunctional telomeres in an ATM-dependent manner (Silverman et al., 2004; Xu and Blackburn, 2004). Furthermore, hRAP1 appears also to regulate homology-driven and telomerase-independent elongation at telomeres (Sfeir et al., 2010). Fission yeast Taz1, which is related to TRF2 but not scRap1, mediates negative regulation of telomerase by independent recruitment of Rap1 and Rif1 (Chikashige and Hiraoka, 2001; Cooper et al., 1997; Kanoh and Ishikawa, 2001). Only few details are known about telomerase inhibition in this organism but spRap1 and spRif1 do not interact with each other and they seem to suppress telomerase via two separate pathways (Kanoh and Ishikawa, 2001; Miller et al., 2005).

Despite the differences between the various model systems, negative regulation of telomerase in *cis* that apparently correlates with the number of ds repeat-bound telomeric proteins appears to be widespread and a well-conserved phenomenon. This mechanism would ensure that telomeres that have reached lengths above a certain average are no longer substrates for further telomerase-mediated elongation. However, if the length of the ds repeats shortens below a certain threshold which assures optimal telomere function, the likelihood of those ends being dealt with, that is elongated, by telomerase must increase sharply. The net result from the negative feedback loop regulation of telomerase in *cis* at individual telomeres would be the maintenance of all telomeres within a narrow length distribution commonly observed in cells.

7.5.2 Regulation by Telomeric ssDNA Binding Proteins

An extensively studied and well-understood example of telomerase regulation by a ss repeat binding protein is that of budding yeast telomerase being regulated by the G-tail-binding protein Cdc13. Cdc13 is a multifunctional protein, containing at least four domains: a N-terminal protein-interaction domain, a recruitment domain (RD) that mediates interactions with the Est1 telomerase subunit, a DNA-binding domain (DBD) responsible for direct association with telomeric ss DNA, and a C-terminal domain that negatively regulates telomerase (Chandra et al., 2001; Hughes et al., 2000; Pennock et al., 2001). The essential function of Cdc13 is to protect telomeres from degradation by nucleases (Garvik et al., 1995), a function that is achieved by association with its binding partners, Stn1 and Ten1, in the CST-complex (Gao et al., 2007; Grandin et al., 1997, 2001; Petreaca et al., 2006, 2007; Xu et al., 2009). Besides its essential role in telomere protection, Cdc13 was also identified as a factor required for yeast telomerase activity *in vivo*. A genetic screen for mutants causing senescence or an "*est*" (ever shorter telomeres) phenotype identified telomerase subunits, as well as a particular allele of Cdc13 (Lendvay et al., 1996). Except for Cdc13, all telomerase subunits encoded by *EST* genes are contained within the telomerase RNP, whether by direct or indirect association with the telomerase RNA, which serves as a flexible scaffold for telomerase holoenzyme assembly (Dandjinou et al., 2004; Zappulla and Cech, 2004). Cdc13 associates with the complex via protein–protein interactions with Est1. The specific mutation in Cdc13, *cdc13-2*, that was identified in the above-mentioned screen by virtue of compromised telomerase function leading to cellular senescence (Nugent et al., 1996), was later mapped as a glutamine to lysine change at position 252 in the Cdc13 RD (Pennock et al., 2001). This amino-acid change disrupts the Cdc13–Est1 interaction which can be compensated for by either expressing a fusion protein formed of Cdc13 and Est1, or by introducing a reciprocal mutation in Est1 protein, *est1-60* (Evans and Lundblad, 1999; Pennock et al., 2001). Interestingly, when Cdc13 was fused directly to the Est2 catalytic subunit, the hybrid protein could bypass the need for Est1 and allowed cell growth in complete absence of Est1 function, suggesting that the G-tail bound Cdc13 functions in telomerase recruitment to its substrate by binding Est1 (Evans and Lundblad, 1999, 2002). Of note, the telomeres of the *est1*Δ strain expressing Cdc13–Est2 fusion were stably maintained at a length slightly below than that of WT telomere length, suggesting that Est1 makes yet another contribution to the optimal telomere repeat addition by telomerase (Evans and Lundblad, 2002). Chromatin immunoprecipitation experiments using cells in different phases of the cell cycle further revealed that the association of both Est2 and Est1 with telomeres peaks in a short time interval in late S phase (Taggart et al., 2002), the time when telomeres are elongated (Marcand et al., 2000). The recruitment of Est2 to telomeres in late S and also to telomeric seed sequence near DSB is impaired in a *cdc13-2* background (Bianchi et al., 2004; Taggart et al., 2002), confirming the role of Cdc13 in telomerase recruitment. Interestingly, similar ChIP experiments also have suggested that Est2 is telomere associated in G1 phase of the cycle, which is not the case for Est1. The G1-specific association of telomerase with telomeres thus is not mediated via a Cdc13–Est1 interaction but instead requires the Yku80 protein (Fisher et al., 2004). Previous work had shown that yeast Ku (YKu) can bind to a specific stem loop in telomerase RNA, *TLC1*, thereby apparently defining a second pathway for telomerase recruitment (Chan et al., 2008; Peterson et al., 2001; Stellwagen et al., 2003). Further work has revealed that YKu–*TLC1* interaction is required for telomerase import and accumulation into the nucleus because *TLC1* RNA is not properly retained in the nucleus in YKu70 null cells or in cells expressing a YKu80 protein lacking the *TLC1*-interacting stem (Gallardo et al., 2008). In addition, a parallel Cdc13-independent, but RPA-dependent pathway operates to facilitate Est1 recruitment but does not affect Est2 association with telomeres (Schramke et al., 2004). It is noteworthy that while the Yku- and RPA-mediated pathways may help Est1 and Est2 proteins attain their optimal levels for telomere extension during S phase, a complete loss of telomere-bound telomerase and induction of senescence are only observed in the absence of Cdc13-dependent recruitment.

The dominant role of the Cdc13-regulated mechanisms amongst the pathways for telomerase recruitment is linked perhaps to its enhancement by the Mre11-complex and the checkpoint kinases Tel1 and Mec1. As discussed earlier, all these factors genetically can be placed in the telomerase pathway and act to promote its action at telomeres. The patterns of Mre11 and Tel1 association with telomeres matches the specific time for telomere elongation (Hector et al., 2007; Sabourin et al., 2007; Viscardi et al., 2007) and they are required for normal levels of Est1 and Est2 recruitment (Goudsouzian et al., 2006; Tsukamoto et al., 2001). As in DSB repair, the Mre11-comoplex functions as a Tel1 recruiter via an interaction between a Xrs2 C-terminal domain and Tel1 (Tsukamoto et al., 2005). In situations in which this interaction is disrupted, the recruitment of both Tel1 and Est2 to telomeres is compromised and the cells exhibit severe telomere shortening (Sabourin et al., 2007; Shima et al., 2005; Tsukamoto et al., 2005). Intriguingly, the RD of Cdc13 contains several potential PI3K-kinase consensus sites and some have been shown to be phosphorylated by Tel1 and Mec1 *in vitro* (Tseng et al., 2006). Point mutations abolishing some of these sites also lead to gradual telomere shortening and senescence phenotype, suggesting that the Tel1-dependent Cdc13 phosphorylation controls the recruitment function of Cdc13 (Tseng et al., 2006). Finally, recent analyses of the temporal events occurring at an induced single short telomeric end have revealed an increased binding of both Est1 and Est2 at short telomere as compared to a wild-type length telomere (Bianchi and Shore, 2007b; Sabourin et al., 2007). Tel1 is also preferentially enriched at eroded telomeres in cells lacking telomerase as well as at artificially shortened telomeres (Bianchi and Shore, 2007b; Hector et al., 2007; Sabourin et al., 2007). Based on these results, it has been proposed that Cdc13 bound to telomeric G-tails is modified by Tel1 and Mec1 kinases and these phosphorylation events are required to facilitate a productive Cdc13–Est1 interaction promoting telomerase association or activation at telomere ends.

The stimulation of telomerase activity by the Cdc13–Est1 interaction is not the only function of Cdc13 in telomerase regulation: cells that contain certain other Cdc13 alleles display a telomere overelongation, suggesting that this telomere protein may also exert an inhibitory effect on telomerase (Chandra et al., 2001; Grandin et al., 2000). This negative regulation appears to be mediated by the Cdc13-interacting partner Stn1 and, perhaps, the assembly or a conformational change in the CST complex. The association between Cdc13 and Stn1 is abolished in *cdc13-2* mutants, suggesting that Stn1 and Est1 compete for the same binding surface on Cdc13 (Chandra et al., 2001). Telomerase repression by CST relies on an interaction between the Cdc13 C-terminus and Stn1–Ten1, which could be stabilized after Stn1 displaces Est1 from binding in RD (Chandra et al., 2001; DeZwaan and Freeman, 2009; Puglisi et al., 2008). It has been proposed that Cdc13–Est1 and Cdc13–Stn1 interactions may reflect the transition between an extendible and nonextendible telomere states and, as it will be discussed in the following section, constitute an important mechanism for coordination of conventional replication with telomerase. Interestingly, the switch between Cdc13-telomerase extendable and CST nonextendable structures at telomeres appears to involve the yeast Hsp90 chaperone, which can modulate CST assembly and Cdc13 binding to telomeric DNA (DeZwaan et al., 2009).

In mammalian cells, telomerase access to the chromosomal end presumably is restrained by adopting the mentioned t-loop configuration, since in this configuration, the 3′-end of the ss telomeric overhang may be tucked back and base-paired with the C-strand. It logically follows that the unfolding of the t-loop structure itself should represent an important regulatory step, necessary to make the 3′ terminus available for binding by ss telomere DNA-binding proteins, such as the POT1–TPP1 subcomplex, and for telomerase. Although the vertebrate POT1–TPP1 complex has a high affinity for ss overhangs, it is primarily recruited to telomeres by bridging interactions with TIN2, which binds both TRF1 and TRF2 (Hockemeyer et al., 2007; Houghtaling et al., 2004; Liu et al., 2004; Loayza and De Lange, 2003; O'Connor et al., 2006; Wang et al., 2007). This higher-order arrangement of the vertebrate telomeric complex has lead to the idea that telomerase action might be regulated by transferring POT1–TPP1 from its shelterin-bound state to an overhang-bound state, in which the complex facilitates telomerase access. The switch between the two states could take place just after telomere replication when the t-loop is unfolded. Alternatively, one could speculate that telomeres that are insufficiently long to form t-loops or to sequester all available POT1–TPP1 in shelterin-bound configuration would also be elongated in a preferential manner. Irrespective, there still is a debate regarding the effect of the G-tail-bound POT1 on telomerase activity. *In vitro* observations made by studying POT1 effects on telomerase activity have shown that POT1 bound very near to the 3′ terminus of the overhang blocks telomerase access while, on the contrary, more internal binding stimulates telomerase activity (Kelleher et al., 2005; Lei et al., 2005). *In vivo*, interference with POT1 levels through expression of shRNAs or via the expression of a dominant-negative POT1 allele, which is proficient in TPP1-mediated telomerase recruitment to telomeres but deficient in ssDNA binding, results in telomere lengthening, suggesting that POT1 is a negative telomerase regulator (Loayza and De Lange, 2003; Veldman et al., 2004; Ye et al., 2004b). In yet other settings, the overexpression of the full-length POT1 yielded results suggesting that it is a positive regulator of telomerase (Armbruster et al., 2001; Colgin et al., 2003; Liu et al., 2004). More recent studies have lent further support to the notion that the POT1–TPP1 complex is a positive regulator of human telomerase. Both sequence alignment and crystallographic analyses of a domain in human TPP1 revealed the presence of an OB fold and structural similarities to the beta-subunit of TEBP, the telomere end-binding protein in ciliated protozoa (Wang et al., 2007; Xin et al., 2007). TPP1 associates with telomerase through its OB-fold and TPP1–POT1 association enhances POT1 affinity for telomeric ssDNA, providing a physical link between telomerase and the telosome/shelterin complex (Xin et al., 2007). Moreover, the interaction between TPP1 and POT1 at telomeric overhangs increases the activity and processivity of the human telomerase core enzyme (Wang et al., 2007). Based on these findings, it has been proposed that POT1–TPP1 switches from inhibiting telomerase access to telomeres, as a component of shelterin, to serving as a processivity factor for telomerase during telomere extension at the 3′ overhangs.

The function of ss telomere DNA-binding proteins being able to mediate a transition between "open" and "closed" states in terms of extension may be well conserved. Studies in ciliates have shown that a cooperative binding of TEBPβ and TEBPα to telomeric DNA maintains the overhangs in a nonextendable, perhaps a G-quadruplex configuration (Paeschke et al., 2005). During telomere replication, TEBPβ gets phosphorylated by the CDK and is displaced from the telomere overhang–TEBPα complex by telomerase, two events needed for unfolding of G-tails and concomitant extension by telomerase (Paeschke et al., 2005, 2008). Whether a similar switch between a permissive and a nonpermissive state for telomerase-mediated extension is also regulated by phosphorylation in mammalian cells is yet to be investigated. However, in human cells, the late S–G2 binding of the activated ATM (the human Tel1 homologue) to telomeres parallels the time of t-loop reorganization (Verdun et al., 2005; Verdun and Karlseder, 2006).

In other organisms, the presence of several POT1 paralogs has added another layer of complexity to these regulatory mechanisms. For example, the conditional loss of one of the POT gene products, POT1a, in *Tetrahymena* and mouse cells caused telomere elongation (Jacob et al., 2007; Wu et al., 2006), while elimination of one of the two POT genes in *Arabidopsis* leads to progressive telomere shortening (Surovtseva et al., 2007). The situation currently emerging for fission yeast bears similarities to other organisms with a shelterin-like complex at their telomeres. The fission yeast orthologue of the mammalian TPP1, called Tpz1, nucleates a four-member complex also containing Poz1, Pot1, and Ccq1, all factors involved in telomerase recruitment (Miyoshi et al., 2008; Tomita and Cooper, 2008). Coimmunoprecipitation experiments show that Tpz1 interacts with telomerase only in the presence of Ccq1 and cells devoid of Ccq1 experience telomere shortening, suggesting a positive role for this interaction in telomerase regulation. On the other hand, removal of Poz1 from the complex results in telomere elongation, suggesting that Poz1-mediated bridging of a Tpz1-dependent complex to Taz1-Rap1 at ds telomeric repeats inhibits telomerase activity (Miyoshi et al., 2008; Tomita and Cooper, 2008). Together, these data support a model, in which the dynamic transition of Poz1–Tpz1–Pot1–Ccq1 complex between Taz1–Rap1-anchored and overhang-bound states can act as a regulatory switch for the fission yeast telomerase. Remarkably, this regulatory mechanism definitively resembles the one proposed to operate at human telomeres and some parallels can also be envisaged between this mechanism and the one relying on Est1-dependent activation of telomere-bound inactive telomerase in budding yeast. Nevertheless, fission yeast and mammalian TPP1 apparently play a very particular role in switching between the two states, with no obvious equivalent uncovered for budding yeast. In this regard, it is noteworthy, that both *S. cerevisiae* and *C. albicans* Est3 proteins have recently been shown to contain OB-fold domains mediating a direct association with the telomerase holoenzyme and these proteins have been proposed to be structural TPP1 homologs, thus raising interesting evolutionary questions (Lee et al., 2008; Yu et al., 2008).

Overall, it appears that depending on the organisms, either a few functionally homologous ss telomere DNA-binding proteins perform the multiple functions required for telomerase regulation which range from telomere protection to positive and negative regulation of telomerase at chromosome tips, or multiple protein paralogs have evolved to fulfill these functions separately, thus providing a testimony for the evolutionary flexibility of telomerase regulation by the telomeric G-tail-binding factors.

7.6 TELOMERASE REGULATION IN THE CONTEXT OF CELL DIVISION

To date, there is very good evidence suggesting that conventional DNA replication and telomerase-mediated telomeric repeat addition normally are temporarily and physically coordinated. Early studies in budding yeast have documented that telomere proximal origins of replication fire late in S phase (Ferguson et al., 1991; McCarroll and Fangman, 1988; Reynolds et al., 1989) and that no replication origin is present in the telomeric repeats (Wellinger and Zakian, 1989). The closest origin residing in the subtelomeric region of the chromosome thus serves as a starting point for replication of chromosome ends. In a number of systems analyzed, telomeres represent fragile sites in terms of DNA replication and telomeric chromatin has been shown to impose various difficulties for the passage of the replication fork. These include potential fork pausing and topological constrains that require the presence of certain telomere components and the activity of specialized proteins, such as helicases and nucleases, to relieve the block and allow full telomere replication (Ivessa et al., 2002; Makovets et al., 2004; Miller et al., 2006; Sfeir et al., 2009; Ye et al., 2010). Presently, it is unclear if these events are somehow interconnected with telomere end processing or telomerase action at telomeres, but some data suggest that this may well be the case (Moser and Nakamura, 2009; Verdun and Karlseder, 2007).

At least in yeast, where telomeres are late replicating, the processing events generating the G-tails needed for telomerase association are confined to late S phase (Wellinger et al., 1993b, 1996). The action of telomerase at telomeres also appears to be tightly linked with conventional replication. This link is highlighted by experiments showing that telomeric repeat synthesis onto the ends of yeast linear plasmids depends on the presence of a functional replication origin in the plasmid (Dionne and Wellinger, 1998; Marcand et al., 2000) and that the elongation of shortened telomeres depends on origin firing (Viscardi et al., 2007). These observations have lead to the hypothesis that replication fork passage through yeast telomeres is required for telomerase-mediated repeat addition by virtue of overhang generation or by transient opening to grant telomerase access to telomere. Conversely, the telomeres themselves may influence the timing of firing of the nearby origin in a manner dependent on the actual length of telomeric repeat tract in *cis*, provoking early firing of origins near short telomeres. An artificially created short telomere leads to an induction of the telomere-proximal origin on the same chromosome, concomitant with lengthening of this particular telomere earlier in S phase, but not of another telomere in the same cell that is of average length (Bianchi and Shore, 2007a). The molecular mechanisms behind these observations and their significance in terms of telomerase regulation are not yet fully resolved. However, given that only a few telomeres are extended by telomerase during a single passage though S-phase (Teixeira et al., 2004) and the augmented ssDNA signals at natural telomeres that are seen in late S (Wellinger et al., 1993b), these data are consistent with the possibility that telomere replication unfolds differently for short versus longer telomeres. Perhaps a few, or even only one, very short telomere induces a strong Tel1-mediated signal such that they are coordinately extended by telomerase and the normal DNA replication machinery prior to the replication of longer telomeres, which really are the bulk of yeast telomeres but are not perceived as needing elongation by telomerase. If true, this situation would also have similarities in human cells, where telomeres are replicated throughout S phase but actual telomere processing apparently only occurs in late S/G2, as discussed below.

Indeed, quite different from the situation in yeast, human telomere replication is asynchronous even for two opposite ends of one chromosome and occurs throughout S phase (Wright et al., 1999; Zou et al., 2004). More recent studies in which ChIP using telomere-specific proteins as well as replication proteins was combined with bromodeoxyuridine incorporation into nascent DNA strands have substantiated that human telomeres undergo two steps in terms of telomeric DNA replication and modification, one occurs during S phase, followed by another occurring in G2 phase (Verdun et al., 2005). Interestingly, this second step is accompanied by a transient association of DDR factors with telomeres, suggesting that in humans, as in yeasts, during every given S-phase the processing events required for completeness of telomere structure may be the last ones occurring in cells before entry into mitosis. However, a direct proof for telomere extension by telomerase within this particular interval in the cell cycle is still missing for higher eukaryotes. On the contrary, data from human cancer cells show that the G-strands on most chromosomal ends are rapidly extended by telomerase immediately after their replication, which occurs randomly throughout the S phase (Zhao et al., 2009). Moreover, the fill-in synthesis of the C-strand was delayed into late S/G2 phase, suggesting that G-strand extension by telomerase and the complimentary C-strand resynthesis by the conventional replication machinery are uncoordinated in tumor cells (Zhao et al., 2009). Nonetheless, both leading and lagging ends were subjected to processing in late S and G2 phases to generate G-tails of normal length irrespectively of the presence of active human telomerase (Dai et al., 2010).

For budding yeast there is compelling evidence supporting the idea that the conventional replication machinery and telomerase cooperate to maintain telomeres at an appropriate length. The possibility that the lagging-strand replication machinery is coupled to telomerase action has first been raised from observations indicating that overall telomere length is deregulated in mutants of two components of the lagging-strand apparatus, namely Cdc17 (encoding DNA polymerase alpha) and Cdc44 (encoding the large subunit of replication factor C) (Adams and Holm, 1996; Carson and Hartwell, 1985). These observations were further strengthened by experiments showing that inhibition with aphidicolin of DNA polymerases alpha and delta, both involved in lagging-strand synthesis, resulted in coordinated changes in G- and C-strand telomere length in *Euplotes* (Fan and Price, 1997). Consistently, interference with DNA polymerase alpha in mouse cells by expressing a temperature-sensitive allele at semipermissive temperature results in telomere elongation (Nakamura et al., 2005). In yeast, the requirement for replicative polymerases has also been demonstrated using an assay which monitors the addition of a telomere onto a DSB end that is marked with a short telomeric seed sequence. In this case, the healing of the break through G-strand polymerization by telomerase is inhibited if conventional DNA polymerases are inactivated, suggesting that telomerase needs to interact with the lagging-strand machinery to actively synthesize new repeats (Diede and Gottschling, 1999).

Because in the majority of cases considerable telomere lengthening ensues after an inhibition of lagging-strand components, one could argue that the lagging-strand DNA-replication machinery negatively regulates telomerase. This idea also received support via studies of the budding yeast CST complex. The significant homology between CST members and the RPA complex subunits has lead to the suggestion that CST represents an RPA-like complex, perhaps with a devoted role in telomere replication (Gao et al., 2007). Both Cdc13 and Stn1 have known physical interactions with separate subunits of DNA polymerase alpha-primase complex, Pol1 and Pol12, respectively (Grossi et al., 2004; Qi and Zakian, 2000). The C terminus of Stn1 binds both Cdc13 and Pol12 and the data indicates that the Stn1–Cdc13 interaction is required to limit continuous telomerase action (Puglisi et al., 2008). The negative effect of Stn1 on telomerase action might be regulated by a modification of the CST activity or structure in *cis* at individual telomeres because Stn1 telomere binding is independent of telomeric repeat tract length (Puglisi et al., 2008). Although the precise mechanism remains to be uncovered, it is likely that post-translational modifications, for example, phosphorylation, may determine the balance between an inhibitory effect of Stn1–Cdc13 binding and activation of telomerase through Est1 recruitment. Thus, Cdc13 can be seen as a platform for recruitment of factors that regulate telomerase activity both positively and negatively at the tips of chromosomes and, at the same time, coordinates telomere elongation, C-strand resynthesis, and telomere capping (Bertuch and Lundblad, 2006).

A new aspect in telomerase control by telomeric proteins is a recently uncovered regulation of telomere structure and telomeric proteins by the activity of mitotic CDK. CDK drives the cell division cycle by pairing with different cyclins and phosphorylating numerous targets to promote cell proliferation. Its activity is low in G1 phase of the cycle but increases rapidly during the transition from G1 to S, remains high throughout S, G2, and M, to be again suppressed in late M phase by the activity of the anaphase promoting complex (APC), which marks the end of a cell cycle (Manchado et al., 2010; Mendenhall and Hodge, 1998). As discussed earlier, one important way by which CDK may modulate telomerase activity, is via a modulation of the mechanisms generating G-tails. In budding yeast, high CDK activity is required both for the addition of a new telomere at a broken DNA end by telomerase and for the acquisition of long G-tails at native telomeres in late S phase (Frank et al., 2006; Vodenicharov and Wellinger, 2006). Thus, the processing of telomeric ends, which presumably creates the suitable substrate for telomerase action, requires both high CDK1 activity and the passage of the replication fork (Vodenicharov and Wellinger, 2007). These studies prompted the search for telomere-specific targets of CDK1 and the first one to be identified was Cdc13. Two groups reported CDK1-dependent phosphorylation of Cdc13 as being essential for efficient recruitment of the yeast telomerase complex to telomeres (Li et al., 2009; Tseng et al., 2009). The telomere-shortening phenotype observed in the absence of the specific phosphorylation sites on Cdc13 in this case however is not that pronounced as when Tel1 phosphorylation sites on Cdc13 are mutated, suggesting that multiple modifications by a number of kinases are needed for optimal telomerase recruitment. It has also been proposed that CDK1-mediated phosphorylation of Cdc13 stimulates telomerase by favoring the interaction of Cdc13 with Est1, rather than the competing Stn1–Ten1 complex. While not all details are clear, one can nevertheless imagine the existence of a direct mechanistic link between cell cycle progression and coordination of telomere elongation.

The end of telomerase-mediated telomere replication is marked by telomerase dissociation from the telomeric DNA end. This is an active process requiring the activity of a specialized helicase, which exerts an additional control on telomerase by means of its physical removal form the G-tails. In budding yeast, this step is mediated by the Pif1 helicase, which unwinds the DNA–RNA duplex between telomerase RNA and the telomeric DNA end (Boule et al., 2005). Pif1 is aided in this process by the Est2 protein itself, which encourages dissociation of the DNA–RNA hybrid by a subdomain in its reverse transcriptase finger (Eugster et al., 2006). If judged from the enormous increase in genome instability due to *de novo* telomere addition events in cells lacking this helicase, Pif1 must be an important inhibitor of telomerase (Myung et al., 2001; Pennaneach et al., 2006; Putnam et al., 2004). Finally, a recent study shows that the main ATR/Mec1 DNA damage signaling pathway regulates telomerase action at DSBs by a Mec1-dependent phosphorylation of Pif1. Curiously, this phosphorylation is specifically required for the Pif1-mediated telomerase inhibition that takes place at DNA breaks, but not for that at telomeres (Makovets and Blackburn, 2009). Perhaps, the inhibitory function at telomeres requires very high levels of Pif1, whose expression is cell cycle regulated and peaks in late S (Vega et al., 2007), the time when telomerase acts. Alternatively, Mec1-dependent phosphorylation and Pph3 dephosphorylation of Cdc13 may regulate specifically its binding at DSB versus telomeres (Zhang and Durocher, 2010).

7.7 CONCLUSIONS

The addition of new telomeric repeats by telomerase is a prevalent, but not universal, way for counteracting the erosion at the ends of eukaryotic chromosomes. Some exceptions are found within the dipteran insects, including the fruit fly *Drosphila melanogaster*, that rely on retrotransposition to maintain their telomeres. Further, in certain yeast and human cells telomerase-independent telomere length changes are based on the utilization of HR. The latter represents the primary mechanism for telomere maintenance in about 15% of human cancer cells; however, the majority of tumors still rely on telomerase reactivation to divide indefinitely. Therefore, this underscores the need for a thorough understanding of the molecular regulatory mechanisms that control telomerase activity. A key regulator of telomerase is the telomere itself and a wealth of recent reports allowed significant progress in our understanding of the importance of telomere complex and telomere structure in the functional regulation of telomerase. Telomere-specific proteins may actually play both positive and negative roles in telomerase regulation (Table 7.1). They assist in the recruitment of telomerase to their substrates, on the one hand, and on the other they

TABLE 7.1 Telomeric Proteins Known to Modulate Telomerase Activity in Various Experimental Systems and their Roles in Telomerase Regulation

Organism/ Protein	Part of Complex	Binds Telomeric DNA	Protein that Tethers it to Telomeric DNA	Affects Telomerase Activity
S. cerevisiae				
Rap1	Rap1–Rif1–Rif2	Yes (ds)	n.a.	Negatively
Rif1, Rif2	Rap1–Rif1–Rif2	No	Rap1	Negatively
Yku70, Yku80	Yeast Ku	Yes (ds)	n.a.	Positively
Cdc13	Cdc13–Stn1–Ten1 (CST)	Yes (ss)	n.a.	Positively
Stn1	Cdc13–Stn1–Ten1 (CST)	Yes (ss)	n.a.	Negatively
Ten1	Cdc13–Stn1–Ten1 (CST)	Yes (ss)	n.a.	Negatively
H. sapiens				
TRF1	Shelterin	Yes (ds)	n.a.	Negatively
TRF2	Shelterin	Yes (ds)	n.a.	Negatively
RAP1	Shelterin	No	TRF2	Negatively
TIN2	Shelterin	No	TRF1, TRF2	Negatively
TPP1	Shelterin	No	POT1	Positively and negatively
POT1	Shelterin	Yes (ss)	n.a.	Positively and negatively
Hku		Yes (ds)	n.a.	Positively
S. pombe				
Taz1	Sheltelin-like	Yes (ds)	n.a.	Negatively
spRap1	Sheltelin-like	No	Taz1	Negatively
spRif1	Sheltelin-like	No	Taz1	Negatively
Pot1	Sheltelin-like	Yes (ss)	n.a.	Positively
Tpz1	Sheltelin-like	No	Pot1	Positively
Poz1	Sheltelin-like	No	Taz1/spRap1	Negatively
Ccq1	Sheltelin-like	No	Pot1	Positively
A. thaliana				
TBP1		Yes (ds)	n.a.	Negatively
TRFL1, TRFL2, TRFL4	Sheltelin-like (family I)	Yes (ds)	n.a.	No effect
POT1A		No	?	Positively

Note: (ds), double-stranded telomeric DNA; (ss), single-stranded telomeric DNA; n.a., not applicable; ?, unknown.

gradually inhibit telomerase activity with increasing repeat length. Transient telomere uncapping and recognition as DSB appears as a prerequisite for telomere extension by telomerase. The elicited transient response is a key event in telomerase activation because it involves enzymatic and structural transactions needed to prepare the chromosome end as a substrate suitable for extension by telomerase. Furthermore, it is an exclusive signal emanating from short telomeres only and switches them into an extendable state required for telomerase action. Last but not least, telomerase-mediated telomere replication is closely linked and regulated in a coordinated manner with conventional replication. All these and, perhaps, more unknown molecular levers converge into a complex mechanism for telomerase regulation by the actual structure of the telomere and/or the telomeric complex, the complete characterization of which should occupy the telomere field for many years to come.

ACKNOWLEDGMENTS

The work in the Vodenicharov lab is supported by grants form National Science and Engineering Research Council of Canada and start-up funds from the Faculty of Sciences of the University of Sherbrooke. Work in the Wellinger lab is supported by grants from the Canadian Institutes of Health Research. The authors apologize to all colleagues whose important contributions could not be cited due to space constrains.

REFERENCES

Adams AK, Holm C. (1996) Specific DNA replication mutations affect telomere length in *Saccharomyces cerevisiae*. *Mol. Cell. Biol.* **16**: 4614–4620.

Ahmed S, Hodgkin J. (2000) MRT-2 checkpoint protein is required for germline immortality and telomere replication in *C. elegans*. *Nature*. **403**: 159–164.

Ancelin K, Brunori M, Bauwens S, Koering CE, Brun C, Ricoul M, Pommier JP, Sabatier L, Gilson E. (2002) Targeting assay to study the *cis* functions of human telomeric proteins: evidence for inhibition of telomerase by TRF1 and for activation of telomere degradation by TRF2. *Mol. Cell. Biol.* **22**: 3474–3487.

Armbruster BN, Banik SS, Guo C, Smith AC, Counter CM. (2001) N-terminal domains of the human telomerase catalytic subunit required for enzyme activity *in vivo*. *Mol. Cell. Biol.* **21**: 7775–7786.

Arneric M, Lingner J. (2007) Tel1 kinase and subtelomere-bound Tbf1 mediate preferential elongation of short telomeres by telomerase in yeast. *EMBO Rep.* **8**: 1080–1085.

Autexier C, Lue NF. (2006) The structure and function of telomerase reverse transcriptase. *Annu. Rev. Biochem.* **75**: 493–517.

Aylon Y, Liefshitz B, Kupiec M. (2004) The CDK regulates repair of double-strand breaks by homologous recombination during the cell cycle. *EMBO J.* **23**: 4868–4875.

Bae NS, Baumann P. (2007) A RAP1/TRF2 complex inhibits nonhomologous end-joining at human telomeric DNA ends. *Mol. Cell.* **26**: 323–334.

Bailey SM, Cornforth MN, Kurimasa A, Chen DJ, Goodwin EH. (2001) Strand-specific postreplicative processing of mammalian telomeres. *Science*. **293**: 2462–2465.

Baumann P, Cech TR. (2001) Pot1, the putative telomere end-binding protein in fission yeast and humans. *Science*. **292**: 1171–1175.

Bertuch AA, Lundblad V. (2006) The maintenance and masking of chromosome termini. *Curr. Opin. Cell. Biol.* **18**: 247–253.

Bianchi A, Negrini S, Shore D. (2004) Delivery of yeast telomerase to a DNA break depends on the recruitment functions of Cdc13 and Est1. *Mol. Cell.* **16**: 139–146.

Bianchi A, Shore D. (2007a) Early replication of short telomeres in budding yeast. *Cell*. **128**: 1051–1062.

Bianchi A, Shore D. (2007b) Increased association of telomerase with short telomeres in yeast. *Genes Dev*. **21**: 1726–1730.

Blackburn EH, Greider CW, Szostak JW. (2006) Telomeres and telomerase: the path from maize, *Tetrahymena* and yeast to human cancer and aging. *Nat. Med.* **12**: 1133–1138.

Bonetti D, Clerici M, Anbalagan S, Martina M, Lucchini G, Longhese MP (2010) Shelterin-like proteins and Yku inhibit nucleolytic processing of *Saccharomyces cerevisiae* telomeres. *PLoS Genet*. **6**: e1000966.

Bonetti D, Martina M, Clerici M, Lucchini G, Longhese MP. (2009) Multiple pathways regulate 3′ overhang generation at *S. cerevisiae* telomeres. *Mol. Cell.* **35**: 70–81.

Boule JB, Vega LR, Zakian VA. (2005) The yeast Pif1p helicase removes telomerase from telomeric DNA. *Nature*. **438**: 57–61.

Boulton SJ, Jackson SP. (1996) Identification of a *Saccharomyces cerevisiae* Ku80 homologue: roles in DNA double strand break rejoining and in telomeric maintenance. *Nucl. Acids Res.* **24**: 4639–4648.

Boulton SJ, Jackson SP. (1998) Components of the Ku-dependent non-homologous end-joining pathway are involved in telomeric length maintenance and telomeric silencing. *EMBO J*. **17**: 1819–1828.

Bryan TM, Englezou A, Gupta J, Bacchetti S, Reddel RR. (1995) Telomere elongation in immortal human cells without detectable telomerase activity. *EMBO J*. **14**: 4240–4248.

Budd ME, Reis CC, Smith S, Myung K, Campbell JL. (2006) Evidence suggesting that Pif1 helicase functions in DNA replication with the DNA2 helicase/nuclease and DNA polymerase delta. *Mol. Cell. Biol.* **26**: 2490–2500.

Carson MJ, Hartwell L. (1985) CDC17: an essential gene that prevents telomere elongation in yeast. *Cell*. **42**: 249–257.

Cech TR. (2004) Beginning to understand the end of the chromosome. *Cell*. **116**: 273–279.

Celli GB, de Lange T (2005) DNA processing is not required for ATM-mediated telomere damage response after TRF2 deletion. *Nat. Cell. Biol.* **7**: 712–718.

Cesare AJ, Griffith JD. (2004) Telomeric DNA in ALT cells is characterized by free telomeric circles and heterogeneous t-loops. *Mol. Cell. Biol.* **24**: 9948–9957.

Cesare AJ, Quinney N, Willcox S, Subramanian D, Griffith JD. (2003) Telomere looping in *P. sativum* (common garden pea). *Plant J*. **36**: 271–279.

Cesare AJ, Reddel RR. (2010) Alternative lengthening of telomeres: models, mechanisms and implications. *Nat. Rev. Genet.* **11**: 319–330.

Chai W, Du Q, Shay JW, Wright WE. (2006a) Human telomeres have different overhang sizes at leading versus lagging strands. *Mol. Cell.* **21**: 427–435.

Chai W, Sfeir AJ, Hoshiyama H, Shay JW, Wright WE. (2006b) The involvement of the Mre11/Rad50/Nbs1 complex in the generation of G-overhangs at human telomeres. *EMBO Rep*. **7**: 225–230.

Chakhparonian M, Wellinger RJ. (2003) Telomere maintenance and DNA replication: how closely are these two connected? *Trends Genet*. **19**: 439–446.

Chan A, Boule JB, Zakian VA. (2008) Two pathways recruit telomerase to *Saccharomyces cerevisiae* telomeres. *PLoS Genet*. **4**: e1000236.

Chan SW, Blackburn EH. (2003) Telomerase and ATM/Tel1p protect telomeres from nonhomologous end joining. *Mol. Cell*. **11**: 1379–1387.

Chan SW, Chang J, Prescott J, Blackburn EH. (2001) Altering telomere structure allows telomerase to act in yeast lacking ATM kinases. *Curr. Biol.* **11**: 1240–1250.

Chandra A, Hughes TR, Nugent CI, Lundblad V. (2001) Cdc13 both positively and negatively regulates telomere replication. *Genes Dev*. **15**: 404–414.

Chang M, Arneric M, Lingner J. (2007) Telomerase repeat addition processivity is increased at critically short telomeres in a Tel1-dependent manner in *Saccharomyces cerevisiae*. *Genes Dev*. **21**: 2485–2494.

Chikashige Y, Hiraoka Y. (2001) Telomere binding of the Rap1 protein is required for meiosis in fission yeast. *Curr. Biol.* **11**: 1618–1623.

Churikov D, Price CM. (2008) Pot1 and cell cycle progression cooperate in telomere length regulation. *Nat. Struct. Mol. Biol.* **15**: 79–84.

Clerici M, Mantiero D, Lucchini G, Longhese MP. (2005) The *Saccharomyces cerevisiae* Sae2 protein promotes resection and bridging of double strand break ends. *J. Biol. Chem.* **280**: 38631–38638.

Colgin LM, Baran K, Baumann P, Cech TR, Reddel RR. (2003) Human POT1 facilitates telomere elongation by telomerase. *Curr. Biol.* **13**: 942–946.

Conrad MN, Wright JH, Wolf AJ, Zakian VA. (1990) RAP1 protein interacts with yeast telomeres *in vivo*: overproduction alters telomere structure and decreases chromosome stability. *Cell*. **63**: 739–750.

Cooper JP, Nimmo ER, Allshire RC, Cech TR. (1997) Regulation of telomere length and function by a Myb-domain protein in fission yeast. *Nature*. **385**: 744–747.

Crabbe L, Verdun RE, Haggblom CI, Karlseder J. (2004) Defective telomere lagging strand synthesis in cells lacking WRN helicase activity. *Science*. **306**: 1951–1953.

Craven RJ, Greenwell PW, Dominska M, Petes TD. (2002) Regulation of genome stability by TEL1 and MEC1, yeast homologs of the mammalian ATM and ATR genes. *Genetics*. **161**: 493–507.

Craven RJ, Petes TD. (1999) Dependence of the regulation of telomere length on the type of subtelomeric repeat in the yeast *Saccharomyces cerevisiae*. *Genetics*. **152**: 1531–1541.

d'Adda di Fagagna F, Hande MP, Tong WM, Roth D, Lansdorp PM, Wang ZQ, Jackson SP. (2001) Effects of DNA nonhomologous end-joining factors on telomere length and chromosomal stability in mammalian cells. *Curr. Biol.* **11**: 1192–1196.

d'Adda di Fagagna F, Teo SH, Jackson SP. (2004) Functional links between telomeres and proteins of the DNA-damage response. *Genes Dev*. **18**: 1781–1799.

Dai X, Huang C, Bhusari A, Sampathi S, Schubert K, Chai W. (2010) Molecular steps of G-overhang generation at human telomeres and its function in chromosome end protection. EMBO J. **29**: 2788–2801.

Dandjinou AT, Levesque N, Larose S, Lucier JF, Abou Elela S, Wellinger RJ. (2004) A phylogenetically based secondary structure for the yeast telomerase RNA. *Curr. Biol.* **14**: 1148–1158.

de Lange T. (2004) T-loops and the origin of telomeres. *Nat. Rev. Mol. Cell. Biol.* **5**: 323–329.

de Lange T. (2005) Shelterin: the protein complex that shapes and safeguards human telomeres. *Genes Dev.* **19**: 2100–2110.

DeZwaan DC, Freeman BC (2009) The conserved Est1 protein stimulates telomerase DNA extension activity. *Proc. Natl. Acad. Sci. USA.* **106**: 17337–17342.

DeZwaan DC, Toogun OA, Echtenkamp FJ, Freeman BC (2009) The Hsp82 molecular chaperone promotes a switch between unextendable and extendable telomere states. *Nat. Struct. Mol. Biol.* **16**: 711–716.

Diede SJ, Gottschling DE. (1999) Telomerase-mediated telomere addition *in vivo* requires DNA primase and DNA polymerases alpha and delta. *Cell.* **99**: 723–733.

Diede SJ, Gottschling DE. (2001) Exonuclease activity is required for sequence addition and Cdc13p loading at a *de novo* telomere. *Curr. Biol.* **11**: 1336–1340.

Dionne I, Wellinger RJ. (1996) Cell cycle-regulated generation of single-stranded G-rich DNA in the absence of telomerase. *Proc. Natl. Acad. Sci. USA.* **93**: 13902–13907.

Dionne I, Wellinger RJ. (1998) Processing of telomeric DNA ends requires the passage of a replication fork. *Nucl. Acids Res.* **26**: 5365–5371.

Eugster A, Lanzuolo C, Bonneton M, Luciano P, Pollice A, Pulitzer JF, Stegberg E, Berthiau AS, Forstemann K, Corda Y, Lingner J, Geli V, Gilson E. (2006) The finger subdomain of yeast telomerase cooperates with Pif1p to limit telomere elongation. *Nat. Struct. Mol. Biol.* **13**: 734–739.

Evans SK, Lundblad V. (1999) Est1 and Cdc13 as comediators of telomerase access. *Science.* **286**: 117–120.

Evans SK, Lundblad V. (2002) The Est1 subunit of *Saccharomyces cerevisiae* telomerase makes multiple contributions to telomere length maintenance. *Genetics.* **162**: 1101–1115.

Fan X, Price CM. (1997) Coordinate regulation of G- and C strand length during new telomere synthesis. *Mol. Biol. Cell.* **8**: 2145–2155.

Faure V, Coulon S, Hardy J, Geli V. (2010) Cdc13 and telomerase bind through different mechanisms at the lagging- and leading-strand telomeres. *Mol. Cell.* **38**: 842–852.

Feeser EA, Wolberger C. (2008) Structural and functional studies of the Rap1 C-terminus reveal novel separation-of-function mutants. *J. Mol. Biol.* **380**: 520–531.

Ferguson BM, Brewer BJ, Reynolds AE, Fangman WL. (1991) A yeast origin of replication is activated late in S phase. *Cell.* **65**: 507–515.

Ferreira MG, Cooper JP. (2001) The fission yeast Taz1 protein protects chromosomes from Ku-dependent end-to-end fusions. *Mol. Cell.* **7**: 55–63.

Fisher TS, Taggart AK, Zakian VA. (2004) Cell cycle-dependent regulation of yeast telomerase by Ku. *Nat. Struct. Mol. Biol.* **11**: 1198–1205.

Fisher TS, Zakian VA. (2005) Ku: a multifunctional protein involved in telomere maintenance. *DNA Repair (Amst).* **4**: 1215–1226.

Frank CJ, Hyde M, Greider CW. (2006) Regulation of telomere elongation by the cyclin-dependent kinase CDK1. *Mol. Cell.* **24**: 423–432.

Gallardo F, Olivier C, Dandjinou AT, Wellinger RJ, Chartrand P. (2008) TLC1 RNA nucleo-cytoplasmic trafficking links telomerase biogenesis to its recruitment to telomeres. *EMBO J.* **27**: 748–757.

Gao G, Walser JC, Beaucher ML, Morciano P, Wesolowska N, Chen J, Rong YS. (2010) HipHop interacts with HOAP and HP1 to protect *Drosophila* telomeres in a sequence-independent manner. *EMBO J.* **29**: 819–829.

Gao H, Cervantes RB, Mandell EK, Otero JH, Lundblad V. (2007) RPA-like proteins mediate yeast telomere function. *Nat. Struct. Mol. Biol.* **14**: 208–214.

Garvik B, Carson M, Hartwell L. (1995) Single-stranded DNA arising at telomeres in cdc13 mutants may constitute a specific signal for the RAD9 checkpoint. *Mol. Cell. Biol.* **15**: 6128–6138.

Gottschling DE, Zakian VA. (1986) Telomere proteins: specific recognition and protection of the natural termini of *Oxytricha* macronuclear DNA. *Cell.* **47**: 195–205.

Goudsouzian LK, Tuzon CT, Zakian VA. (2006) S. cerevisiae Tel1p and Mre11p are required for normal levels of Est1p and Est2p telomere association. *Mol. Cell.* **24**: 603–610.

Grandin N, Damon C, Charbonneau M. (2000) Cdc13 cooperates with the yeast Ku proteins and Stn1 to regulate telomerase recruitment. *Mol. Cell. Biol.* **20**: 8397–8408.

Grandin N, Damon C, Charbonneau M. (2001) Ten1 functions in telomere end protection and length regulation in association with Stn1 and Cdc13. *EMBO J.* **20**: 1173–1183.

Grandin N, Reed SI, Charbonneau M. (1997) Stn1, a new *Saccharomyces cerevisiae* protein, is implicated in telomere size regulation in association with Cdc13. *Genes Dev.* **11**: 512–527.

Gravel S, Chapman JR, Magill C, Jackson SP. (2008) DNA helicases Sgs1 and BLM promote DNA double-strand break resection. *Genes Dev.* **22**: 2767–2772.

Gravel S, Larrivee M, Labrecque P, Wellinger RJ. (1998) Yeast Ku as a regulator of chromosomal DNA end structure. *Science.* **280**: 741–744.

Griffith JD, Comeau L, Rosenfield S, Stansel RM, Bianchi A, Moss H, de Lange T. (1999) Mammalian telomeres end in a large duplex loop. *Cell.* **97**: 503–514.

Grossi S, Puglisi A, Dmitriev PV, Lopes M, Shore D. (2004) Pol12, the B subunit of DNA polymerase alpha, functions in both telomere capping and length regulation. *Genes Dev.* **18**: 992–1006.

Hanaoka S, Nagadoi A, Yoshimura S, Aimoto S, Li B, de Lange T, Nishimura Y. (2001) NMR structure of the hRap1 Myb motif reveals a canonical three-helix bundle lacking the positive surface charge typical of Myb DNA-binding domains. *J. Mol. Biol.* **312**: 167–175.

Hardy CF, Balderes D, Shore D. (1992a) Dissection of a carboxy-terminal region of the yeast regulatory protein RAP1 with effects on both transcriptional activation and silencing. *Mol. Cell. Biol.* **12**: 1209–1217.

Hardy CF, Sussel L, Shore D. (1992b) A RAP1-interacting protein involved in transcriptional silencing and telomere length regulation. *Genes Dev.* **6**: 801–814.

Hector RE, Shtofman RL, Ray A, Chen BR, Nyun T, Berkner KL, Runge KW. (2007) Tel1p preferentially associates with short telomeres to stimulate their elongation. *Mol. Cell.* **27**: 851–858.

Hemann MT, Greider CW. (1999) G-strand overhangs on telomeres in telomerase-deficient mouse cells. *Nucl. Acids Res.* **27**: 3964–3969.

Hirano Y, Fukunaga K, Sugimoto K. (2009) Rif1 and rif2 inhibit localization of tel1 to DNA ends. *Mol. Cell.* **33**: 312–322.

Hockemeyer D, Daniels JP, Takai H, de Lange T. (2006) Recent expansion of the telomeric complex in rodents: two distinct POT1 proteins protect mouse telomeres. *Cell.* **126**: 63–77.

Hockemeyer D, Palm W, Else T, Daniels JP, Takai KK, Ye JZ, Keegan CE, de Lange T, Hammer GD. (2007) Telomere protection by mammalian Pot1 requires interaction with Tpp1. *Nat. Struct. Mol. Biol.* **14**: 754–761.

Hockemeyer D, Sfeir AJ, Shay JW, Wright WE, de Lange T. (2005) POT1 protects telomeres from a transient DNA damage response and determines how human chromosomes end. *EMBO J.* **24**: 2667–2678.

Houghtaling BR, Cuttonaro L, Chang W, Smith S. (2004) A dynamic molecular link between the telomere length regulator TRF1 and the chromosome end protector TRF2. *Curr. Biol.* **14**: 1621–1631.

Hsu HL, Gilley D, Blackburn EH, Chen DJ. (1999) Ku is associated with the telomere in mammals. *Proc. Natl. Acad. Sci. USA.* **96**: 12454–12458.

Huertas P, Cortes-Ledesma F, Sartori AA, Aguilera A, Jackson SP. (2008) CDK targets Sae2 to control DNA-end resection and homologous recombination. *Nature.* **455**: 689–692.

Huertas P, Jackson SP. (2009) Human CtIP mediates cell cycle control of DNA end resection and double strand break repair. *J. Biol. Chem.* **284**: 9558–9565.

Hug N, Lingner J. (2006) Telomere length homeostasis. *Chromosoma.* **115**: 413–425.

Hughes TR, Weilbaecher RG, Walterscheid M, Lundblad V. (2000) Identification of the single-strand telomeric DNA binding domain of the *Saccharomyces cerevisiae* Cdc13 protein. *Proc. Natl. Acad. Sci. USA.* **97**: 6457–6462.

Ira G, Pellicioli A, Balijja A, Wang X, Fiorani S, Carotenuto W, Liberi G, Bressan D, Wan L, Hollingsworth NM, Haber JE, Foiani M. (2004) DNA end resection, homologous recombination and DNA damage checkpoint activation require CDK1. *Nature.* **431**: 1011–1017.

Ivessa AS, Zhou JQ, Schulz VP, Monson EK, Zakian VA. (2002) *Saccharomyces* Rrm3p, a 5′ to 3′ DNA helicase that promotes replication fork progression through telomeric and subtelomeric DNA. *Genes Dev.* **16**: 1383–1396.

Jacob NK, Kirk KE, Price CM. (2003) Generation of telomeric G strand overhangs involves both G and C strand cleavage. *Mol. Cell.* **11**: 1021–1032.

Jacob NK, Lescasse R, Linger BR, Price CM. (2007) *Tetrahymena* POT1a regulates telomere length and prevents activation of a cell cycle checkpoint. *Mol. Cell. Biol.* **27**: 1592–1601.

Jacob NK, Skopp R, Price CM. (2001) G-overhang dynamics at *Tetrahymena* telomeres. *EMBO J.* **20**: 4299–4308.

Jazayeri A, Falck J, Lukas C, Bartek J, Smith GC, Lukas J, Jackson SP. (2006) ATM- and cell cycle-dependent regulation of ATR in response to DNA double-strand breaks. *Nat. Cell. Biol.* **8**: 37–45.

Jeong HS, Backlund PS, Chen HC, Karavanov AA, Crouch RJ. (2004) RNase H2 of *Saccharomyces cerevisiae* is a complex of three proteins. *Nucl. Acids Res.* **32**: 407–414.

Kanoh J, Ishikawa F. (2001) spRap1 and spRif1, recruited to telomeres by Taz1, are essential for telomere function in fission yeast. *Curr. Biol.* **11**: 1624–1630.

Karamysheva ZN, Surovtseva YV, Vespa L, Shakirov EV, Shippen DE. (2004) A C-terminal Myb extension domain defines a novel family of double-strand telomeric DNA-binding proteins in *Arabidopsis*. *J. Biol. Chem.* **279**: 47799–47807.

Karlseder J. (2009) Chromosome end protection becomes even more complex. *Nat. Struct. Mol. Biol.* **16**: 1205–1206.

Karlseder J, Broccoli D, Dai Y, Hardy S, de Lange T. (1999) p53- and ATM-dependent apoptosis induced by telomeres lacking TRF2. *Science.* **283**: 1321–1325.

Kelleher C, Kurth I, Lingner J. (2005) Human protection of telomeres 1 (POT1) is a negative regulator of telomerase activity *in vitro*. *Mol. Cell. Biol.* **25**: 808–818.

Kibe T, Ono Y, Sato K, Ueno M. (2007) Fission yeast Taz1 and RPA are synergistically required to prevent rapid telomere loss. *Mol. Biol. Cell.* **18**: 2378–2387.

Kim SH, Beausejour C, Davalos AR, Kaminker P, Heo SJ, Campisi J. (2004) TIN2 mediates functions of TRF2 at human telomeres. *J. Biol. Chem.* **279**: 43799–43804.

Kim SH, Kaminker P, Campisi J. (1999) TIN2, a new regulator of telomere length in human cells. *Nat. Genet.* **23**: 405–412.

Kyrion G, Boakye KA, Lustig AJ. (1992) C-terminal truncation of RAP1 results in the deregulation of telomere size, stability, and function in *Saccharomyces cerevisiae*. *Mol. Cell. Biol.* **12**: 5159–5173.

Lam YC, Akhter S, Gu P, Ye J, Poulet A, Giraud-Panis MJ, Bailey SM, Gilson E, Legerski RJ, Chang S. (2010) SNMIB/Apollo protects leading-strand telomeres against NHEJ-mediated repair. *EMBO J.* **29**: 2230–2241.

Laroche T, Martin SG, Gotta M, Gorham HC, Pryde FE, Louis EJ, Gasser SM. (1998) Mutation of yeast Ku genes disrupts the subnuclear organization of telomeres. *Curr. Biol.* **8**: 653–656.

Larrivee M, LeBel C, Wellinger RJ. (2004) The generation of proper constitutive G-tails on yeast telomeres is dependent on the MRX complex. *Genes Dev.* **18**: 1391–1396.

LeBel C, Wellinger RJ. (2005) Telomeres: what's new at your end? *J. Cell. Sci.* **118**: 2785–2788.

Lee J, Mandell EK, Tucey TM, Morris DK, Lundblad V. (2008) The Est3 protein associates with yeast telomerase through an OB-fold domain. *Nat. Struct. Mol. Biol.* **15**: 990–997.

Lei M, Zaug AJ, Podell ER, Cech TR. (2005) Switching human telomerase on and off with hPOT1 protein *in vitro*. *J. Biol. Chem.* **280**: 20449–20456.

Lenain C, Bauwens S, Amiard S, Brunori M, Giraud-Panis MJ, Gilson E. (2006) The Apollo 5′ exonuclease functions together with TRF2 to protect telomeres from DNA repair. *Curr. Biol.* **16**: 1303–1310.

Lendvay TS, Morris DK, Sah J, Balasubramanian B, Lundblad V. (1996) Senescence mutants of *Saccharomyces cerevisiae* with a defect in telomere replication identify three additional EST genes. *Genetics.* **144**: 1399–1412.

Levy DL, Blackburn EH. (2004) Counting of Rif1p and Rif2p on *Saccharomyces cerevisiae* telomeres regulates telomere length. *Mol. Cell. Biol.* **24**: 10857–10867.

Li B, de Lange T. (2003) Rap1 affects the length and heterogeneity of human telomeres. *Mol. Biol. Cell.* **14**: 5060–5068.

Li B, Oestreich S, de Lange T. (2000) Identification of human Rap1: implications for telomere evolution. *Cell.* **101**: 471–483.

Li S, Makovets S, Matsuguchi T, Blethrow JD, Shokat KM, Blackburn EH. (2009) Cdk1-dependent phosphorylation of Cdc13 coordinates telomere elongation during cell-cycle progression. *Cell.* **136**: 50–61.

Limbo O, Chahwan C, Yamada Y, de Bruin RA, Wittenberg C, Russell P. (2007) Ctp1 is a cell-cycle-regulated protein that functions with Mre11 complex to control double-strand break repair by homologous recombination. *Mol. Cell.* **28**: 134–146.

Lingner J, Cooper JP, Cech TR. (1995) Telomerase and DNA end replication: no longer a lagging strand problem? *Science.* **269**: 1533–1534.

Lipps HJ, Rhodes D. (2009) G-quadruplex structures: *in vivo* evidence and function. *Trends Cell. Biol.* **19**: 414–422.

Liu D, Safari A, O'Connor MS, Chan DW, Laegeler A, Qin J, Songyang Z. (2004) PTOP interacts with POT1 and regulates its localization to telomeres. *Nat. Cell. Biol.* **6**: 673–680.

Loayza D, De Lange T. (2003) POT1 as a terminal transducer of TRF1 telomere length control. *Nature.* **423**: 1013–1018.

Longhese MP. (2008) DNA damage response at functional and dysfunctional telomeres. *Genes Dev.* **22**: 125–140.

Lue NF. (2010) Plasticity of telomere maintenance mechanisms in yeast. *Trends Biochem. Sci.* **35**: 8–17.

Lundblad V. (2003) Telomere replication: an Est fest. *Curr. Biol.* **13**: R439–R441.

Lundblad V, Blackburn EH. (1993) An alternative pathway for yeast telomere maintenance rescues est1-senescence. *Cell.* **73**: 347–360.

Lundblad V, Szostak JW. (1989) A mutant with a defect in telomere elongation leads to senescence in yeast. *Cell.* **57**: 633–643.

Lustig AJ, Kurtz S, Shore D. (1990) Involvement of the silencer and UAS binding protein RAP1 in regulation of telomere length. *Science.* **250**: 549–553.

Lustig AJ, Petes TD. (1986) Identification of yeast mutants with altered telomere structure. *Proc. Natl. Acad. Sci. USA.* **83**: 1398–1402.

Lydall D. (2003) Hiding at the ends of yeast chromosomes: telomeres, nucleases and checkpoint pathways. *J. Cell. Sci.* **116**: 4057–4065.

Lydall D. (2009) Taming the tiger by the tail: modulation of DNA damage responses by telomeres. *EMBO J.* **28**: 2174–2187.

Lydall D, Weinert T. (1995) Yeast checkpoint genes in DNA damage processing: implications for repair and arrest. *Science.* **270**: 1488–1491.

Makarov VL, Hirose Y, Langmore JP. (1997) Long G tails at both ends of human chromosomes suggest a C strand degradation mechanism for telomere shortening. *Cell.* **88**: 657–666.

Makovets S, Blackburn EH. (2009) DNA damage signalling prevents deleterious telomere addition at DNA breaks. *Nat. Cell. Biol.* **11**: 1383–1386.

Makovets S, Herskowitz I, Blackburn EH. (2004) Anatomy and dynamics of DNA replication fork movement in yeast telomeric regions. *Mol. Cell. Biol.* **24**: 4019–4031.

Mallory JC, Bashkirov VI, Trujillo KM, Solinger JA, Dominska M, Sung P, Heyer WD, Petes TD. (2003) Amino acid changes in Xrs2p, Dun1p, and Rfa2p that remove the preferred targets of the ATM family of protein kinases do not affect DNA repair or telomere length in *Saccharomyces cerevisiae. DNA Repair (Amst).* **2**: 1041–1064.

Manchado E, Eguren M, Malumbres M. (2010) The anaphase-promoting complex/cyclosome (APC/C): cell-cycle-dependent and -independent functions. *Biochem. Soc. Trans.* **38**: 65–71.

Marcand S, Brevet V, Gilson E. (1999) Progressive *cis*-inhibition of telomerase upon telomere elongation. *EMBO J.* **18**: 3509–3519.

Marcand S, Brevet V, Mann C, Gilson E. (2000) Cell cycle restriction of telomere elongation. *Curr. Biol.* **10**: 487–490.

Marcand S, Gilson E, Shore D. (1997) A protein-counting mechanism for telomere length regulation in yeast. *Science.* **275**: 986–990.

Marcand S, Pardo B, Gratias A, Cahun S, Callebaut I. (2008) Multiple pathways inhibit NHEJ at telomeres. *Genes Dev.* **22**: 1153–1158.

McCarroll RM, Fangman WL. (1988) Time of replication of yeast centromeres and telomeres. *Cell.* **54**: 505–513.

McClintock B. (1938) The production of homozygous deficient tissues with mutant characteristics by means of the aberrant mitotic behavior of ring-shaped chromosomes. *Genetics.* **23**: 315–376.

McClintock B. (1941) The stability of broken ends of chromosomes in *Zea mays*. *Genetics*. **26**: 234–282.

McElligott R, Wellinger RJ. (1997) The terminal DNA structure of mammalian chromosomes. *EMBO J*. **16**: 3705–3714.

Mendenhall MD, Hodge AE. (1998) Regulation of Cdc28 cyclin-dependent protein kinase activity during the cell cycle of the yeast *Saccharomyces cerevisiae*. *Microbiol. Mol. Biol. Rev*. **62**: 1191–1243.

Mieczkowski PA, Mieczkowska JO, Dominska M, Petes TD. (2003) Genetic regulation of telomere-telomere fusions in the yeast *Saccharomyces cerevisae*. *Proc. Natl. Acad. Sci. USA*. **100**: 10854–10859.

Miller KM, Ferreira MG, Cooper JP. (2005) Taz1, Rap1 and Rif1 act both interdependently and independently to maintain telomeres. *EMBO J*. **24**: 3128–3135.

Miller KM, Rog O, Cooper JP. (2006) Semi-conservative DNA replication through telomeres requires Taz1. *Nature*. **440**: 824–828.

Mimitou EP, Symington LS. (2008) Sae2, Exo1 and Sgs1 collaborate in DNA double-strand break processing. *Nature*. **455**: 770–774.

Mishra K, Shore D. (1999) Yeast Ku protein plays a direct role in telomeric silencing and counteracts inhibition by rif proteins. *Curr. Biol*. **9**: 1123–1126.

Miyake Y, Nakamura M, Nabetani A, Shimamura S, Tamura M, Yonehara S, Saito M, Ishikawa F. (2009) RPA-like mammalian Ctc1–Stn1–Ten1 complex binds to single-stranded DNA and protects telomeres independently of the Pot1 pathway. *Mol. Cell*. **36**: 193–206.

Miyoshi T, Kanoh J, Saito M, Ishikawa F. (2008) Fission yeast Pot1–Tpp1 protects telomeres and regulates telomere length. *Science*. **320**: 1341–1344.

Moretti P, Freeman K, Coodly L, Shore D. (1994) Evidence that a complex of SIR proteins interacts with the silencer and telomere-binding protein RAP1. *Genes Dev*. **8**: 2257–2269.

Moretti P, Shore D. (2001) Multiple interactions in Sir protein recruitment by Rap1p at silencers and telomeres in yeast. *Mol. Cell. Biol*. **21**: 8082–8094.

Moser BA, Nakamura TM. (2009) Protection and replication of telomeres in fission yeast. *Biochem. Cell. Biol*. **87**: 747–758.

Moser BA, Subramanian L, Khair L, Chang YT, Nakamura TM. (2009) Fission yeast Tel1 (ATM) and Rad3(ATR) promote telomere protection and telomerase recruitment. *PLoS Genet*. **5**: e1000622.

Muller HJ. (1938) The remaking of chromosomes. *Collecting Net*. **13**: 181–198.

Munoz-Jordan JL, Cross GA, de Lange T, Griffith JD. (2001) t-loops at trypanosome telomeres. *EMBO J*. **20**: 579–588.

Murnane JP, Sabatier L, Marder BA, Morgan WF. (1994) Telomere dynamics in an immortal human cell line. *EMBO J*. **13**: 4953–4962.

Murti KG, Prescott DM. (1999) Telomeres of polytene chromosomes in a ciliated protozoan terminate in duplex DNA loops. *Proc. Natl. Acad. Sci. USA*. **96**: 14436–14439.

Myung K, Chen C, Kolodner RD. (2001) Multiple pathways cooperate in the suppression of genome instability in *Saccharomyces cerevisiae*. *Nature*. **411**: 1073–1076.

Naito T, Matsuura A, Ishikawa F. (1998) Circular chromosome formation in a fission yeast mutant defective in two ATM homologues. *Nat. Genet*. **20**: 203–206.

Nakamura M, Nabetani A, Mizuno T, Hanaoka F, Ishikawa F. (2005) Alterations of DNA and chromatin structures at telomeres and genetic instability in mouse cells defective in DNA polymerase alpha. *Mol. Cell. Biol.* **25**: 11073–11088.

Nakamura TM, Moser BA, Russell P. (2002) Telomere binding of checkpoint sensor and DNA repair proteins contributes to maintenance of functional fission yeast telomeres. *Genetics.* **161**: 1437–1452.

Negrini S, Ribaud V, Bianchi A, Shore D. (2007) DNA breaks are masked by multiple Rap1 binding in yeast: implications for telomere capping and telomerase regulation. *Genes Dev.* **21**: 292–302.

Nimonkar AV, Ozsoy AZ, Genschel J, Modrich P, Kowalczykowski SC. (2008) Human exonuclease 1 and BLM helicase interact to resect DNA and initiate DNA repair. *Proc. Natl. Acad. Sci. USA.* **105**: 16906–16911.

Nugent CI, Bosco G, Ross LO, Evans SK, Salinger AP, Moore JK, Haber JE, Lundblad V. (1998) Telomere maintenance is dependent on activities required for end repair of double-strand breaks. *Curr. Biol.* **8**: 657–660.

Nugent CI, Hughes TR, Lue NF, Lundblad V. (1996) Cdc13p: a single-strand telomeric DNA-binding protein with a dual role in yeast telomere maintenance. *Science.* **274**: 249–252.

O'Connor MS, Safari A, Liu D, Qin J, Songyang Z. (2004) The human Rap1 protein complex and modulation of telomere length. *J. Biol. Chem.* **279**: 28585–28591.

O'Connor MS, Safari A, Xin H, Liu D, Songyang Z. (2006) A critical role for TPP1 and TIN2 interaction in high-order telomeric complex assembly. *Proc. Natl. Acad. Sci. USA.* **103**: 11874–11879.

O'Sullivan RJ, Karlseder J. (2010) Telomeres: protecting chromosomes against genome instability. *Nat. Rev. Mol. Cell. Biol.* **11**: 171–181.

Ohki R, Tsurimoto T, Ishikawa F. (2001) *In vitro* reconstitution of the end replication problem. *Mol. Cell. Biol.* **21**: 5753–5766.

Olovnikov AM. (1971) [Principle of marginotomy in template synthesis of polynucleotides]. *Dokl. Akad. Nauk. SSSR.* **201**: 1496–1499.

Olovnikov AM. (1973) A theory of marginotomy. The incomplete copying of template margin in enzymic synthesis of polynucleotides and biological significance of the phenomenon. *J. Theor. Biol.* **41**: 181–190.

Ono Y, Tomita K, Matsuura A, Nakagawa T, Masukata H, Uritani M, Ushimaru T, Ueno M. (2003) A novel allele of fission yeast rad11 that causes defects in DNA repair and telomere length regulation. *Nucl. Acids Res.* **31**: 7141–7149.

Opresko PL, Otterlei M, Graakjaer J, Bruheim P, Dawut L, Kolvraa S, May A, Seidman MM, Bohr VA. (2004) The Werner syndrome helicase and exonuclease cooperate to resolve telomeric D loops in a manner regulated by TRF1 and TRF2. *Mol. Cell.* **14**: 763–774.

Opresko PL, von Kobbe C, Laine JP, Harrigan J, Hickson ID, Bohr VA. (2002) Telomere-binding protein TRF2 binds to and stimulates the Werner and Bloom syndrome helicases. *J. Biol. Chem.* **277**: 41110–41119.

Osterhage JL, Friedman KL. (2009) Chromosome end maintenance by telomerase. *J. Biol. Chem.* **284**: 16061–16065.

Paeschke K, Juranek S, Simonsson T, Hempel A, Rhodes D, Lipps HJ. (2008) Telomerase recruitment by the telomere end binding protein-beta facilitates G-quadruplex DNA unfolding in ciliates. *Nat. Struct. Mol. Biol.* **15**: 598–604.

Paeschke K, Simonsson T, Postberg J, Rhodes D, Lipps HJ. (2005) Telomere end-binding proteins control the formation of G-quadruplex DNA structures *in vivo*. *Nat. Struct. Mol. Biol.* **12**: 847–854.

Palm W, de Lange T. (2008) How shelterin protects mammalian telomeres. *Annu. Rev. Genet.* **42**: 301–334.

Pardo B, Marcand S. (2005) Rap1 prevents telomere fusions by nonhomologous end joining. *EMBO J.* **24**: 3117–3127.

Parenteau J, Wellinger RJ. (2002) Differential processing of leading- and lagging-strand ends at *Saccharomyces cerevisiae* telomeres revealed by the absence of Rad27p nuclease. *Genetics.* **162**: 1583–1594.

Pennaneach V, Putnam CD, Kolodner RD. (2006) Chromosome healing by *de novo* telomere addition in *Saccharomyces cerevisiae*. *Mol. Microbiol.* **59**: 1357–1368.

Pennock E, Buckley K, Lundblad V. (2001) Cdc13 delivers separate complexes to the telomere for end protection and replication. *Cell.* **104**: 387–396.

Peterson SE, Stellwagen AE, Diede SJ, Singer MS, Haimberger ZW, Johnson CO, Tzoneva M, Gottschling DE. (2001) The function of a stem-loop in telomerase RNA is linked to the DNA repair protein Ku. *Nat. Genet.* **27**: 64–67.

Petreaca RC, Chiu HC, Eckelhoefer HA, Chuang C, Xu L, Nugent CI. (2006) Chromosome end protection plasticity revealed by Stn1p and Ten1p bypass of Cdc13p. *Nat. Cell. Biol.* **8**: 748–755.

Petreaca RC, Chiu HC, Nugent CI. (2007) The role of Stn1p in *Saccharomyces cerevisiae* telomere capping can be separated from its interaction with Cdc13p. *Genetics.* **177**: 1459–1474.

Pina B, Fernandez-Larrea J, Garcia-Reyero N, Idrissi FZ. (2003) The different (sur)faces of Rap1p. *Mol. Genet. Genomics.* **268**: 791–798.

Polotnianka RM, Li J, Lustig AJ. (1998) The yeast Ku heterodimer is essential for protection of the telomere against nucleolytic and recombinational activities. *Curr. Biol.* **8**: 831–834.

Porter SE, Greenwell PW, Ritchie KB, Petes TD. (1996) The DNA-binding protein Hdf1p (a putative Ku homologue) is required for maintaining normal telomere length in *Saccharomyces cerevisiae*. *Nucl. Acids Res.* **24**: 582–585.

Price CM, Cech TR. (1987) Telomeric DNA–protein interactions of *Oxytricha* macronuclear DNA. *Genes Dev.* **1**: 783–793.

Puglisi A, Bianchi A, Lemmens L, Damay P, Shore D. (2008) Distinct roles for yeast Stn1 in telomere capping and telomerase inhibition. *EMBO J.* **27**: 2328–2339.

Putnam CD, Pennaneach V, Kolodner RD. (2004) Chromosome healing through terminal deletions generated by *de novo* telomere additions in *Saccharomyces cerevisiae*. *Proc. Natl. Acad. Sci. USA.* **101**: 13262–13267.

Qi H, Zakian VA. (2000) The *Saccharomyces* telomere-binding protein Cdc13p interacts with both the catalytic subunit of DNA polymerase alpha and the telomerase-associated est1 protein. *Genes Dev.* **14**: 1777–1788.

Qiu J, Qian Y, Frank P, Wintersberger U, Shen B. (1999) *Saccharomyces cerevisiae* RNase H (35) functions in RNA primer removal during lagging-strand DNA synthesis, most efficiently in cooperation with Rad27 nuclease. *Mol. Cell. Biol.* **19**: 8361–8371.

Raffa GD, Raimondo D, Sorino C, Cugusi S, Cenci G, Cacchione S, Gatti M, Ciapponi L. (2010) Verrocchio, a *Drosophila* OB fold-containing protein, is a component of the terminin telomere-capping complex. *Genes Dev.* **24**: 1596–1601.

Raffa GD, Siriaco G, Cugusi S, Ciapponi L, Cenci G, Wojcik E, Gatti M. (2009) The *Drosophila* modigliani (moi) gene encodes a HOAP-interacting protein required for telomere protection. *Proc. Natl. Acad. Sci. USA.* **106**: 2271–2276.

Raices M, Verdun RE, Compton SA, Haggblom CI, Griffith JD, Dillin A, Karlseder J. (2008) *C. elegans* telomeres contain G-strand and C-strand overhangs that are bound by distinct proteins. *Cell.* **132**: 745–757.

Ranganathan V, Heine WF, Ciccone DN, Rudolph KL, Wu X, Chang S, Hai H, Ahearn IM, Livingston DM, Resnick I, Rosen F, Seemanova E, Jarolim P, DePinho RA, Weaver DT. (2001) Rescue of a telomere length defect of Nijmegen breakage syndrome cells requires NBS and telomerase catalytic subunit. *Curr. Biol.* **11**: 962–966.

Ray A, Runge KW. (1999) Varying the number of telomere-bound proteins does not alter telomere length in tel1Delta cells. *Proc. Natl. Acad. Sci. USA.* **96**: 15044–15049.

Reynolds AE, McCarroll RM, Newlon CS, Fangman WL. (1989) Time of replication of ARS elements along yeast chromosome III. *Mol. Cell. Biol.* **9**: 4488–4494.

Riha K, Shippen DE. (2003) Ku is required for telomeric C-rich strand maintenance but not for end-to-end chromosome fusions in *Arabidopsis*. *Proc. Natl. Acad. Sci. USA.* **100**: 611–615.

Riha K, Watson JM, Parkey J, Shippen DE. (2002) Telomere length deregulation and enhanced sensitivity to genotoxic stress in *Arabidopsis* mutants deficient in Ku70. *EMBO J.* **21**: 2819–2826.

Ritchie KB, Mallory JC, Petes TD. (1999) Interactions of TLC1 (which encodes the RNA subunit of telomerase), TEL1, and MEC1 in regulating telomere length in the yeast *Saccharomyces cerevisiae*. *Mol. Cell. Biol.* **19**: 6065–6075.

Ritchie KB, Petes TD. (2000) The Mre11p/Rad50p/Xrs2p complex and the Tel1p function in a single pathway for telomere maintenance in yeast. *Genetics.* **155**: 475–479.

Rog O, Miller KM, Ferreira MG, Cooper JP. (2009) Sumoylation of RecQ helicase controls the fate of dysfunctional telomeres. *Mol. Cell.* **33**: 559–569.

Rong YS. (2008) Telomere capping in *Drosophila*: dealing with chromosome ends that most resemble DNA breaks. *Chromosoma.* **117**: 235–242.

Sabourin M, Tuzon CT, Zakian VA. (2007) Telomerase and Tel1p preferentially associate with short telomeres in *S. cerevisiae*. *Mol. Cell.* **27**: 550–561.

Sarthy J, Bae NS, Scrafford J, Baumann P. (2009) Human RAP1 inhibits non-homologous end joining at telomeres. *EMBO J.* **28**: 3390–3399.

Sartori AA, Lukas C, Coates J, Mistrik M, Fu S, Bartek J, Baer R, Lukas J, Jackson SP. (2007) Human CtIP promotes DNA end resection. *Nature.* **450**: 509–514.

Schramke V, Luciano P, Brevet V, Guillot S, Corda Y, Longhese MP, Gilson E, Geli V. (2004) RPA regulates telomerase action by providing Est1p access to chromosome ends. *Nat. Genet.* **36**: 46–54.

Sfeir A, Kabir S, van Overbeek M, Celli GB, de Lange T. (2010) Loss of Rap1 induces telomere recombination in the absence of NHEJ or a DNA damage signal. *Science.* **327**: 1657–1661.

Sfeir A, Kosiyatrakul ST, Hockemeyer D, MacRae SL, Karlseder J, Schildkraut CL, de Lange T. (2009) Mammalian telomeres resemble fragile sites and require TRF1 for efficient replication. *Cell.* **138**: 90–103.

Sfeir AJ, Chai W, Shay JW, Wright WE. (2005) Telomere-end processing the terminal nucleotides of human chromosomes. *Mol. Cell.* **18**: 131–138.

Shakirov EV, Surovtseva YV, Osbun N, Shippen DE. (2005) The *Arabidopsis* Pot1 and Pot2 proteins function in telomere length homeostasis and chromosome end protection. *Mol. Cell. Biol.* **25**: 7725–7733.

Shima H, Suzuki M, Shinohara M. (2005) Isolation and characterization of novel xrs2 mutations in *Saccharomyces cerevisiae*. *Genetics*. **170**: 71–85.

Shippen-Lentz D, Blackburn EH. (1990) Functional evidence for an RNA template in telomerase. *Science*. **247**: 546–552.

Shore D. (2001) Telomeric chromatin: replicating and wrapping up chromosome ends. *Curr. Opin. Genet. Dev.* **11**: 189–198.

Silverman J, Takai H, Buonomo SB, Eisenhaber F, de Lange T. (2004) Human Rif1, ortholog of a yeast telomeric protein, is regulated by ATM and 53BP1 and functions in the S-phase checkpoint. *Genes Dev*. **18**: 2108–2119.

Smogorzewska A, Karlseder J, Holtgreve-Grez H, Jauch A, de Lange T. (2002) DNA ligase IV-dependent NHEJ of deprotected mammalian telomeres in G1 and G2. *Curr. Biol.* **12**: 1635–1644.

Smogorzewska A, van Steensel B, Bianchi A, Oelmann S, Schaefer MR, Schnapp G, de Lange T. (2000) Control of human telomere length by TRF1 and TRF2. *Mol. Cell. Biol.* **20**: 1659–1668.

Stellwagen AE, Haimberger ZW, Veatch JR, Gottschling DE. (2003) Ku interacts with telomerase RNA to promote telomere addition at native and broken chromosome ends. *Genes Dev*. **17**: 2384–2395.

Sugawara N, Szostak JW. (1986) Telomeres of *Schizosaccaromyces pombe*. *Yeast (Suppl.)*. **2**: S373.

Surovtseva YV, Churikov D, Boltz KA, Song X, Lamb JC, Warrington R, Leehy K, Heacock M, Price CM, Shippen DE. (2009) Conserved telomere maintenance component 1 interacts with STN1 and maintains chromosome ends in higher eukaryotes. *Mol. Cell.* **36**: 207–218.

Surovtseva YV, Shakirov EV, Vespa L, Osbun N, Song X, Shippen DE. (2007) *Arabidopsis* POT1 associates with the telomerase RNP and is required for telomere maintenance. EMBO J. **26**: 3653–3661.

Sussel L, Shore D. (1991) Separation of transcriptional activation and silencing functions of the RAP1-encoded repressor/activator protein 1: isolation of viable mutants affecting both silencing and telomere length. *Proc. Natl. Acad. Sci. USA*. **88**: 7749–7753.

Taggart AK, Teng SC, Zakian VA. (2002) Est1p as a cell cycle-regulated activator of telomere-bound telomerase. *Science*. **297**: 1023–1026.

Takai H, Smogorzewska A, de Lange T. (2003) DNA damage foci at dysfunctional telomeres. *Curr. Biol.* **13**: 1549–1556.

Takai KK, Hooper S, Blackwood S, Gandhi R, de Lange T. (2010) *In vivo* stoichiometry of shelterin components. *J. Biol. Chem.* **285**: 1457–1467.

Takata H, Tanaka Y, Matsuura A. (2005) Late S phase-specific recruitment of Mre11 complex triggers hierarchical assembly of telomere replication proteins in *Saccharomyces cerevisiae*. *Mol. Cell.* **17**: 573–583.

Tarsounas M, Munoz P, Claas A, Smiraldo PG, Pittman DL, Blasco MA, West SC. (2004) Telomere maintenance requires the RAD51D recombination/repair protein. *Cell*. **117**: 337–347.

Teixeira MT, Arneric M, Sperisen P, Lingner J. (2004) Telomere length homeostasis is achieved via a switch between telomerase- extendible and -nonextendible states. *Cell.* **117**: 323–335.

Tomita K, Cooper JP. (2008) Fission yeast Ccq1 is telomerase recruiter and local checkpoint controller. *Genes Dev.* **22**: 3461–3474.

Tomita K, Kibe T, Kang HY, Seo YS, Uritani M, Ushimaru T, Ueno M. (2004) Fission yeast Dna2 is required for generation of the telomeric single-strand overhang. *Mol. Cell. Biol.* **24**: 9557–9567.

Tseng SF, Lin JJ, Teng SC. (2006) The telomerase-recruitment domain of the telomere binding protein Cdc13 is regulated by Mec1p/Tel1p-dependent phosphorylation. *Nucl. Acids Res.* **34**: 6327–6336.

Tseng SF, Shen ZJ, Tsai HJ, Lin YH, Teng SC. (2009) Rapid Cdc13 turnover and telomere length homeostasis are controlled by Cdk1-mediated phosphorylation of Cdc13. *Nucl. Acids Res.* **37**: 3602–3611.

Tsukamoto Y, Mitsuoka C, Terasawa M, Ogawa H, Ogawa T. (2005) Xrs2p regulates Mre11p translocation to the nucleus and plays a role in telomere elongation and meiotic recombination. *Mol. Biol. Cell.* **16**: 597–608.

Tsukamoto Y, Taggart AK, Zakian VA. (2001) The role of the Mre11–Rad50–Xrs2 complex in telomerase-mediated lengthening of *Saccharomyces cerevisiae* telomeres. *Curr. Biol.* **11**: 1328–1335.

van Overbeek M, de Lange T. (2006) Apollo, an Artemis-related nuclease, interacts with TRF2 and protects human telomeres in S phase. *Curr. Biol.* **16**: 1295–1302.

van Steensel B, Smogorzewska A, de Lange T. (1998) TRF2 protects human telomeres from end-to-end fusions. *Cell.* **92**: 401–413.

Vaziri H, West MD, Allsopp RC, Davison TS, Wu YS, Arrowsmith CH, Poirier GG, Benchimol S. (1997) ATM-dependent telomere loss in aging human diploid fibroblasts and DNA damage lead to the post-translational activation of p53 protein involving poly(ADP-ribose) polymerase. *EMBO J.* **16**: 6018–6033.

Vega LR, Mateyak MK, Zakian VA. (2003) Getting to the end: telomerase access in yeast and humans. *Nat. Rev. Mol. Cell. Biol.* **4**: 948–959.

Vega LR, Phillips JA, Thornton BR, Benanti JA, Onigbanjo MT, Toczyski DP, Zakian VA. (2007) Sensitivity of yeast strains with long G-tails to levels of telomere-bound telomerase. *PLoS Genet.* **3**: e105.

Veldman T, Etheridge KT, Counter CM. (2004) Loss of hPot1 function leads to telomere instability and a cut-like phenotype. *Curr. Biol.* **14**: 2264–2270.

Verdun RE, Crabbe L, Haggblom C, Karlseder J. (2005) Functional human telomeres are recognized as DNA damage in G2 of the cell cycle. *Mol. Cell.* **20**: 551–561.

Verdun RE, Karlseder J. (2006) The DNA damage machinery and homologous recombination pathway act consecutively to protect human telomeres. *Cell.* **127**: 709–720.

Verdun RE, Karlseder J. (2007) Replication and protection of telomeres. *Nature.* **447**: 924–931.

Vespa L, Couvillion M, Spangler E, Shippen DE. (2005) ATM and ATR make distinct contributions to chromosome end protection and the maintenance of telomeric DNA in *Arabidopsis*. *Genes Dev.* **19**: 2111–2115.

Vespa L, Warrington RT, Mokros P, Siroky J, Shippen DE. (2007) ATM regulates the length of individual telomere tracts in *Arabidopsis*. *Proc. Natl. Acad. Sci. USA.* **104**: 18145–18150.

Viscardi V, Bonetti D, Cartagena-Lirola H, Lucchini G, Longhese MP. (2007) MRX-dependent DNA damage response to short telomeres. *Mol. Biol. Cell.* **18**: 3047–3058.

Vodenicharov MD, Laterreur N, Wellinger RJ. (2010) Telomere capping in non-dividing yeast cells requires Yku and Rap1. *EMBO J*. **29**: 3007–3019.

Vodenicharov MD, Wellinger RJ. (2006) DNA degradation at unprotected telomeres in yeast is regulated by the CDK1 (Cdc28/Clb) cell-cycle kinase. *Mol. Cell.* **24**: 127–137.

Vodenicharov MD, Wellinger RJ. (2007) The cell division cycle puts up with unprotected telomeres: cell cycle regulated telomere uncapping as a means to achieve telomere homeostasis. *Cell Cycle*. **6**: 1161–1167.

Wang F, Podell ER, Zaug AJ, Yang Y, Baciu P, Cech TR, Lei M. (2007) The POT1–TPP1 telomere complex is a telomerase processivity factor. *Nature*. **445**: 506–510.

Wang Y, Ghosh G, Hendrickson EA. (2009) Ku86 represses lethal telomere deletion events in human somatic cells. *Proc. Natl. Acad. Sci. USA*. **106**: 12430–12435.

Watson JD. (1972) Origin of concatemeric T7 DNA. *Nat. New Biol.* **239**: 197–201.

Wellinger RJ. (2009) The CST complex and telomere maintenance: the exception becomes the rule. *Mol. Cell.* **36**: 168–169.

Wellinger RJ. (2010) When the caps fall off: responses to telomere uncapping in yeast. *FEBS Lett*. **584**: 3734–3740.

Wellinger RJ, Ethier K, Labrecque P, Zakian VA. (1996) Evidence for a new step in telomere maintenance. *Cell.* **85**: 423–433.

Wellinger RJ, Wolf AJ, Zakian VA. (1993a) Origin activation and formation of single-strand TG1-3 tails occur sequentially in late S phase on a yeast linear plasmid. *Mol. Cell. Biol.* **13**: 4057–4065.

Wellinger RJ, Wolf AJ, Zakian VA. (1993b) *Saccharomyces* telomeres acquire single-strand TG1-3 tails late in S phase. *Cell.* **72**: 51–60.

Wellinger RJ, Zakian VA. (1989) Lack of positional requirements for autonomously replicating sequence elements on artificial yeast chromosomes. *Proc. Natl. Acad. Sci. USA*. **86**: 973–977.

Wotton D, Shore D. (1997) A novel Rap1p-interacting factor, Rif2p, cooperates with Rif1p to regulate telomere length in *Saccharomyces cerevisiae*. *Genes Dev*. **11**: 748–760.

Wright WE, Tesmer VM, Liao ML, Shay JW. (1999) Normal human telomeres are not late replicating. *Exp. Cell. Res.* **251**: 492–499.

Wu L, Multani AS, He H, Cosme-Blanco W, Deng Y, Deng JM, Bachilo O, Pathak S, Tahara H, Bailey SM, Deng Y, Behringer RR, Chang S. (2006) Pot1 deficiency initiates DNA damage checkpoint activation and aberrant homologous recombination at telomeres. *Cell.* **126**: 49–62.

Wu P, van Overbeek M, Rooney S, de Lange T. (2010) Apollo contributes to G overhang maintenance and protects leading-end telomeres. *Mol Cell*. **39**: 606–617.

Xin H, Liu D, Wan M, Safari A, Kim H, Sun W, O'Connor MS, Songyang Z. (2007) TPP1 is a homologue of ciliate TEBP-beta and interacts with POT1 to recruit telomerase. *Nature*. **445**: 559–562.

Xu L, Blackburn EH. (2004) Human Rif1 protein binds aberrant telomeres and aligns along anaphase midzone microtubules. *J. Cell. Biol.* **167**: 819–830.

Xu L, Petreaca RC, Gasparyan HJ, Vu S, Nugent CI. (2009) TEN1 is essential for CDC13-mediated telomere capping. *Genetics*. **183**: 793–810.

Ye J, Lenain C, Bauwens S, Rizzo A, Saint-Leger A, Poulet A, Benarroch D, Magdinier F, Morere J, Amiard S, Verhoeyen E, Britton S, Calsou P, Salles B, Bizard A, Nadal M, Salvati

E, Sabatier L, Wu Y, Biroccio A, et al. (2010) TRF2 and apollo cooperate with topoisomerase 2alpha to protect human telomeres from replicative damage. *Cell.* **142**: 230–242.

Ye JZ, Donigian JR, van Overbeek M, Loayza D, Luo Y, Krutchinsky AN, Chait BT, de Lange T. (2004a) TIN2 binds TRF1 and TRF2 simultaneously and stabilizes the TRF2 complex on telomeres. *J. Biol. Chem.* **279**: 47264–47271.

Ye JZ, Hockemeyer D, Krutchinsky AN, Loayza D, Hooper SM, Chait BT, de Lange T. (2004b) POT1-interacting protein PIP1: a telomere length regulator that recruits POT1 to the TIN2/TRF1 complex. *Genes Dev.* **18**: 1649–1654.

Yu EY, Wang F, Lei M, Lue NF. (2008) A proposed OB-fold with a protein-interaction surface in *Candida albicans* telomerase protein Est3. *Nat. Struct. Mol. Biol.* **15**: 985–989.

Zappulla DC, Cech TR. (2004) Yeast telomerase RNA: a flexible scaffold for protein subunits. *Proc. Natl. Acad. Sci. USA.* **101**: 10024–10029.

Zaug AJ, Podell ER, Cech TR. (2005) Human POT1 disrupts telomeric G-quadruplexes allowing telomerase extension *in vitro*. *Proc. Natl. Acad. Sci. USA.* **102**: 10864–10869.

Zhang W, Durocher D. (2010) De novo telomere formation is suppressed by the Mec1-dependent inhibition of Cdc13 accumulation at DNA breaks. *Genes Dev.* **24**: 502–515.

Zhao Y, Sfeir AJ, Zou Y, Buseman CM, Chow TT, Shay JW, Wright WE. (2009) Telomere extension occurs at most chromosome ends and is uncoupled from fill-in in human cancer cells. *Cell.* **138**: 463–475.

Zhu XD, Kuster B, Mann M, Petrini JH, de Lange T. (2000) Cell-cycle-regulated association of RAD50/MRE11/NBS1 with TRF2 and human telomeres. *Nat. Genet.* **25**: 347–352.

Zhu XD, Niedernhofer L, Kuster B, Mann M, Hoeijmakers JH, de Lange T. (2003) ERCC1/XPF removes the 3′ overhang from uncapped telomeres and represses formation of telomeric DNA-containing double minute chromosomes. *Mol. Cell.* **12**: 1489–1498.

Zhu Z, Chung WH, Shim EY, Lee SE, Ira G. (2008) Sgs1 helicase and two nucleases Dna2 and Exo1 resect DNA double-strand break ends. *Cell.* **134**: 981–994.

Zou Y, Gryaznov SM, Shay JW, Wright WE, Cornforth MN. (2004) Asynchronous replication timing of telomeres at opposite arms of mammalian chromosomes. *Proc. Natl. Acad. Sci. USA.* **101**: 12928–12933.

8

OFF-TELOMERE FUNCTIONS OF TELOMERASE

Kenkichi Masutomi and William C. Hahn

As described in prior chapters, the core components of telomerase include the catalytic protein subunit TERT and a noncoding RNA *TERC* that provides the template for the telomere-synthesis reaction. This reverse transcriptase ribonucleoprotein complex maintains telomeres when expressed constitutively. In addition to the well-described function in telomere maintenance, accumulating evidence indicates that TERT has additional functions. In this chapter, we review these functions and recent biochemical and genetic experiments that define additional TERT complexes.

8.1 EVIDENCE FOR NONTELOMERE DIRECTED TERT FUNCTIONS IN MALIGNANT TRANSFORMATION

As described in prior chapters, the majority of human tumors express telomerase constitutively and show stable telomere length with extended passage in culture. Suppression of TERT (Allsopp et al., 2003; Masutomi et al., 2003; Yuan et al., 1999) or *TERC* (Blasco et al., 1997; Herbert et al., 1999), expression of dominantly interfering TERT mutants (Hahn et al., 1999b; Zhang et al., 1999), or pharmacologic inhibition of TERT (Herbert et al., 2002; Marian et al., 2010; Pascolo et al., 2002; Seimiya et al., 2002, 2005) induces telomere shortening and limits malignant potential in human cancer cell lines or under certain genetic backgrounds, particularly in the setting of intact p53 function (Chin et al., 1999), in genetically engineered mice (Blasco et al., 1997; Gonzalez-Suarez et al., 2000; Greenberg et al., 1999). Conversely

Telomerases: Chemistry, Biology, and Clinical Applications, First Edition.
Edited by Neal F. Lue and Chantal Autexier.

overexpression of TERT facilitates the immortalization of human cells (Bodnar et al., 1998) and cooperates with other oncogenes to induce cell transformation in a wide range of primary human cells (Elenbaas et al., 2001; Hahn et al., 1999a; MacKenzie et al., 2002; Rich et al., 2001). Taken together, these findings confirm that TERT plays an important role in cell transformation.

In addition, accumulating evidence indicates that hTERT contributes to the transformation of human cells through both telomere-dependent and telomere-independent mechanisms (Artandi et al., 2002; Flores et al., 2005; Gonzalez-Suarez et al., 2001; Imamura et al., 2008; Masutomi et al., 2005; Stewart et al., 2002). Although most cancer cells maintain stable telomere lengths by constitutive expression of telomerase, cells that lack TERT and maintain their telomeres through the alternative lengthening of telomeres (ALT) recombination-based mechanism exhibit long and heterogeneous telomeres, including some chromosomes that lack detectable telomeres (Bryan et al., 1997a,b). In such cells, Stewart et al. demonstrated that expression of TERT-enhanced tumorigenicity in a manner that depends on the catalytic activity of TERT. Specifically, in a subset of immortal but nontumorigenic human cell lines expressing the SV40 large and small t antigens, the expression of the oncogenic version of the H-RAS oncogene led to anchorage-independent colony formation but failed to render these cells tumorigenic. However, the ectopic expression of wild-type TERT in ALT cells expressing the SV40 proteins and oncogenic H-RAS allowed these cells to form tumors. In contrast, expression of either a carboxyterminal epitope hemagluttinin (HA)-tagged TERT mutant unable to elongate telomeres (hTERT-HA) (Stewart et al., 2002) or a catalytically inactive mutant (DN-hTERT) (S.A. Stewart and W.C. Hahn, unpublished observations) failed to induce tumorigenicity. Although one cannot eliminate the possibility that hTERT forms a cap that protects short telomeres (Zhu et al., 1999), these observations suggest that telomere elongation alone is not sufficient to explain the contribution of this additional function of hTERT in transformation.

Similarly, expression of TERT in murine tumors that also use ALT-like recombination-based mechanisms rendered such tumors more tumorigenic (Chang et al., 2003). Chang et al. showed that serial transplantation of subcutaneous tumors derived from late passage doubly deficient genetically engineered mice lacking *mTERC* and *Ink4a/Arf* resulted in the outgrowth of tumors that showed evidence of telomere maintenance via ALT. Significantly, despite a marked increase in telomere reserve, cells derived from the ALT+ subcutaneous tumors were unable to generate lung metastases while cells derived from *mTERC*+/−, *Ink4a/Arf*−/− mice formed such metastatic lesions, indicating functional differences in these principal mechanisms of telomere maintenance *in vivo*. These observations suggest that although telomere dysfunction plays a key role in tumor initiation, tumors that harbor active telomerase exhibit differences in malignant progression compared to tumors that maintain telomeres through ALT. Consistent with these observations, the presence of ALT use in osteosarcomas correlates with a better prognosis (Henson et al., 2005; Johnson et al., 2005; Sanders et al., 2004).

In addition, transgenic mice that overexpress TERT exhibit increased tumor susceptibility even when telomere length is not limiting. Specifically, expression of TERT under the control of the β-actin promoter and cytomegalovirus (CMV) enhancer elements led to increased incidence of breast tumors (Artandi et al., 2002). Similarly, overexpression of TERT in basal keratinocytes using the bovine keratin 5 promoter led to increased responsiveness of the epidermis to the mitogenic effects of phorbol esters, increased wound healing, and susceptibility to the development of skin papillomas (Gonzalez-Suarez et al., 2001). Since telomere lengths are not limiting in these mice, these observations provide further evidence that TERT contributes to malignant transformation beyond its effects on telomere maintenance.

Mechanistically, overexpression of hTERT in normal human cells renders such cells more resistant to apoptotic signals (Forsythe et al., 2002; Hahn et al., 1999b; Holt et al., 1999; Lee et al., 2008) and increases the rate of wound healing (Gonzalez-Suarez et al., 2001), even when telomere length is not limiting. Inhibition of telomerase activity inhibits the long-term proliferation of human cancer cell lines. In some cells, cell death ensues in a manner that depends on telomere length (Forsythe et al., 2002; Hahn et al., 1999b; Holt et al., 1999; Zhang et al., 1999); however, in other cell lines, cell death occurs well before significant telomere shortening (Lee et al., 2008; Li et al., 2004), suggesting either a nontelomere elongation directed function or a telomere-capping function independent of telomere elongation for TERT (Zhu et al., 1999).

8.2 EVIDENCE FOR NONTELOMERE DIRECTED TERT FUNCTIONS IN STEM-CELL FUNCTION

In addition to cancer cell lines, TERT is expressed in normal and malignant stem cells (Allsopp et al., 2003; Hiyama et al., 1995; Marian et al., 2009, 2010; Marian and Shay, 2009). The expression of telomerase has been proposed to maintain telomere length in these cell populations. Consistent with this notion, mutants in TERT have been found in the subsets of patients with the stem-cell disorder dyskeratosis congenita (Mochizuki et al., 2004; Ruggero et al., 2003) as well as subsets of patients with idiopathic pulmonary fibrosis (Armanios et al., 2007; Tsakiri et al., 2007) and aplastic anemia (Yamaguchi et al., 2005).

Specifically, Yamaguchi et al. studied the link between mutations of TERT and aplastic anemia. They screened blood or marrow cells from patients with acquired aplastic anemia and identified five heterozygous, nonsynonymous mutations in TERT among seven unrelated patients. When they coexpressed wild-type TERT and mutant TERT containing the mutations found in these aplastic anemia patients in a telomerase-deficient cell line, they found that each of these mutations showed decreased telomerase activity and concluded that haploinsufficiency of TERT contributed to the telomere shortening phenotypes observed in such patients. However, since these patients did not exhibit stem-cell failure in other tissue compartments, it remains possible that such mutations, particularly those distant from the catalytic center of TERT affect other TERT activities.

Indeed, although telomere maintenance is one function for TERT in stem cells, several laboratories have shown that telomere lengths shorten when such cells are propagated in culture or *in vivo* (Allsopp et al., 2001; Son et al., 2000). Moreover, some have reported that telomere homeostasis and self-renewal potential are not necessarily linked (Allsopp et al., 2003). Specifically, Allsopp et al. developed a transgenic mouse strain in which mouse TERT is overexpressed in hematopoietic stem cells (HSCs) and assessed the effect on telomere length and transplantation capacity of HSCs during serial transplantation. Although telomere length remained stable during serial transplantation of whole bone marrow or HSCs from transgenic mice, this telomere-length maintenance failed to confer an extension of HSC transplantation capacity. For both transgenic and nontransgenic animals, HSCs could be serially transplanted no more than four times.

In support of these observations, when TERT is expressed in normal skin stem cells with an inducible promoter, such transgenic mice show abnormal hair growth due to defects in the cycling of hair-follicle stem cells (Flores et al., 2005; Sarin et al., 2005). Sarin et al. showed that conditional transgenic induction of TERT in mouse skin epithelium induced a rapid transition from telogen to anagen, thereby facilitating robust hair growth. TERT overexpression promotes this developmental transition by causing proliferation of quiescent, multipotent stem cells in the hair follicle bulge region. Importantly, overexpression of TERT in mice that are deficient for *TERC* resulted in similar results, confirming that this TERT function did not require catalytic activity of TERT on telomeres. Similarly, Flores et al. (2005) also showed that TERT overexpression in the absence of changes in telomere length promoted stem-cell mobilization, hair growth, and stem-cell proliferation *in vitro*.

In some reports, these phenotypes are also observed in mice lacking *mTERC* or when TERT mutants lacking telomerase activity are expressed. Recent work suggests that through interactions with the SWI/SNF component BRG1, TERT modifies chromatin to activate the pathways regulated by Myc and Wnt (Wingless integration) (Choi et al., 2008). Choi et al. demonstrated that a TERT mutant lacking reverse transcriptase function retained the full activities of wild-type TERT in enhancing keratinocyte proliferation in skin and in activating resting hair follicle stem cells. TERT transcriptional response strongly resembles those mediated by Myc and Wnt, two proteins intimately associated with stem-cell function and cancer. These data show that TERT controls tissue progenitor cells via transcriptional regulation of a developmental program converging on the Myc and Wnt pathways. Together these observations provide strong evidence that TERT contributes to stem-cell function through mechanisms beyond telomere maintenance.

Moreover, recent work has identified a second noncoding RNA that binds TERT, the *RNA component of mitochondrial RNA polymerase* (*RMRP*) and forms a RNA-dependent RNA polymerase (Maida et al., 2009). *RMRP* is mutated in the pleiotropic syndrome cartilage hair hypoplasia (CHH), a syndrome characterized by short stature, hair and nail abnormalities, and immunodeficiency (Calado and Young, 2008; Guggenheim et al., 2006; Liu and Ellis, 2006; Ridanpaa et al., 2001)—phenotypes predicted to occur due to defects in stem-cell function. Further work is necessary to determine whether the TERT–*RMRP* complex plays a role in the pathogenesis of this disorder.

8.3 MULTIPLE TERT COMPLEXES AND BIOCHEMICAL ACTIVITIES

The observations described above suggest that TERT has several activities important for cell transformation and stem-cell function. Moreover, several lines of evidence suggest that hTERT shuttles among multiple intracellular complexes (Fu and Collins, 2007; Tycowski et al., 2009; Venteicher et al., 2008, 2009). In this section, we will review the biochemical activities linked to TERT beyond its role in telomere synthesis.

8.3.1 Terminal Transferase

Similar to other polymerases, TERT requires divalent cations for activity. Lue et al. examined the effects of altering cations on TERT activity and found that the addition of manganese to both yeast and mammalian TERT altered the activity of TERT (Lue et al., 2005). Specifically, in the presence of manganese, TERT switched to a template- and RNA-independent mode of DNA synthesis. Even as a terminal transferase, TERT showed a preference for G-rich substrates. Although it remains unclear when TERT may function as a terminal transferase, these observations confirm that TERT carries this activity.

8.3.2 RNA-Dependent RNA Polymerase

In addition to its role as a telomere-specific reverse transcriptase and a terminal transferase, TERT has recently been shown to form a complex with the noncoding RNA *RMRP*, which acts as a RdRP (Maida et al., 2009). As an RdRP, TERT contributes to the production of double-stranded RNAs *in vitro* and *in vivo* that can be processed into small interfering RNA.

Although the finding that hTERT forms a second enzymatic complex was surprising, TERT is evolutionarily related to both viral reverse transcriptases and viral RdRPs (Nakamura et al., 1997) and recent structural work confirms that TERT is a closed right-handed polymerase (Gillis et al., 2008; Mitchell et al., 2010). Several primer-dependent RdRPs found in viruses are also right-handed polymerases (Salgado et al., 2006). In *S. pombe*, the RDRC (RNA-directed RNA polymerase complex) together with RITS (RNA-induced transcriptional gene silencing) complex play an important role in heterochromatin formation (Motamedi et al., 2004; Sugiyama et al., 2005). The observation that TERT forms an RdRP with *RMRP* suggests that this mammalian RdRP may play a similar role in post-transcriptional gene silencing. However, further work is necessary to characterize the role of the TERT–*RMRP* complex in these activities as well as to determine the function of this complex in stem-cell function and cell transformation.

8.3.3 Polymerase-Independent Activities

As described earlier, the conditional expression TERT in murine skin alters the normal cycling of hair follicle stem cells. Importantly, these phenotypes do not require an active telomerase holoenzymes as similar effects were observed in mice also lacking *mTERC* (Sarin et al., 2005) and when a catalytically inactive (DN-TERT) was expressed (Choi et al., 2008). Since the catalytic domain of TERT, including that required for telomerase (reverse transcriptase), RdRP, and terminal transferase activities, is shared, these data suggest that phenotype mediated by the conditional overexpression is independent of the polymerase activity of TERT. In a recent work, TERT was shown to interact with the chromatin remodeling protein BRG1, which binds β-catenin, the central transactivator in the Wnt pathway (Park et al., 2009). Given the important role that Wnt signaling plays in stem-cell function, these observations suggest an additional TERT function that is independent of its activity at telomeres.

In addition, several reports suggest that TERT is present in mitochondria (Santos et al., 2004, 2006). Specifically, Santos and her colleagues found that hTERT is targeted to the mitochondria by an N-terminal leader sequence, and that mitochondrial extracts contain telomerase activity. Although the specific complexes formed by this mitochondrial TERT remain undefined, some work suggests that mitochondrially located TERT plays a role in protecting cells from apoptosis (Chakraborty et al., 2006; Del Bufalo et al., 2005; Fu et al., 2000; Kang et al., 2004; Massard et al., 2006; Sahin and Depinho, 2010; Santos et al., 2004, 2006; Zhang et al., 2003). Although the mechanism by which mitochondrial TERT protects cells from cell death remain undefined, these observations merit further investigation since it is unlikely that mitochondrial TERT acts on telomeres.

8.4 CONCLUSIONS

Thus, these observations indicate that hTERT exhibits at least three enzymatic activities: a telomere-specific RNA-dependent DNA polymerase, a terminal trans-

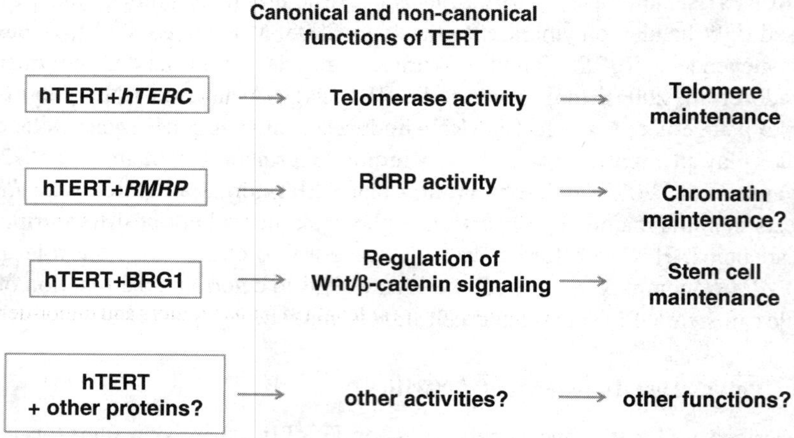

FIGURE 8.1 Canonical and noncanonical functions of TERT. TERT forms several distinct complexes, which have different biological functions.

ferase, and a RNA-dependent RNA polymerase, and exists as at least three complexes: one composed of hTERT and *hTERC* that is involved in telomere maintenance, a RNA-dependent RNA polymerase composed of hTERT and *RMRP*, and a third complex composed of BRG1 that regulates Wnt/β-catenin signaling in normal stem cells (Kalani et al., 2008; Park et al., 2009; Fig. 8.1).

Each of these complexes and enzymatic activities appears to contribute to both normal and malignant cell physiology, and it is likely that TERT affects these phenotypes through both telomere-dependent and telomere-independent mechanisms. Further work will be necessary to dissect the relative contributions of each of these complexes. Indeed, since it is clear that TERT shuttles among different intracellular compartments (Fu and Collins, 2007; Wong et al., 2002), defining not only the roles of each of these complexes but their locations in specific types of tissues and cells will provide new insights into the roles TERT play in both normal and malignant cells.

REFERENCES

Allsopp RC, Cheshier S, Weissman IL. (2001) Telomere shortening accompanies increased cell cycle activity during serial transplantation of hematopoietic stem cells. *J. Exp. Med.* **193**(8): 917–924.

Allsopp RC, Morin GB, Horner JW, DePinho R, Harley CB, Weissman IL. (2003) Effect of TERT over-expression on the long-term transplantation capacity of hematopoietic stem cells. *Nat. Med.* **9**(4): 369–371.

Armanios MY, Chen JJ, Cogan JD, Alder JK, Ingersoll RG, Markin C, Lawson WE, Xie M, Vulto I, Phillips JA 3rd, Lansdorp PM, Greider CW, Loyd JE. (2007) Telomerase mutations in families with idiopathic pulmonary fibrosis. *N. Eng. J. Med.* **356**(13): 1317–1326.

Artandi SE, Alson S, Tietze MK, Sharpless NE, Ye S, Greenberg RA, Castrillon DH, Horner JW, Weiler SR, Carrasco RD, DePinho RA. (2002) Constitutive telomerase expression promotes mammary carcinomas in aging mice. *Proc. Natl. Acad. Sci. USA.* **99**(12): 8191–8196.

Blasco MA, Lee HW, Hande MP, Samper E, Lansdorp PM, DePinho RA, Greider CW. (1997) Telomere shortening and tumor formation by mouse cells lacking telomerase RNA. *Cell.* **91**(1): 25–34.

Bodnar AG, Ouellette M, Frolkis M, Holt SE, Chiu CP, Morin GB, Harley CB, Shay JW, Lichtsteiner S, Wright WE. (1998) Extension of life-span by introduction of telomerase into normal human cells. *Science.* **279**(5349): 349–352.

Bryan TM, Englezou A, Dalla-Pozza L, Dunham MA, Reddel RR. (1997a) Evidence for an alternative mechanism for maintaining telomere length in human tumors and tumor-derived cell lines. *Nat. Med.* **3**(11): 1271–1274.

Bryan TM, Marusic L, Bacchetti S, Namba M, Reddel RR. (1997b) The telomere lengthening mechanism in telomerase-negative immortal human cells does not involve the telomerase RNA subunit. *Hum. Mol. Genet.* **6**(6): 921–926.

Calado RT, Young NS. (2008) Telomere maintenance and human bone marrow failure. *Blood.* **111**(9): 4446–4455.

Chakraborty S, Ghosh U, Bhattacharyya NP, Bhattacharya RK, Roy M. (2006) Inhibition of telomerase activity and induction of apoptosis by curcumin in K-562 cells. *Mutat. Res.* **596** (1–2): 81–90.

Chang S, Khoo CM, Naylor ML, Maser RS, DePinho RA. (2003) Telomere-based crisis: functional differences between telomerase activation and ALT in tumor progression. *Genes Dev.* **17**(1): 88–100.

Chin L, Artandi SE, Shen Q, Tam A, Lee SL, Gottlieb GJ, Greider CW, DePinho RA. (1999) p53 deficiency rescues the adverse effects of telomere loss and cooperates with telomere dysfunction to accelerate carcinogenesis. *Cell.* **97**(4): 527–538.

Choi J, Southworth LK, Sarin KY, Venteicher AS, Ma W, Chang W, Cheung P, Jun S, Artandi MK, Shah N, Kim SK, Artandi SE. (2008) TERT promotes epithelial proliferation through transcriptional control of a Myc- and Wnt-related developmental program. *PLoS Genet.* **4** (1): e10.

Del Bufalo D, Rizzo A, Trisciuoglio D, Cardinali G, Torrisi MR, Zangemeister-Wittke U, Zupi G, Biroccio A. (2005) Involvement of hTERT in apoptosis induced by interference with Bcl-2 expression and function. *Cell Death Differ.* **12**(11): 1429–1438.

Elenbaas B, Spirio L, Koerner F, Fleming MD, Zimonjic DB, Donaher JL, Popescu NC, Hahn WC, Weinberg RA. (2001) Human breast cancer cells generated by oncogenic transformation of primary mammary epithelial cells. *Genes Dev.* **15**(1): 50–65.

Flores I, Cayuela ML, Blasco MA. (2005) Effects of telomerase and telomere length on epidermal stem cell behavior. *Science* **309**(5738): 1253–1256.

Forsythe HL, Elmore LW, Jensen KO, Landon MR, Holt SE. (2002) Retroviral-mediated expression of telomerase in normal human cells provides a selective growth advantage. *Int. J. Oncol.* **20**(6): 1137–1143.

Fu D, Collins K. (2007) Purification of human telomerase complexes identifies factors involved in telomerase biogenesis and telomere length regulation. *Mol. Cell.* **28**(5): 773–785.

Fu W, Killen M, Culmsee C, Dhar S, Pandita TK, Mattson MP. (2000) The catalytic subunit of telomerase is expressed in developing brain neurons and serves a cell survival-promoting function. *J. Mol. Neurosci.* **14**(1–2): 3–15.

Gillis AJ, Schuller AP, Skordalakes E. (2008) Structure of the *Tribolium castaneum* telomerase catalytic subunit TERT. *Nature.* **455**(7213): 633–637.

Gonzalez-Suarez E, Samper E, Flores JM, Blasco MA. (2000) Telomerase-deficient mice with short telomeres are resistant to skin tumorigenesis. *Nat. Genet.* **26**(1): 114–117.

Gonzalez-Suarez E, Samper E, Ramirez A, Flores JM, Martin-Caballero J, Jorcano JL, Blasco MA. (2001) Increased epidermal tumors and increased skin wound healing in transgenic mice overexpressing the catalytic subunit of telomerase, mTERT, in basal keratinocytes. *EMBO J.* **20**(11): 2619–2630.

Greenberg RA, Chin L, Femino A, Lee KH, Gottlieb GJ, Singer RH, Greider CW, DePinho RA. (1999) Short dysfunctional telomeres impair tumorigenesis in the INK4a(delta2/3) cancer-prone mouse. *Cell.* **97**(4): 515–525.

Guggenheim R, Somech R, Grunebaum E, Atkinson A, Roifman CM. (2006) Bone marrow transplantation for cartilage-hair-hypoplasia. *Bone Marrow Transpl.* **38**(11): 751–756.

Hahn WC, Counter CM, Lundberg AS, Beijersbergen RL, Brooks MW, Weinberg RA. (1999a) Creation of human tumour cells with defined genetic elements. *Nature.* **400**(6743): 464–468.

Hahn WC, Stewart SA, Brooks MW, York SG, Eaton E, Kurachi A, Beijersbergen RL, Knoll JH, Meyerson M, Weinberg RA. (1999b) Inhibition of telomerase limits the growth of human cancer cells. *Nat. Med.* **5**(10): 1164–1170.

Henson JD, Hannay JA, McCarthy SW, Royds JA, Yeager TR, Robinson RA, Wharton SB, Jellinek DA, Arbuckle SM, Yoo J, Robinson BG, Learoyd DL, Stalley PD, Bonar SF, Yu D, Pollock RE, Reddel RR. (2005) A robust assay for alternative lengthening of telomeres in tumors shows the significance of alternative lengthening of telomeres in sarcomas and astrocytomas. *Clin. Cancer Res.* **11**(1): 217–225.

Herbert B, Pitts AE, Baker SI, Hamilton SE, Wright WE, Shay JW, Corey DR. (1999) Inhibition of human telomerase in immortal human cells leads to progressive telomere shortening and cell death. *Proc. Natl. Acad. Sci. USA.* **96**(25): 14276–14281.

Herbert BS, Pongracz K, Shay JW, Gryaznov SM. (2002) Oligonucleotide N3′−>P5′ phosphoramidates as efficient telomerase inhibitors. *Oncogene.* **21**(4): 638–642.

Hiyama K, Hirai Y, Kyoizumi S, Akiyama M, Hiyama E, Piatyszek MA, Shay JW, Ishioka S, Yamakido M. (1995) Activation of telomerase in human lymphocytes and hematopoietic progenitor cells. *J. Immunol.* **155**(8): 3711–3715.

Holt SE, Glinsky VV, Ivanova AB, Glinsky GV. (1999) Resistance to apoptosis in human cells conferred by telomerase function and telomere stability. *Mol. Carcinog.* **25**(4): 241–248.

Imamura S, Uchiyama J, Koshimizu E, Hanai J, Raftopoulou C, Murphey RD, Bayliss PE, Imai Y, Burns CE, Masutomi K, Gagos S, Zon LI, Roberts TM, Kishi S. (2008) A non-canonical function of zebrafish telomerase reverse transcriptase is required for developmental hematopoiesis. *PLoS One.* **3**(10): e3364.

Johnson JE, Varkonyi RJ, Schwalm J, Cragle R, Klein-Szanto A, Patchefsky A, Cukierman E, von Mehren M, Broccoli D. (2005) Multiple mechanisms of telomere maintenance exist in liposarcomas. *Clin. Cancer Res.* **11**(15): 5347–5355.

Kalani MY, Cheshier SH, Cord BJ, Bababeygy SR, Vogel H, Weissman IL, Palmer TD, Nusse R. (2008) Wnt-mediated self-renewal of neural stem/progenitor cells. *Proc. Natl. Acad. Sci. USA.* **105**(44): 16970–16975.

Kang HJ, Choi YS, Hong SB, Kim KW, Woo RS, Won SJ, Kim EJ, Jeon HK, Jo SY, Kim TK, Bachoo R, Reynolds IJ, Gwag BJ, Lee HW. (2004) Ectopic expression of the catalytic subunit of telomerase protects against brain injury resulting from ischemia and NMDA-induced neurotoxicity. *J. Neurosci.* **24**(6): 1280–1287.

Lee J, Sung YH, Cheong C, Choi YS, Jeon HK, Sun W, Hahn WC, Ishikawa F, Lee HW. (2008) TERT promotes cellular and organismal survival independently of telomerase activity. *Oncogene.* **27**(26): 3754–3760.

Li S, Rosenberg JE, Donjacour AA, Botchkina IL, Hom YK, Cunha GR, Blackburn EH. (2004) Rapid inhibition of cancer cell growth induced by lentiviral delivery and expression of mutant-template telomerase RNA and anti-telomerase short-interfering RNA. *Cancer Res.* **64**(14): 4833–4840.

Liu JM, Ellis SR. (2006) Ribosomes and marrow failure: coincidental association or molecular paradigm? *Blood.* **107**(12): 4583–4588.

Lue NF, Bosoy D, Moriarty TJ, Autexier C, Altman B, Leng S. (2005) Telomerase can act as a template- and RNA-independent terminal transferase. *Proc. Natl. Acad. Sci. USA.* **102**(28): 9778–9783.

MacKenzie KL, Franco S, Naiyer AJ, May C, Sadelain M, Rafii S, Moore MA. (2002) Multiple stages of malignant transformation of human endothelial cells modelled by co-expression of telomerase reverse transcriptase, SV40 T antigen and oncogenic N-ras. *Oncogene*. **21**(27): 4200–4211.

Maida Y, Yasukawa M, Furuuchi M, Lassmann T, Possemato R, Okamoto N, Kasim V, Hayashizaki Y, Hahn WC, Masutomi K. (2009) An RNA-dependent RNA polymerase formed by TERT and the RMRP RNA. *Nature*. **461**(7261): 230–235.

Marian CO, Cho SK, McEllin BM, Maher EA, Hatanpaa KJ, Madden CJ, Mickey BE, Wright WE, Shay JW, Bachoo RM. (2010) The telomerase antagonist, imetelstat, efficiently targets glioblastoma tumor-initiating cells leading to decreased proliferation and tumor growth. *Clin. Cancer Res.* **16**(1): 154–163.

Marian CO, Shay JW. (2009) Prostate tumor-initiating cells: a new target for telomerase inhibition therapy? *Biochim. Biophys. Acta.* **1792**(4): 289–296.

Marian CO, Wright WE, Shay JW. (2009) The effects of telomerase inhibition on prostate tumor-initiating cells. *Int. J. Cancer*

Massard C, Zermati Y, Pauleau AL, Larochette N, Metivier D, Sabatier L, Kroemer G, Soria JC. (2006) hTERT: a novel endogenous inhibitor of the mitochondrial cell death pathway. *Oncogene*. **25**(33): 4505–4514.

Masutomi K, Possemato R, Wong JM, Currier JL, Tothova Z, Manola JB, Ganesan S, Lansdorp PM, Collins K, Hahn WC. (2005) The telomerase reverse transcriptase regulates chromatin state and DNA damage responses. *Proc. Natl. Acad. Sci. USA*. **102**(23): 8222–8227.

Masutomi K, Yu EY, Khurts S, Ben-Porath I, Currier JL, Metz GB, Brooks MW, Kaneko S, Murakami S, DeCaprio JA, Weinberg RA, Stewart SA, Hahn WC. (2003) Telomerase maintains telomere structure in normal human cells. *Cell*. **114**(2): 241–253.

Mitchell M, Gillis A, Futahashi M, Fujiwara H, Skordalakes E. (2010) Structural basis for telomerase catalytic subunit TERT binding to RNA template and telomeric DNA. *Nat. Struct. Molec. Biol.* **17**(4): 513–518.

Mochizuki Y, He J, Kulkarni S, Bessler M, Mason PJ. (2004) Mouse dyskerin mutations affect accumulation of telomerase RNA and small nucleolar RNA, telomerase activity, and ribosomal RNA processing. *Proc. Natl. Acad. Sci. USA*. **101**(29): 10756–10761.

Motamedi MR, Verdel A, Colmenares SU, Gerber SA, Gygi SP, Moazed D. (2004) Two RNAi complexes, RITS and RDRC, physically interact and localize to noncoding centromeric RNAs. *Cell*. **119**(6): 789–802.

Nakamura TM, Morin GB, Chapman KB, Weinrich SL, Andrews WH, Lingner J, Harley CB, Cech TR. (1997) Telomerase catalytic subunit homologs from fission yeast and human. *Science*. **277**(5328): 955–959.

Park JI, Venteicher AS, Hong JY, Choi J, Jun S, Shkreli M, Chang W, Meng Z, Cheung P, Ji H, McLaughlin M, Veenstra TD, Nusse R, McCrea PD, Artandi SE. (2009) Telomerase modulates Wnt signalling by association with target gene chromatin. *Nature*. **460**(7251): 66–72.

Pascolo E, Wenz C, Lingner J, Hauel N, Priepke H, Kauffmann I, Garin-Chesa P, Rettig WJ, Damm K, Schnapp A. (2002) Mechanism of human telomerase inhibition by BIBR1532, a synthetic, non-nucleosidic drug candidate. *J. Biol. Chem.* **277**(18): 15566–15572.

Rich JN, Guo C, McLendon RE, Bigner DD, Wang XF, Counter CM. (2001) A genetically tractable model of human glioma formation. *Cancer Res.* **61**(9): 3556–3560.

Ridanpaa M, van Eenennaam H, Pelin K, Chadwick R, Johnson C, Yuan B, vanVenrooij W, Pruijn G, Salmela R, Rockas S, Makitie O, Kaitila I, de la Chapelle A. (2001) Mutations in the RNA component of RNase MRP cause a pleiotropic human disease, cartilage–hair hypoplasia. *Cell.* **104**(2): 195–203.

Ruggero D, Grisendi S, Piazza F, Rego E, Mari F, Rao PH, Cordon-Cardo C, Pandolfi P.P. (2003) Dyskeratosis congenita and cancer in mice deficient in ribosomal RNA modification. *Science* **299**(5604): 259–262.

Sahin E, Depinho RA. (2010) Linking functional decline of telomeres, mitochondria and stem cells during ageing. *Nature.* **464**(7288): 520–528.

Salgado PS, Koivunen MR, Makeyev EV, Bamford DH, Stuart DI, Grimes JM. (2006) The structure of an RNAi polymerase links RNA silencing and transcription. *PLoS Biol.* **4**(12): e434.

Sanders RP, Drissi R, Billups CA, Daw NC, Valentine MB, Dome JS. (2004) Telomerase expression predicts unfavorable outcome in osteosarcoma. *J. Clin. Oncol.* **22**(18): 3790–3797.

Santos JH, Meyer JN, Skorvaga M, Annab LA, Van Houten B. (2004) Mitochondrial hTERT exacerbates free-radical-mediated mtDNA damage. *Aging Cell.* **3**(6): 399–411.

Santos JH, Meyer JN, Van Houten B. (2006) Mitochondrial localization of telomerase as a determinant for hydrogen peroxide-induced mitochondrial DNA damage and apoptosis. *Hum. Mol. Genet.* **15**(11): 1757–1768.

Sarin KY, Cheung P, Gilison D, Lee E, Tennen RI, Wang E, Artandi MK, Oro AE, Artandi SE. (2005) Conditional telomerase induction causes proliferation of hair follicle stem cells. *Nature.* **436**(7053): 1048–1052.

Seimiya H, Muramatsu Y, Ohishi T, Tsuruo T. (2005) Tankyrase 1 as a target for telomere-directed molecular cancer therapeutics. *Cancer Cell.* **7**(1): 25–37.

Seimiya H, Oh-hara T, Suzuki T, Naasani I, Shimazaki T, Tsuchiya K, Tsuruo T. (2002) Telomere shortening and growth inhibition of human cancer cells by novel synthetic telomerase inhibitors MST-312, MST-295, and MST-1991. *Mol. Cancer Ther.* **1**(9): 657–665.

Son NH, Murray S, Yanovski J, Hodes RJ, Weng N. (2000) Lineage-specific telomere shortening and unaltered capacity for telomerase expression in human T and B lymphocytes with age. *J. Immunol.* **165**(3): 1191–1196.

Stewart SA, Hahn WC, O'Connor BF, Banner EN, Lundberg AS, Modha P, Mizuno H, Brooks MW, Fleming M, Zimonjic DB, Popescu NC, Weinberg RA. (2002) Telomerase contributes to tumorigenesis by a telomere length-independent mechanism. *Proc. Natl. Acad. Sci. USA.* **99**(20): 12606–12611.

Sugiyama T, Cam H, Verdel A, Moazed D, Grewal SI. (2005) RNA-dependent RNA polymerase is an essential component of a self-enforcing loop coupling heterochromatin assembly to siRNA production. *Proc. Natl. Acad. Sci. USA.* **102**(1): 152–157.

Tsakiri KD, Cronkhite JT, Kuan PJ, Xing C, Raghu G, Weissler JC, Rosenblatt RL, Shay JW, Garcia CK. (2007) Adult-onset pulmonary fibrosis caused by mutations in telomerase. *Proc. Natl. Acad. Sci. USA.* **104**(18): 7552–7557.

Tycowski KT, Shu MD, Kukoyi A, Steitz JA. (2009) A conserved WD40 protein binds the Cajal body localization signal of scaRNP particles. *Mol. Cell.* **34**(1): 47–57.

Venteicher AS, Abreu EB, Meng Z, McCann KE, Terns RM, Veenstra TD, Terns MP, Artandi SE. (2009) A human telomerase holoenzyme protein required for Cajal body localization and telomere synthesis. *Science.* **323**(5914): 644–648.

Venteicher AS, Meng Z, Mason PJ, Veenstra TD, Artandi SE. (2008) Identification of ATPases pontin and reptin as telomerase components essential for holoenzyme assembly. *Cell.* **132**: 945–971.

Wong JM, Kusdra L, Collins K. (2002) Subnuclear shuttling of human telomerase induced by transformation and DNA damage. *Nat. Cell. Biol.* **4**(9): 731–736.

Yamaguchi H, Calado RT, Ly H, Kajigaya S, Baerlocher GM, Chanock SJ, Lansdorp PM, Young NS. (2005) Mutations in TERT, the gene for telomerase reverse transcriptase, in aplastic anemia. *N Engl J Med.* **352**(14): 1413–1424.

Yuan X, Ishibashi S, Hatakeyama S, Saito M, Nakayama J, Nikaido R, Haruyama T, Watanabe Y, Iwata H, Iida M, Sugimura H, Yamada N, Ishikawa F. (1999) Presence of telomeric G-strand tails in the telomerase catalytic subunit TERT knockout mice. *Genes Cells.* **4**(10): 563–572.

Zhang P, Chan SL, Fu W, Mendoza M, Mattson MP. (2003) TERT suppresses apoptotis at a premitochondrial step by a mechanism requiring reverse transcriptase activity and 14-3-3 protein-binding ability. *FASEB J.* **17**(6): 767–769.

Zhang X, Mar V, Zhou W, Harrington L, Robinson MO. (1999) Telomere shortening and apoptosis in telomerase-inhibited human tumor cells. *Genes Dev.* **13**(18): 2388–2399.

Zhu J, Wang H, Bishop JM, Blackburn EH. (1999) Telomerase extends the lifespan of virus-transformed human cells without net telomere lengthening. *Proc. Natl. Acad. Sci. USA.* **96**(7): 3723–3728.

9

MURINE MODELS OF DYSFUNCTIONAL TELOMERES AND TELOMERASE

YIE LIU AND LEA HARRINGTON

9.1 INTRODUCTION

Telomeres are specialized structures consisting of tandem G-rich repeats of the sequence 5′-(TTAGGG)$_n$-3′ in mammals, and together with telomeric protein complexes such as shelterin serve to cap the ends of linear chromosomes (Fig. 9.1). Telomeres prevent the recognition of chromosome termini as broken DNA ends and are critical to maintain genomic stability. Telomere dysfunction can trigger the recruitment of several double-strand breakage response factors, including γ-H2AX, 53BP1, Mre11/Rad50/NBS1, and phosphorylated ATM. These resultant telomere-dysfunction induced foci (TIF) coincide with the activation of the checkpoint kinases ATR and/or ATM, which in turn phosphorylate Chk1 and Chk2 and lead to p53- and p21-dependent cell apoptosis and cycle arrest (Celli and de Lange, 2005; d'Adda di Fagagna et al., 2003; Herbig et al., 2004; Karlseder et al., 1999; Takai et al., 2003; Fig. 9.1). Telomere dysfunction has been linked to premature aging, bone marrow failure syndromes, and tumor formation (reviewed in Sahin and Depinho, 2010). Accumulating evidence suggests that telomere integrity depends on the ability to maintain a critical reservoir of telomere length and/or the ability to mask chromosome ends from being recognized as damaged DNA. In addition to shelterin, the telomerase reverse transcriptase also plays an essential role in telomere length maintenance and

Telomerases: Chemistry, Biology, and Clinical Applications, First Edition.
Edited by Neal F. Lue and Chantal Autexier.

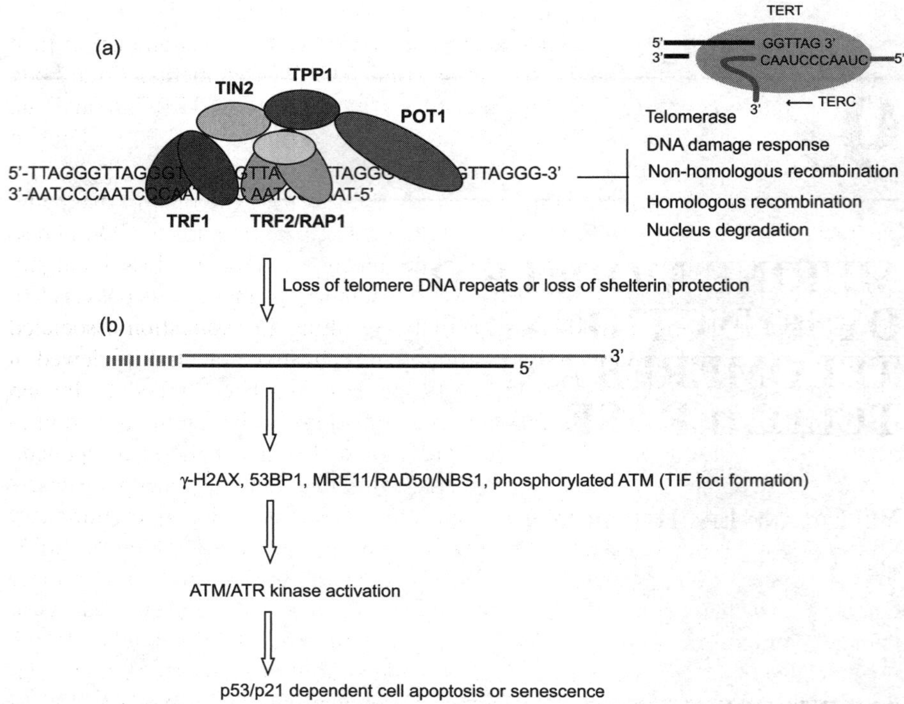

FIGURE 9.1 Telomere maintenance by telomerase and shelterin, and the consequences of telomere dysfunction. (a) Telomere DNA, telomerase, and shelterin. Telomeres cap the chromosome ends and protect against NHEJ, HR, DNA damage signaling, and nucleolytic degradation. The access of telomerase to the telomere is limited by telomere-bound POT1 and TRF1. (b) Dysfunctional telomeres arise via loss of telomere DNA repeats or loss of protection of shelterin, resulting in the induction of DNA damage foci at telomeres (TIF) and activation of ATM–ATR kinase pathways. These signaling cascades in turn can lead to p53/p21 dependent cell apoptosis, cell cycle arrest, and cellular senescence. (See the color version of this figure in Color Plates section.)

telomere capping (reviewed in Blackburn, 2001; Palm and de Lange, 2008). In this chapter, we will present key advances that have employed knockout or transgenic murine models to elucidate the role of telomerase and shelterin *in vivo*.

9.2 TELOMERASE

Mammalian telomerase is a large ribonucleoprotein complex whose enzymatic activity depends on two core components; TERT (telomerase reverse transcriptase) and TERC (telomerase RNA) that contains the template for reverse transcription of telomeric repeats (reviewed in Kelleher et al., 2002 and Chapters 2 and 3). However, stable and active telomerase requires other components. Dyskerin and its interacting proteins NHP2, NOP10, and GAR1 bind to the TERC H/ACA box and are critical for telomerase assembly and TERC stability, and TCAB1 (telomerase Cajal body protein 1) interacts with dyskerin and recognizes the CAB box in TERC to facilitate telomerase assembly and trafficking to Cajal bodies (Cohen et al., 2007; Dragon et al., 2000; Fu and Collins, 2007; Pogacic et al., 2000; Venteicher et al., 2009).

During DNA replication, telomerase is recruited to the 3′ telomeric overhang, where TERT extends telomere DNA repeats using TERC as a template. Telomerase activity is essential in preventing replication-dependent telomere loss in highly proliferative cells and cancer cells. However, most human somatic cells possess low or undetectable levels of telomerase activity, resulting in replication-associated telomere shortening and progressive restriction in replicative potential (reviewed in Shay and Wright, 2005). Telomere shortening has also been linked to human syndromes with early mortality. For example, mutations in the telomerase components, TERT, TERC, DKC, NHP2, and NOP10 (and also in a shelterin component, TIN2; see Section 2 below) are associated with rare human genetic disorders including dyskeratosis congenita (DC), aplastic anemia, and idiopathic pulmonary fibrosis (IPF) (Armanios et al., 2005; Marrone et al., 2007; Savage et al., 2008; Vulliamy et al., 2001; Walne et al., 2007). Patients harboring mutations in these components display telomere shortening and, in some instances, an increased risk of hematopoietic cancers, further underscoring the importance of telomere length maintenance and integrity in genome stability (reviewed in Armanios, 2009; Calado, 2009; Savage and Alter, 2008). Furthermore, mutations in the WRN and ATM genes cause inherited premature-aging syndromes, namely Werner (WRN) and Ataxia telangiectasia (AT). Cells derived from WRN and AT patients exhibit marked telomere attrition, chromosome instability, and premature senescence, and these defects are rescued *in vitro* by enforced TERT expression (Crabbe et al., 2007; Metcalfe et al., 1996), thus suggesting a further link between telomere attrition and the pathogenesis of premature-aging syndromes.

Unicellular (e.g., yeasts, ciliates) and multicellular model organisms (worms, plants, fishes, and sea urchins) have been instrumental in dissecting the role of telomerase and its associated components in telomere length maintenance and genome integrity. This chapter will focus on the use of *TERT* and *TERC* null or transgenic murine models to evaluate telomerase function and telomere attrition in the content of mammalian tissue function and organismal aging. Key findings summarized herein include: (1) TERT and TERC are the essential genetic components of mammalian telomerase and are essential for telomere length maintenance and genome stability *in vivo*, (2) telomere dysfunction arises from critically short telomeres and triggers a DNA damage response that affects the function of stem and progenitor cells, (3) telomere dysfunction can promote or suppress cancer development, (4) telomere dysfunction leads to human pathologies and is required for phenotypic penetration in *Atm*, *Wrn*, and *Blm* deficient mice, (5) the dosage of telomerase components is critical for telomere maintenance, (6) and telomere homeostasis impinges on murine aging and development.

9.2.1 TERT and TERC in Telomere Length Maintenance and Genome Stability

Telomerase activity can be detected in several murine tissues, but *TERT* and *TERC* null mice possess no detectable telomerase activity, which provided unequivocal evidence that mice possess one telomerase reverse transcriptase and one telomerase RNA template and both are required for telomerase catalysis (Blasco et al., 1997; Liu et al., 2000; Yuan et al., 1999). Early-generation *TERC* or *TERT* null mice do not exhibit telomere dysfunction in a genetic background with intrinsically long telomere reserves. Successive generations of telomerase null mice gradually exhaust their telomere reserves and exhibit increased chromosome ends without detectable telomere signals and with chromosome end-to-end fusions. These results confirm that TERC and TERT are essential for telomere length maintenance in mammals. Critically shortened telomeres in late-generation *TERC* mice activate a DNA damage response, leading to γH2AX detection at telomeres (Hao et al., 2004) and p53/p21 activation (Chin et al., 1999). The phenotypic consequences of telomere loss appear indistinguishable between *TERT* and *TERC* null mice (Vidal-Cardenas and Greider, 2010), and are collectively referred to as telomerase null mice in the following sections.

9.2.2 Critically Short Telomeres are Responsible for Telomere Dysfunction and Primarily Affect the Function of Stem and Progenitor Cells in a Variety of Tissue Types

Human patients harboring mutations in telomerase components display telomere erosion and suffer ailments in highly proliferative tissues such as the blood, skin, intestine, and nails (reviewed in Vulliamy and Dokal, 2008). In mice, successive interbreeding of telomerase knockout mice leads to gradual telomere loss and acquisition of similar defects in hematopoietic, intestinal, epidermal progenitor, and stem cells. As a result, late-generation telomerase knockout mice display pleiotropic phenotypes including infertility, an abnormal hematological profile, and reduced long-term repopulation of bone marrow progenitor cells, reduced B and T cell proliferation, impaired germinal centre function, and tissue atrophy in spleen, small intestine, skin, and testis (Blasco, 2007; Erdmann et al., 2004; Herrera et al., 1999; Lee et al., 1998; Rudolph et al., 1999; Sahin and Depinho, 2010). In addition, they possess a shortened life span, hair graying, alopecia, hematopoietic ablation, and decreased wound healing in response to stress, reduced weight, and an increased incidence of spontaneous tumors (reviewed in Blasco, 2005). In addition, late-generation null mice also exhibit impairment in the replicative capacity of insulin-producing beta-cells, and possess alterations in glucose metabolism and insulin secretion that mimics age-associated type II diabetes (Guo et al., 2011; Kuhlow et al., 2010). The murine telomerase knockout model enabled the determination that the shortest telomeres, not average telomere length, is the critical variable in susceptibility to tissue dysfunction (Erdmann et al., 2004; Hemann et al., 2001b; Samper et al., 2001). The reintroduction of telomerase into late-generation telomerase null mice is sufficient to elongate critically short telomeres and prevent end-to-end fusions, which correlates with the rescue of phenotypes associated with telomere loss (Jaskelioff et al., 2010; Siegl-Cachedenier et al., 2007a). These findings support the assertion that the loss of telomere integrity is primarily responsible for the reduced long-term viability of tissue that undergoes constant renewal in the knockout mice.

Mounting evidence supports the notion that these phenotypes arise because critically short telomeres trigger a DNA damage response, leading to cell apoptosis, cell-cycle arrest, and defective renewal and differentiation in stem, progenitor, and highly proliferative cell types. Critically short telomeres serve to activate and stabilize the p53 protein family and their target p21, which play prominent roles in growth arrest and/or apoptosis of affected tissues in late-generation telomerase null mice (Chin et al., 1999). Although inactivation of p53 and p21 does not rescue telomere length or the end-to-end fusions observed in late-generation telomerase null mice, loss of p53 checkpoint function can rescue cell death, while deletion of p21 does not rescue cell apoptosis but improves cell-proliferation defects and overall lifespan (Chin et al., 1999; Choudhury et al., 2007; Flores and Blasco, 2009). Deletion of the mismatch repair genes *Pms2* and *Msh2* attenuate short telomere-induced p21 activation, and these deletions elicit a similar effect as p21 deficiency on organism fitness and survival of late-generation telomerase null mice (Martinez et al., 2009a; Siegl-Cachedenier et al., 2007b). Furthermore, disruption of the exonuclease *Exo1* impairs cell-cycle arrest and the DNA damage response, leading to an improvement in organ maintenance and lifespan of late-generation telomerase null mice (Schaetzlein et al., 2007). Taken together, these observations demonstrate that p53, p21, MMR, and Exo1 contribute to short telomere-induced DNA damage signaling, and that disruption of these genes can, in some instances, improve overall fitness of late-generation telomerase null mice.

Notably, different laboratory murine strains influence the temporal onset of phenotypes associated with telomerase deficiency (*TERC* or *TERT*). In inbred C57BL/6 or mixed C57BL/6 and 129sv strains, with initially long telomere reserves, 4–6 knockout generations are required to achieve critically short telomeres (Blasco et al., 1997; Erdmann et al., 2004; Hathcock et al., 2002; Herrera et al., 1999). When the mice are backcrossed into a CAST/EiJ background with short telomere reserves, the early-generation *TERC* null mice almost immediately exhibit reduced survival and tissue-renewal defects (Hao et al., 2005). These studies demonstrate that initial telomere length influences the temporal onset of phenotypes associated with telomerase deficiency.

9.2.3 Telomere Dysfunction in Cancer Development

Although late-generation telomerase knockout mice can develop spontaneous tumors (Rudolph et al., 1999), in general these mice exhibit a decreased incidence of tumors with increasing generations, suggesting that short telomeres can also act to suppress tumor progression. The suppressive function of short telomeres occurs via induction of p53 and p21-dependent cell-cycle arrest, cellular senescence, and/or cell apoptosis. For instance, the rescue of cell apoptosis by deletion of p53 confers a survival advantage to late-generation telomerase null mice, yet it further increases genome instability. As a result, p53 deficiency cooperates with telomere dysfunction to accelerate carcinogenesis (Chin et al., 1999; Farazi et al., 2006). In mice with loss-of-function in the tumor suppressors p16 and p19ARF (but retaining intact p53), short telomeres are associated with a severe defect in growth and inability to escape senescence/crisis, concomitant with impaired tumorigenesis (Greenberg et al., 1999; Khoo et al., 2007). Moreover, targeting *Cdkn1a* (p21) and *Exo1* genes rescues DNA damage signal induction and p53/p21 dependent cell-cycle arrest in late-generation telomerase null mice (Choudhury et al., 2007; Schaetzlein et al., 2007) and yet does not further increase chromosomal instability or cancer formation of late-generation telomerase null animals. Thus, loss of telomere function in cancer-prone mice may impair or promote tumor formation, which is influenced by whether the strain possesses intact p53 function. However, other factors may also contribute to the impact of short telomeres on tumorigenesis. Both mismatch repair proteins MSH2 and PMS2 signal cell-cycle arrest in a p53/p21-dependent manner in response to short telomeres, and deletion of *Msh2* and *Pms2* rescues cell-cycle arrest, but not apoptosis. Although *Msh2* and *Pms2* null mice are cancer-prone, successive generations of telomerase and *Pms2* double null mice exhibit a decreased incidence of tumors, compared with single *Pms2* null mice (Siegl-Cachedenier et al., 2007b). In contrast to loss of PMS2, *Msh2* deficiency abolishes the tumor-suppressor activity of short telomeres (Martinez et al., 2009a; Siegl-Cachedenier et al., 2007a). Thus, *Msh2* deficiency can overcome its attenuation of downstream p53/p21 activation and lead to tumorigenesis.

Short telomeres can also influence tumorigenesis when combined with other defective tumor suppressors or dysregulated oncogenes. Short telomeres promote genome instability and appear to drive early carcinogenesis, but reduce the size and incidence of macroscopic adenomas in mice carrying an APC mutation (Rudolph et al., 2001). Short telomeres also delay tumor formation in mice that express K-ras, but further introduction of a heterozygous p53 mutation allows the development of aggressive tumors with more chromosomal instability and high metastatic potential (Perera et al., 2008). Short telomeres suppress tumor formation in Emu-myc transgenic mice, an effect that is abrogated by loss of p53 function (Feldser and Greider, 2007). These studies reinforce the notion that p53 status regulates the impact of short telomeres on tumorigenesis. Myc has also been shown to activate telomerase in the skin and induce skin papillomatosis, and inactivation of telomerase or the presence of short telomeres reduces the incidence of these skin lesions (Flores et al., 2006). Conversely, short telomeres appear to have minimal influence on carcinogenesis elicited by the viral oncoproteins SV40 or HPV16 (Argilla et al., 2004).

Short telomeres and telomerase deficiency can also impact tumorigenesis when combined with backgrounds defective in telomere protection. Overexpression of the telomere repeat binding protein, TRF2, in murine stratified epithelia (*K5TRF2* mice) results in XPF nuclease-dependent telomere attrition and chromosomal instability, including the development of preneoplastic and neoplastic lesions in stratified epithelia and UV-induced skin cancer. The onset of spontaneous or UV-induced epithelial carcinogenesis is further increased in successive generations of *K5TRF2* animals in which telomerase is absent (Blanco et al., 2007). ATM protects short telomeres, but is not required for short telomere-induced cell death (Feldser et al., 2006; Qi et al., 2003; Wong et al., 2003). Late-generation *Atm* and telomerase compound null mice display increased end-to-end fusions and apoptosis and a reduced incidence of lymphoma. Additional p53 deficiency allows development of lymphoma with shorter latency and higher penetrance (Maser et al., 2007). Thus, short telomeres in late-generation *Atm* and telomerase compound null mice suppress the development of lymphoma, which may relate to p53 activation. Both Ku86 and DNA-Pkcs are involved in telomere protection (Bailey et al., 1999; Goytisolo et al., 2001; Samper et al., 2000). Deletion of Ku86 or DNA-Pkcs in combination with progressive telomere shortening accelerates loss of organism viability (Espejel et al., 2004; Wong et al., 2007), but does not affect the incidence of cancer (Espejel et al., 2004).

Taken together, these studies illustrate that telomere dysfunction can have dual effects on tumorigenesis, which is dictated by different genetic contexts, and that the p53 checkpoint plays a dominant role in determining the outcome of short telomere-associated tumorigenesis.

9.2.4 Telomere Dysfunction in Human Premature Aging Syndromes

In humans, loss of WRN, BLM, or ATM can lead to segmental progeroid syndromes (Werner, Bloom, and AT, respectively), characterized by accelerated aging and cancer susceptibility. None of these degenerative pathologies has been recapitulated in *Wrn*, *Blm*, or *Atm* gene mutant murine strains. However, when these mutant mice are crossed into a telomerase null background, partial features of the pathologies associated with accelerated ageing are precipitated by exhaustion of telomere reserves (Chang et al., 2004; Du et al., 2004; Wong et al., 2003). For example, late-generation mice deficient in *Wrn* and *TERC* display accelerated telomere loss and increased genome instability, accompanied by higher rates of apoptosis in highly proliferative tissues and accelerated age-related phenotypes including premature death, hair graying, alopecia, osteoporosis, type II diabetes, cataracts, glucose intolerance, wound healing defects, and increased malignancies. Similar to WRN-deficient human fibroblasts, the compound null primary murine embryonic fibroblasts (MEFs) exhibit profound premature replicative senescence. In this setting, *Wrn* deficiency promotes telomere sister-chromatid exchange, allowing MEFs to escape from senescence and engage in telomerase-independent telomere maintenance and tumor formation (Laud et al., 2005). These studies suggest that telomere attrition may contribute a critical role to the pathogenesis of these genetic disorders in humans.

The human syndrome DC is characterized by skin pigmentation, leukoplasia, nail dystrophy, and bone marrow failure, which are tissues that require constant cell turnover and tissue replacement. Inactivating mutations in one allele of TERT or TERC (autosomal dominant), or dyskerin (X-linked), have been found in patients affected by DC; in some cases, the disease displays genetic anticipation (earlier disease onset) in successive generations (reviewed in Armanios, 2009; Calado, 2009; Savage and Alter, 2008). Dyskerin possesses several distinctive cellular functions in ribosome biogenesis, snRNA maturation, and telomerase assembly and TERC stability (Meier, 2005). It is unclear if all or some of these cellular functions could contribute to the pathogenesis of the disease. Some pathogenic dyskerin mutations affect telomerase activity and telomere length, but can also affect pseudouridylation or activate a DNA damage response independent of telomere length in mice (Gu et al., 2008; Mochizuki et al., 2004). Nevertheless, *TERC* null mice with short telomeres exhibit partial features of pathologies associated with DC, including an abnormal hematological profile and reduced renewal of bone marrow progenitor cells (Hao et al., 2005; Herrera et al., 1999; Lee et al., 1998; Rudolph et al., 1999). Moreover, after backcrossing into a CAST/EiJ background, *TERC* heterozygous mice also display haploinsufficiency for telomere maintenance, leading to hematopoietic and immunologic defects (Hao et al., 2005). Although the complete spectra of DC epithelial phenotypes are not observed in *TERC* null mice, additional deletion of *Pot1b* in this background results in a nearly complete penetration of typical DC features (see below). Similarly, mice expressing the mutation in dystrophin responsible for Duchenne muscular dystrophy (DMD) recapitulate the human disorder only in the context of shorter telomeres (Sacco et al., 2010). Thus, it appears that age-associated disorders in humans exhibit more complete penetrance in mice when telomeres approach the shorter lengths observed in humans.

9.2.5 Consequence of Altered Dosage of TERT and TERC in Telomere Length Maintenance

Human germline cells possess high telomerase activity and their telomeres are maintained about an equilibrium length, but quiescent stem cells express low levels of telomerase activity and undergo telomere shortening within an individual's lifetime (Rufer et al., 1999; Vaziri et al., 1994; Wright et al., 1996). Thus, low telomerase activity has been proposed to be limiting for telomere elongation during stem-cell renewal. Several mouse genetic studies support that dosage of telomerase does indeed affect telomere length.

The first indication that telomerase is haploinsufficient in mammals came from the discovery of telomere erosion in embryonic stem cells heterozygous for *mTERT* (Liu et al., 2000). Subsequently, telomere erosion in heterozygous animals was established in successive generations of *mTERT*$^{+/-}$ (Liu et al., 2002b) and *mTERC*$^{+/-}$ animals (Hao et al., 2005). In further support of haploinsufficiency, in the progeny of crosses between C57BL/6 (a strain that possesses a long telomere reserve) and CAST/Ei or SPRET/Ei strains (shorter telomere reserve), the ability to elongate the shorter telomeres to a new equilibrium length in the offspring was compromised in *mTERC*$^{+/-}$ mice (Hathcock et al., 2002). Even wild-type littermates of heterozygous progeny exhibited shorter telomeres (Hao et al., 2005; Liu et al., 2002b). In fact, the ability to reset *mTERT*$^{+/+}$ littermate telomeres to a wild-type equilibrium length requires at least two successive generations of crosses to C57BL/6, and in some instances telomeres remain short despite outbreeding to wild-type animals (Chiang et al., 2010; Erdmann et al., 2004). These genetic experiments support the notion that one copy of m*TERT* and *mTERC* is haploinsufficient for the maintenance of long telomeres.

Although telomerase appears haploinsufficient for the maintenance of long telomeres, mice heterozygous for telomerase nonetheless appear proficient to rescue critically short chromosome ends from a complete loss of telomeric DNA. Although average telomere length in *mTERT*$^{+/-}$ ES cells and mice approaches that of *mTERT*$^{-/-}$ mice, *mTERT*$^{+/-}$ ES cells retain a minimal telomere DNA signal at all chromosome ends and, unlike their null littermates, do not exhibit chromosome end-to-end fusions (Liu et al., 2002b). This study implies that limiting telomerase can protect short telomeres from becoming critically short. In support of the notion that telomerase RNA levels are more limiting than TERT, *mTERC*$^{+/-}$ offspring of C57BL6 and CAST/Ei and SPRET/Ei intercrosses are less proficient in their ability to elongate critically short telomeres (Hathcock et al., 2002). Despite this observation, crosses between late-generation heterozygous (for either TERT or TERC) and null animals (both with short telomeres) result in a preferential rescue of critically short ends in the heterozygous progeny (Erdmann et al., 2004; Hathcock et al., 2005; Hemann et al., 2001b; Meznikova et al., 2009; Samper et al., 2001). This evidence supports the notion that limiting telomerase can selectively add telomere repeats to the shortest telomeres, although it is unable to maintain overall telomere length. This selective elongation does not appear to be mediated by ATM or ATR (Feldser et al., 2006; McNees et al., 2010). It is not yet clear whether a reduction in telomere-bound factors that inhibit telomere elongation may promote elongation even when telomerase levels are reduced, as has been observed in yeast (Bianchi and Shore, 2008; Teixeira et al., 2004). In human cancer cells with short average telomere lengths, all telomeres appear to be substrates for elongation during successive cell divisions (Zhao et al., 2009).

Initial telomere length or strain background may also influence the response to telomerase dosage. Whereas defects in highly proliferative tissues are observed after only five successive crosses between CAST/Ei *mTERC*$^{+/-}$ mice (Hao et al., 2005), these defects are not evident in up to 10 generations of crosses between C57BL/6 *mTERT*$^{+/-}$ and wild-type mice (Erdmann et al., 2004). In fact, after 10 heterozygote × wild-type crosses, *mTERT*$^{+/-}$ mice do not exhibit further telomere erosion, and telomere-length equilibrates (Meznikova et al., 2009). This apparent equilibration depends on *mTERT*, since telomere erosion and its phenotypic consequences resume if null littermates are subsequently interbred (Meznikova et al., 2009). In a separately generated *mTERT* knockout, successive heterozygous breeding for 17 generations also resulted in mice with stably short but functional telomeres, and no bone marrow defects were observed. However, short telomeres in this strain remained short, even in wild-type littermates after a further six crosses to wild-type C57BL/6 mice (Chiang et al., 2010). These results indicate that *mTERT* heterozygous animals in a C57BL/6 background do not suffer an eventual loss in the ability to maintain critically short ends. This clearly distinct response (compared with CAST/Ei *mTERC*$^{+/-}$ mice) to prolonged telomerase heterozygosity underscores the influence of different genetic backgrounds. Notably, the phenotypes of *mTERC*$^{+/-}$ and *mTERT*$^{+/-}$ mice in CAST/Ei are the same (Strong et al., 2011).

TERT overexpression appears to play a protective role in murine disease and ageing. Enforced overexpression of mTERT in stratified epithelia (K5-mTERT) increased telomerase activity (Gonzalez-Suarez et al., 2001) and delayed telomere loss with age (Tomas-Loba et al., 2008). K5-mTERT mice possess a lower incidence of age-related degenerative diseases, but a higher incidence of both induced and spontaneous tumors (Gonzalez-Suarez et al., 2001, 2005). To overcome the potential cancer-prone consequences of TERT overexpression, K5-mTERT was introduced into a genetic background in which the tumor suppressors p53, p16, and p19ARF were overexpressed. The resultant offspring exhibited improved fitness and an increased median lifespan (Tomas-Loba et al., 2008).

9.2.6 Telomerase and Telomeres in Murine Stem-Cell Function in Aging and Immunity

Stem cells have the capability to self-renew, migrate out of the stem-cell niche, and differentiate into different cell lineages. Critically short telomeres impair the survival of tissue stem and progenitor cells. Making an example of the hematopoietic lineage, late-generation telomerase null mice exhibit reduced blood lymphocyte counts and B cells in the spleen, reflecting a decreased number of follicles in the spleen and impaired germinal center formation. The proliferative response of B and T cells upon mitogen stimulation is also decreased (Herrera et al., 1999, 2000; Lee et al., 1998). Critically short telomeres also affect the microenvironment that promotes the self-renewal capacity of stem cells. For instance, telomere dysfunction leads to reduced epidermal stem-cell proliferation capacity and impaired mobilization of stem cells out of their niche (Flores et al., 2005). On the other hand, overexpression of TERT promotes epidermal and hair follicle stem-cell proliferation and epidermal stem-cell mobilization (Flores et al., 2005; Sarin et al., 2005). Telomere dysfunction also limits self-renewal of bone marrow hematopoietic stem cells in serial transplantation experiments (Allsopp et al., 2003a; Samper et al., 2002) and alters stem-cell microenvironments that normally sustain the proper function of transplanted wild-type stem cells (Ju et al., 2007). Furthermore, telomerase is involved in maintaining the undifferentiated state of adult murine bone marrow mesenchymal stem cells (mMSCs), and loss of telomere integrity results in the failure of mMSCs to differentiate into a variety of lineage cell types (Liu et al., 2004b).

Mounting evidence also suggests that telomere shortening can contribute to stem-cell dysfunction with age. Telomere shortening occurs in normal murine aging (Coviello-McLaughlin and Prowse, 1997; Flores et al., 2008), although this observation does not establish whether short telomeres contribute to physiologic ageing *per se*. Telomere length varies within tissues, with the longest telomeres occurring in stem-cell compartments where age-associated telomere erosion occurs and is correlated with a decline in stem-cell function (Flores et al., 2008). Overexpression of TERT in epithelial cells delays telomere loss with age, promotes hair growth, improves skin and intestine barrier functions and wound healing, and increases lifespan (Flores et al., 2005; Gonzalez-Suarez et al., 2005; Tomas-Loba et al., 2008). Experiments involving the transient reintroduction of *mTERT* in adult *mTERT* null mice with telomere-induced stem-cell defects suggest that telomere-induced damage can be averted or reversed even in mature animals (Jaskelioff et al., 2010).

Interestingly, the stabilization of telomere length observed in older animals upon TERT overexpression is not sufficient to increase the number of HSC serial transplantations that still allow the rescue of viability in irradiated recipient animals (Allsopp et al., 2003b). In some genetic analyses, TERT appears to play a role in stem-cell function independent of its activity at the telomere. For example, the induction of hair follicle stem-cell proliferation by TERT overexpression does not require *TERC*, and *mTERT* influences the transcriptional regulation of a MYC and WNT-related developmental program even in early-generation *mTERT* null animals (Choi et al., 2008; Park et al., 2009; Sarin et al., 2005) (see below). These results imply that TERT may promote the proliferation of stem cells independently from its role in telomere elongation.

9.2.7 Telomerase and Telomere Length in Murine Development

Telomere integrity is vital for germline function. Successive generations of telomerase-null mice exhibit reduced litter sizes (Blasco et al., 1997; Erdmann et al., 2004; Herrera et al., 1999; Meznikova et al., 2009) because dysfunctional telomeres trigger male germ-cell apoptosis (Hemann et al., 2001a; Liu et al., 2004c) or female germ-cell arrest in early meiosis (Liu et al., 2004c). Furthermore, telomere dysfunction can severely impair chromosome pairing and synapsis, and reduces meiotic recombination (Liu et al., 2004c). Moreover, telomere dysfunction contributes to aberrant cleavage and preimplantation development of embryos derived from both *in vivo* and *in vitro* fertilization of sperm with oocytes (Liu et al., 2002a).

At different stages of murine development, telomere length is regulated by either telomerase or telomere recombination. In the early embryo cleavages, telomere lengthening occurs in oocytes as a result of increased telomere recombination. At the blastocyst stage, telomere recombination decreases and telomerase is activated (Liu et al., 2002a). An independent study illustrates that telomere lengthening occurs only at the transition from morula to blastocyst, which is mediated by telomerase (Schaetzlein et al., 2004). Nevertheless, a small population (5%) of murine ES cells derived from the inner cell mass of blastocysts retain a high incidence of telomere recombination (Wang et al., 2005a). Coincidentally, 5% of murine ES cells express Zscan4 (a zinc finger and SCAN domain containing 4) at a given time, and its expression influences telomerase-independent and telomere recombination-dependent telomere elongation (Zalzman et al., 2010). In the adult mouse, telomerase activity can be detected in several tissues (Blasco et al., 1997; Coviello-McLaughlin and Prowse, 1997; Liu et al., 2000) where telomere recombination remains low or undetectable (Morrish and Greider, 2009; Rhee et al., 2010; Wang et al., 2005a,b). However, short telomeres can initiate telomere-recombination events even in the presence of telomerase, and these events participate in telomere maintenance without significantly increasing telomere length in mice (Morrish and Greider, 2009). Similar findings are also observed in telomerase positive human cells with telomere dysfunction (Brault and Autexier, 2011).

Not all development roles of TERT depend on telomere maintenance. TERT interacts with BRG1, a SWI/SNF-related chromatin remodeling protein that modulates expression of Wnt-dependent genes (Park et al., 2009). Although early-generation *TERT* null mice appear normal, they nonetheless display homeotic transformation, characterized by the loss of the thirteenth rib on one or both sides. This phenotype resembles *Wnt* mutant mice (Park et al., 2009), suggesting that TERT contributes to murine development by regulating the WNT pathway. Furthermore, inducible TERT overexpression enhances kerotinocyte proliferation and activates resting hair follicle stem cells, even when TERT is catalytically incapable of telomere addition (Choi et al., 2008). This transcriptional response to TERT overexpression resembles Myc and Wnt-induced gene-expression profiles (Choi et al., 2008). Collectively, these findings indicate that TERT may regulate Myc- or Wnt- pathways independent of telomere addition.

9.2.8 Telomerase and Telomeres in Nuclear Reprogramming

Recent technological advances have permitted the reprogramming of adult differentiated cells to a pluripotent state by somatic cell nuclear transfer (SCNT) or via the overexpression of specific transcription factors (reviewed in Yamanaka and Blau, 2010). Compared to somatic cells, reprogrammed pluripotent cells possess a low density of trimethylated histones H3K9 and H4K20, a high level of telomere-associated transcripts (TERRA), high levels of telomerase activity, and longer telomeres (reviewed in Marion and Blasco, 2010). During reprogramming via SCNT, somatic cells from several species have exhibited telomere lengthening (Betts et al., 2001; Clark et al., 2003; Lanza et al., 2000; Schaetzlein et al., 2004; Tian et al., 2000; Wakayama et al., 2000). Telomerase upregulation is associated with telomere lengthening during both murine and cattle embryogenesis (Schaetzlein et al., 2004). Similarly, telomerase is activated in murine induced pluripotent stem (iPS) cells and is responsible for telomere elongation (Marion et al., 2009b). A recent encouraging report demonstrates that telomere elongation and *TERC* upregulation can be achieved in iPS cells from human DC patients (Agarwal et al., 2010), which opens the possibility of iPS cells as an autologous therapy for DC patients.

The reprogramming efficiency of cells derived from late-generation telomerase null mice is dramatically decreased, but reintroduction of telomerase or blockage of the p53-mediated DNA-damage response can restore this defect (Marion et al., 2009a,b). These observations support the notion that telomere integrity and intact p53 function is essential for iPS cell generation and genome integrity. Besides alterations in telomerase activity and telomere length, murine iPS cells also show reduced heterochromatic marks and increased telomere transcription (Marion et al., 2009b). Thus, telomere status in murine iPS cells resembles that of murine ES cells.

9.3 SHELTERIN

In mammals, telomeres are bound by shelterin, a six subunit complex composed of the telomere repeat binding factors TRF1, TRF2, POT1, and their associated proteins

RAP1, TPP1, and TIN2 (Fig. 9.1 and Palm and de Lange, 2008; Xin et al., 2008). TRF1 and TRF2 bind to duplex telomeric DNA and anchor shelterin along the telomere repeats (Bilaud et al., 1997; Broccoli et al., 1997; Chong et al., 1995). POT1 binds to the single-stranded G-rich DNA overhang (Baumann and Cech, 2001; Yang et al., 2005; Ye et al., 2004b). TIN2 serves as the hub of the complex linking TRF1 and TRF2 (Kim et al., 2004; Liu et al., 2004a; Ye et al., 2004a) while also recruiting POT1 to the complex via TPP1 (Houghtaling et al., 2004; Liu et al., 2004a; Ye et al., 2004b). RAP1 associates with the telomere protein complex through its association with TRF2 (Li et al., 2000; O'Connor et al., 2004). Telomere protein complexes and protein components are also found in other organisms, demonstrating the importance of these telomere-specific proteins to telomere function (Longhese, 2008; see also Chapters 6 and 7).

In humans, the shelterin proteins regulate telomere length and telomere capping. POT1 and TPP1 form a heterodimer and modulate telomerase function by negatively regulating telomerase access to the 3′-overhang or serving as a telomerase processivity factor for telomere extension (Latrick and Cech, 2010; Wang et al., 2007; Xin et al., 2007; Zaug et al., 2010). Although TRF1 and TRF2 coat double-stranded telomere DNA and are closely related, structural analyses argue that they recruit different proteins to telomeres to facilitate distinctive functions (Chen et al., 2008; Kim et al., 2009). TRF1 negatively controls telomere length *in cis* via a length-dependent "counting" mechanism, in which an interaction of POT1/TPP1 with TRF1 allows communication between the double-stranded telomeres and telomerase (Loayza and De Lange, 2003). TRF2 and POT1 are vital for the formation or regulation of the telomeric t-loop structure and mask chromosome ends from evoking a DNA damage response or undergoing recombination (reviewed in de Lange, 2005).

Murine TRF1, TRF2, POT1a (one of two murine POT1 paralogs), TIN2, and TPP1 are essential for early embryonic development (Chiang et al., 2004; Hockemeyer et al., 2006; Karlseder et al., 2003; Vlangos et al., 2009; Wu et al., 2006), demonstrating the vital roles of shelterin proteins in murine development. Several important advances in delineating shelterin proteins function have been achieved using murine conditional knockout models. These genetic models allow the identification of shelterin components in: (1) the regulation of telomere length by ensuring efficient telomere replication; (2) the protection of ends from an ATM and ATR-dependent DNA damage response; (3) restraining telomere recombination, NHEJ and HR; (4) protecting telomeres from damage-induced bypass of mitosis and consequent aneuploidy, (5) an association with human premature aging syndromes, (6) recruitment of telomerase to telomeres for telomere addition, and (7) the regulation of subtelomeric silencing, genomewide transcription, and NF-kB signaling.

9.3.1 Shelterin Regulates Telomere Replication

Unusual DNA structures at telomeres (i.e., t-loop and the potential for G-quadruplexes) may represent an obstacle for DNA replication at telomeres. *In vitro*, telomeric DNA is a poor substrate for replication and Okazaki fragment synthesis (Ohki et al., 2001). Thus, telomeres pose a challenge to the DNA-replication machinery. In *S. pombe*, deletion of the telomere-binding protein Taz1 leads to replication fork stalling at the telomere (Miller et al., 2006). In mice, efficient telomere replication also requires shelterin components.

Upon loss of *Trf1* in MEFs, telomeres activate an ATM or S-phase-dependent ATR response (Martinez et al., 2009b; Sfeir et al., 2009). *Trf1*-deficient MEFs and ES cells exhibit an increased level of aberrant telomeres with multitelomeric signals, referred to as fragile telomeres (Iwano et al., 2004; Martinez et al., 2009b; Okamoto et al., 2008; Sfeir et al., 2009). ATR knockdown or partial inhibition of DNA synthesis by treatment with low doses of aphidicolin further induce an increase in the frequency of fragile telomeres (Martinez et al., 2009b; Sfeir et al., 2009). These properties resemble interstitial fragile sites (Murga et al., 2009), which are genomic regions where replication forks also stall and collapse (Glover et al., 2007). Indeed, telomeres in *Trf1*-deficient MEFs display telomere-replication defects. Normally, the replication fork progresses from subtelomeric sites into the telomeric DNA, and occasionally replication initiation occurs within telomere repeats; however, in *Trf1*-deficient cells the fork has a greater tendency to stall when it encounters telomeric DNA, and the replication efficiency of telomeric DNA is also diminished (Sfeir et al., 2009). Collectively, these data indicate that mammalian telomeres resemble fragile sites and require TRF1 for efficient replication. The fact that BLM and REL1 knockdown further increase the frequency of aberrant telomeres in *Trf11* deficient cells implies that these helicases may function to remove aberrant telomere structures that can otherwise hamper telomere replication (Sfeir et al., 2009). The mammalian homolog of the RecQ helicase Pif1, which plays a critical role in DNA replication and telomere homeostasis in yeast, is dispensable for these functions in mice, suggesting functional redundancy or divergence in the role of helicases during replication of the mammalian telomere (Snow et al., 2007).

Tpp1 and *Rap1* deficiency also lead to telomere fragility in MEFs, characterized by telomeres with multitelomere signals that are enhanced upon aphidicolin treatment (Martinez et al., 2010; Tejera et al., 2010). In addition, *Tpp1*-deficient MEFs display increased colocalization of the replication factor PCNA and BrdU at telomeres, suggesting perturbed DNA replication at telomeres. These results support the role of TPP1 and RAP1 in preventing telomere fragility and regulating telomere replication.

9.3.2 Shelterin Shields Ends from an ATM and/or ATR-Dependent DNA Damage Response

In human cells, loss of TRF2 induces an ATM-dependent DNA damage response, involving the phosphorylation of Chk2, p53, and p21-dependent cell-cycle arrest, as well as the localization of γH2AX and 53BP1 to telomeres (Celli and de Lange, 2005; Karlseder et al., 1999; Takai et al., 2003). Induction of a DNA damage response is also observed in MEFs upon loss of *Trf2*, and this response is abolished when *Trf2* is disrupted in *Atm* null MEFs (Denchi and de Lange, 2007). These results show that ATM, but not ATR, is involved in the damage response induced at telomeres upon *Trf2* loss.

POT1 is highly conserved among different species, including fission yeast, plants, *C. elegans*, and vertebrates. Humans possess a single POT1 gene, and diminished POT1 leads to a transient accumulation of γ-H2AX and 53BP1 at telomeres in the G1 phase of the cell cycle (Hockemeyer et al., 2005) and an ATR-dependent DNA damage response (Denchi and de Lange, 2007). Rodents, on the other hand, possess two POT1 orthologs, POT1a and POT1b. Deletion of *Pot1a*, not *Pot1b*, results in embryonic lethality (He et al., 2006; Hockemeyer et al., 2006; Palm et al., 2009; Wu et al., 2006). MEFs deficient in *Pot1a*/*Pot1b* display a high frequency of γ-H2AX and 53BP1 foci at telomeres (Hockemeyer et al., 2006; Wu et al., 2006), including induction of ATR kinase and phosphorylation of Chk2 (Denchi and de Lange, 2007). Upon ATR knockdown, the formation of telomere-induced foci (TIFs) and phosphorylation of Chk2 in *Pot1a*/*Pot1b* null MEFs is significantly reduced (Denchi and de Lange, 2007). Taken together, these data demonstrate that TRF2 and POT1 function independently to repress the activation of the ATM and ATR DNA damage response, respectively.

Trf1 and *Tpp1* deletion can also induce p53 and pRb-mediated senescence and activation of the ATM and ATR checkpoint kinases, CHK1 and CHK2 respectively, leading to γ-H2AX and 53BP1 telomeric foci (Martinez et al., 2009b; Okamoto et al., 2008; Tejera et al., 2010). Upon treatment with an ATM inhibitor, telomeric foci are significantly reduced, and an additive suppressive effect was observed after ATM/ATR inhibition (Martinez et al., 2009b; Tejera et al., 2010). These results suggest that TRF1 or TPP1 deletion can activate both ATM and ATR DNA damage signaling pathways.

9.3.3 Shelterin Restrains Telomere Recombination, NHEJ and HR

Trf1-deficient murine cells exhibit fewer telomere fusions (Iwano et al., 2004; Martinez et al., 2009b; Sfeir et al., 2009), while deletion of *Trf2* gene leads to an increased incidence of telomere fusions that can be rescued by *Atm* deficiency (Denchi and de Lange, 2007). Although *Pot1a*/*Pot1b* null MEFs have a weak telomere fusion phenotype, *Pot1a* knockdown in *Trf2*/*Atm* null MEFs results in frequent telomere fusions, but further inhibition of ATR partially suppresses the fusion phenotype (Denchi and de Lange, 2007). Thus, NHEJ of telomeres after *Trf2* deletion involves signaling by ATM or ATR. Disruption of *Trf2* in *Ku70*-deficient MEFs or deletion of *Pot1a*/*Pot1b* in MEFs promotes HR between telomere sister chromatids, referred to telomere sister chromatid exchange (T-SCE) (Celli et al., 2006; He et al., 2006; Palm et al., 2009; Wu et al., 2006).

In shelterin, RAP1 interacts with TRF2 directly. However, deletion of *Rap1* did not induce TIF or telomere fusions in MEFs. In addition, a TRF2 mutant that does not bind RAP1 ($TRF2^{\Delta Rap1}$) can repress TIF and telomere fusions in TRF2 deficient MEFs (Sfeir et al., 2010). Thus, RAP1 does not appear to be required for the repression of the ATM signal or NHEJ via TRF2. On the other hand, disruption of *Rap1* in *Ku70* null MEFs induces T-SCEs despite the absence of TIF. Introduction of $TRF2^{\Delta Rap1}$ into *Trf2*/*Ku70* deficient MEFs fails to repress T-SCEs (Sfeir et al., 2010). In addition, a conditional knockout strain that eliminates the telomere localization domain and the nuclear localization signal of Rap1 exhibits increased T-SCEs (Martinez et al., 2010). Collectively, these data indicate that TRF2 predominantly represses ATM signaling and NHEJ at telomeres, while RAP1 and POT1a/POT1b repress HR at telomeres. Further, TRF2 and POT1a/b protect telomeres from triggering DNA damage signaling and telomere recombination, whereas RAP1 suppresses HR at telomeres in the absence of damage signaling.

Another shelterin component, TPP1, interacts with POT1 and bridges POT1 and the other shelterin components at double-stranded telomeric DNA. *Tpp1* conditional knockout MEFs or mice harboring a hypomorphic *Tpp1* allele exhibit TIFs and increased SCE (Else et al., 2007; Hockemeyer et al., 2007; Tejera et al., 2010). Thus, in addition to its role in telomerase recruitment and processivity (see below), TPP1 is also required for telomere capping.

9.3.4 Persistent Telomere Damage and p53 Deficiency Induces Endoreplication

One hallmark of tumors is genomic instability, including chromosome aneuploidy. Normally, human somatic cells possess a limited proliferative capacity in culture, which correlates with telomere attrition (reviewed in Shay and Wright, 2005). Telomere erosion commonly limits cell survival following extensive proliferation and thereby suppresses malignant transformation. However, in exceptional cases, such as in combination with p53 deficiency, loss of telomere integrity facilitates chromosomal instability and promotes tumor formation (Chin et al., 1999). Recently, Davoli et al. (2010) explored the possible mechanism of how persistent telomere damage can lead to polyploidy in cells with defective p53 function.

Telomere damage due to *Pot1a/Pot1b* deficiency can induce a persistent ATR kinase response (Denchi and de Lange, 2007). Polyploidy is increased upon deletion of *Pot1a/Pot1b* in SV40 large T-expressing MEFs (thus defective in p53, which would otherwise block the entry of tetraploid cells into S phase). Inhibition of ATM and ATR diminishes the activation of CHK1 and CHK2 and decreases the fraction of polyploid cells (Davoli et al., 2010). Similarly, *Tpp1* deletion in murine epidermis can also lead to a persistent telomere damage response and increased polyploidy (Tejera et al., 2010). Thus, DNA damage signals involving an ATM/ATR kinase response are required for increased polyploidy in response to telomere dysfunction.

Pot1a/Pot1b deficient cells become polyploid by entering S phase without progression through mitosis, referred as endoreduplication. In *Pot1a/Pot1b*-deficient cells, persistent ATM/ATM kinase signaling prevents activation of Cdk1/CyclinB, which is otherwise required for entry into mitosis. In addition, the DNA replication inhibitor geminin is degraded, while Cdt1 is reexpressed; the former prevents rereplication in G2 and the latter is required for origin licensing. Consequently *Pot1a/Pot1b* deficient cells reenter S phase and become tetraploid (Davoli et al., 2010). Thus, telomere dysfunction may contribute to tumorigenesis by permitting polyploidization.

9.3.5 Murine Models with Abnormal Shelterin Function Mimic Human Diseases

Mutations in the telomerase components and associated proteins, TERT, TERC, DKC, NHP2, NOP20, and the shelterin protein TIN2 are linked to the pathogenesis of DC (Armanios et al., 2005; Marrone et al., 2007; Savage et al., 2008; Vulliamy et al., 2001; Walne et al., 2007). In mice, deficiencies in other shelterin components such as Pot1b, Trf1, Rap1, Tpp1, and Trf2 also cause phenotypes similar to human DC patients.

Unlike other telomere repeat-binding factors, *Pot1b* deficiency does not result in murine embryonic lethality and infertility, but results in an increased 3′-overhang and telomere shortening, possibly due to extensive C-strand degradation (Hockemeyer et al., 2008) and diminished recruitment of telomerase to telomeres (He et al., 2009). Telomerase RNA disruption in a *Pot1b*-deficient background exacerbates telomere shortening and dysfunction, resulting in an ATR-dependent DNA damage response and telomere fusions. *Pot1b/mTERC* null pups die a few days after birth, however $Pot1b^{-/-}$ $mTERC^{+/-}$ mice can survive to adulthood, but exhibit increased cell apoptosis, testicular atrophy, and germ-cell depletion (Hockemeyer et al., 2008). In addition, these mice have characteristic symptoms of DC, including anemia, leucopenia, thrombocytopenia, progressive bone marrow failure, cutaneous phenotypes, for example, hyperpigmentation and abnormal nails, and a marked reduction in life span (Hockemeyer et al., 2008). Thus, it may be that shorter telomeres and loss of end protection are required in order to achieve full DC disease penetrance. Similarly, transgenic mice with increased TRF1 or TRF2 expression in epithelia (namely K5-TRF1 or K5-TRF2 mice) display XPF nuclease-dependent severe telomere shortening, a DNA damage response at telomeres, and telomere fusions. The skin exhibits premature dermal deterioration, hyperpigmentation, and an increased incidence of skin cancer (Munoz et al., 2005, 2009; Stout and Blasco, 2009). In combination with telomerase deficiency, telomere-related abnormalities, and TRF2-induced epithelia carcinogenesis are markedly accelerated (Blanco et al., 2007). Similar to the murine models of telomerase haploinsufficiency, these results demonstrate that exposure of telomeres to extensive degradation can lead to pathological defects that mimic human disorders, and underscore the importance of telomere repeat binding factors in protecting telomeres against nuclease-mediated telomere degradation.

Mice with a conditionally targeted *Trf1* gene in stratified epithelia (namely $Trf1^{\Delta/\Delta}$ K5-Cre mice) are viable, but show a severe perinatal mortality (Martinez et al., 2009b). Unlike $Pot1b^{-/-}/mTERC^{+/-}$, K5-TRF1, or K5-TRF2 mice, $Trf1^{\Delta/\Delta}$ K5-Cre mice exhibit normal telomere lengths, yet the mutant mice exhibit persistent activation of a telomere-induced DNA damage response, including γ-H2AX and 53BP1 foci formation at telomeres and induction of p53/Rb-dependent senescence. In addition, the mice display severe skin hyperpigmentation and epithelial atrophies. *p53* deficiency rescues these proliferation defects, but longer-lived $Trf1^{\Delta/\Delta}$ K5-Cre $p53^{-/-}$ mice develop additional epithelial abnormalities, e.g. nail dystrophy and oral leukoplakia. Long-lived $Trf1^{\Delta/\Delta}$ K5-Cre $p53^{-/-}$ mice also develop a higher incidence of squamous cell carcinomas. Thus, TRF1 dysfunction can lead to the activation of a DNA damage response at telomeres and the development of skin neoplasms even in the absence of telomere shortening. It is possible that TRF1 may act as a tumor suppressor by preventing telomere-induced genome instability.

Mice with conditionally targeted *Tpp1* and *Rap1* in stratified epithelia (*Tpp1* and *Rap1*$^{\Delta/\Delta}$ K5-Cre mice, respectively) possess short telomere lengths and persistent activation of a telomere-induced DNA damage response (Martinez et al., 2010; Tejera et al., 2010). These mutant mice also display skin hyperpigmentation and/or severe defects in hair follicle morphogenesis. Abolishing p53 function can rescue these abnormalities in *Tpp1*$^{\Delta/\Delta}$ K5-Cre animals (Tejera et al., 2010). Thus, TPP1 and RAP1 play an essential role in telomere capping and hair follicle development in mice.

9.3.6 Shelterin in Telomerase Recruitment

TPP1 interacts with telomerase to recruit telomerase in human cells (Abreu et al., 2010; Xin et al., 2007). *In vitro*, TPP1 stimulates telomerase processivity in complex with POT1, and via an interaction between TPP1 and the N-terminal domain of TERT (Latrick and Cech, 2010; Zaug et al., 2010). *Tpp1* deletion does not affect the binding of other shelterin components to telomeres, but results in reduced TERT binding at telomeres in MEFs (Abreu et al., 2010; Xin et al., 2007). *Tpp1*-deficient MEFs and epidermis also exhibit shorter telomeres than the corresponding wild-type tissues. Although telomeres undergo a net elongation during iPS cell generation, telomere elongation is abolished in iPS cell clones generated from *Tpp1* null MEFs (Tejera et al., 2010). These data support a role for TPP1 in promoting telomere addition via the recruitment of telomerase to telomeres.

9.3.7 RAP1 Regulates Subtelomeric Silencing, Genome-Wide Transcription, and NF-kB Signaling

The predominant roles of shelterin are to cap and protect telomeres from DNA damage and to regulate telomerase access, however, not all its roles are telomere-restricted. In budding yeast, ScRap1p directly binds to telomeres and regulates telomere length and function (Conrad et al., 1990; Ray and Runge, 1999). ScRap1p is also found at nontelomere sites where it regulates genome-wide gene transcription and gene silencing at subtelomeric region (Kurtz and Shore, 1991; Sussel and Shore, 1991). Unlike its yeast homolog, mammalian RAP1 does not directly bind to telomere DNA, but is recruited to telomeres via its interaction with TRF2 (Li et al., 2000). Nevertheless, RAP1 inhibits telomere recombination in mice (Martinez et al., 2010; Sfeir et al., 2010). Besides telomeres, RAP1–DNA binding sites are also found throughout the genome, but predominately at subtelomeric regions and near gene-coding regions similar to those of known transcription factors. The recruitment of RAP1 to these sites is dependent upon TRF2. *Rap1* deficiency elicits a more pronounced overexpression of genes within the 3′ subtelomeric region, and also leads to dysregulation of genes involved in cell adhesion, malignant transformation, and metabolism (Martinez et al., 2010). Thus, mammalian RAP1 appears to control both telomere function and gene expression via its binding to telomeric and nontelomeric sites.

RAP1 also modulates NF-kB-mediated signaling (Teo et al., 2010). A significant level of RAP1 can be detected in the cytoplasm where TRF2 is undetectable. In this context, RAP1 activates NF-kB by interacting with IKKs and participates in the recruitment of IKKs to the p65 subunit of NF-kB and p65 phosphorylation by IKKs. Rap1 heterozygous mice exhibit a lower induction of NF-kB target-genes in response to increasing doses of lipopolysaccharide (LPS) and are more resistant to endotoxic shock, reflecting a defective activation of NF-kB-dependent cytokines (Teo et al., 2010). These results support an extratelomeric role of RAP1 in the regulation of stress-induced cell signaling.

9.4 CONCLUDING REMARKS

Many inbred murine strains have considerably longer telomeres and higher levels of telomerase activity than humans. In addition, murine and human cells can respond differently to genome instability and loss of checkpoint function (Smogorzewska and de Lange, 2002). Nevertheless, numerous studies of mice deficient in telomere- or telomerase-associated proteins have firmly established the relevance of murine models to human disease, and have contributed seminal discoveries concerning the role of telomere homeostasis in mammalian development, premature aging, and cancer. Studies of transgenic murine models have also revealed tissue-specific consequences of perturbation in telomerase or shelterin, which would not have been possible in unicellular models. Moreover, these genetic models have shown that telomerase and shelterin components may function at extratelomeric sites. Challenges and opportunities that remain include targeting other as yet unexplored tissues for their response to telomere attrition, and the use of ever-more precise and targeted genetic approaches to knock in or conditionally express specific alleles of telomerase or shelterin components implicated in human diseases. We can be confident that the mouse as a model for telomere research has a long future.

ACKNOWLEDGMENT

Yie Liu is supported by the Intramural Research Program of the NIA, National Institutes of Health. Lea Harrington acknowledges past and present support from the Howard Hughes Medical Institute, NIA/NIH, Wellcome Trust, and Medical Research Council UK.

REFERENCES

Abreu E, Aritonovska E, Reichenbach P, Cristofari G, Culp B, Terns RM, Lingner J, Terns MP. (2010) TIN2-tethered TPP1 recruits human telomerase to telomeres *in vivo*. *Mol. Cell, Biol.* **30**(12): 2971–2982.

Agarwal S, Loh YH, McLoughlin EM, Huang J, Park IH, Miller JD, Huo H, Okuka M, Dos Reis RM, Loewer S, Ng HH, Keefe DL, Goldman FD, Klingelhutz AJ, Liu L, Daley GQ. (2010) Telomere elongation in induced pluripotent stem cells from dyskeratosis congenita patients. *Nature*. **464**(7286): 292–296.

Allsopp RC, Morin GB, DePinho R, Harley CB, Weissman IL. (2003a) Telomerase is required to slow telomere shortening and extend replicative lifespan of HSCs during serial transplantation. *Blood*. **102**(2): 517–520.

Allsopp RC, Morin GB, Horner JW, DePinho R, Harley CB, Weissman IL. (2003b) Effect of TERT over-expression on the long-term transplantation capacity of hematopoietic stem cells. *Nat. Med.* **9**(4): 369–371.

Argilla D, Chin K, Singh M, Hodgson JG, Bosenberg M, de Solorzano CO, Lockett S, DePinho RA, Gray J, Hanahan D. (2004) Absence of telomerase and shortened telomeres have minimal effects on skin and pancreatic carcinogenesis elicited by viral oncogenes. *Cancer Cell*. **6**(4): 373–385.

Armanios M. (2009) Syndromes of telomere shortening. *Annu. Rev. Genomics Hum. Genet.* **10**: 45–61.

Armanios M, Chen JL, Chang YP, Brodsky RA, Hawkins A, Griffin CA, Eshleman JR, Cohen AR, Chakravarti A, Hamosh A, Greider CW. (2005) Haploinsufficiency of telomerase reverse transcriptase leads to anticipation in autosomal dominant dyskeratosis congenita. *Proc. Natl. Acad. Sci. USA*. **102**(44): 15960–15964.

Bailey SM, Meyne J, Chen DJ, Kurimasa A, Li GC, Lehnert BE, Goodwin EH. (1999) DNA double-strand break repair proteins are required to cap the ends of mammalian chromosomes. *Proc. Natl. Acad. Sci. USA*. **96**(26): 14899–14904.

Baumann P, Cech TR. (2001) Pot1, the putative telomere end-binding protein in fission yeast and humans. *Science*. **292**(5519): 1171–1175.

Betts D, Bordignon V, Hill J, Winger Q, Westhusin M, Smith L, King W. (2001) Reprogramming of telomerase activity and rebuilding of telomere length in cloned cattle. *Proc. Natl. Acad. Sci. USA*. **98**(3): 1077–1082.

Bianchi A, Shore D. (2008) How telomerase reaches its end: mechanism of telomerase regulation by the telomeric complex. *Mol. Cell.* **31**(2): 153–165.

Bilaud T, Brun C, Ancelin K, Koering CE, Laroche T, Gilson E. (1997) Telomeric localization of TRF2, a novel human telobox protein. *Nat. Genet.* **17**(2): 236–239.

Blackburn EH. (2001) Switching and signaling at the telomere. *Cell*. **106**(6): 661–673.

Blanco R, Munoz P, Flores JM, Klatt P, Blasco MA. (2007) Telomerase abrogation dramatically accelerates TRF2-induced epithelial carcinogenesis. *Genes Dev*. **21**(2): 206–220.

Blasco MA. (2005) Mice with bad ends: mouse models for the study of telomeres and telomerase in cancer and aging. *EMBO J*. **24**(6): 1095–1103.

Blasco MA. (2007) Telomere length, stem cells and aging. *Nat. Chem. Biol.* **3**(10): 640–649.

Blasco MA, Lee HW, Hande MP, Samper E, Lansdorp PM, DePinho RA, Greider CW. (1997) Telomere shortening and tumor formation by mouse cells lacking telomerase RNA. *Cell*. **91**(1): 25–34.

Brault ME, Autexier C. (2011) Telomeric recombination induced by dysfunctional telomeres. *Mol. Biol. Cell.* **22**(2): 179–188.

Broccoli D, Smogorzewska A, Chong L, de Lange T. (1997) Human telomeres contain two distinct Myb-related proteins, TRF1 and TRF2. *Nat. Genet.* **17**(2): 231–235.

Calado RT. (2009) Telomeres and marrow failure. *Hematol. Am. Soc. Hematol. Educ. Program*: 338–343.

Celli GB, de Lange T. (2005) DNA processing is not required for ATM-mediated telomere damage response after TRF2 deletion. *Nat. Cell Biol.* **7**(7): 712–718.

Celli GB, Denchi EL, de Lange T. (2006) Ku70 stimulates fusion of dysfunctional telomeres yet protects chromosome ends from homologous recombination. *Nat. Cell Biol.* **8**(8): 885–890.

Chang S, Multani AS, Cabrera NG, Naylor ML, Laud P, Lombard D, Pathak S, Guarente L, DePinho RA. (2004) Essential role of limiting telomeres in the pathogenesis of Werner syndrome. *Nat. Genet.* **36**(8): 877–882.

Chen Y, Yang Y, van Overbeek M, Donigian JR, Baciu P, de Lange T, Lei M. (2008) A shared docking motif in TRF1 and TRF2 used for differential recruitment of telomeric proteins. *Science*. **319**(5866): 1092–1096.

Chiang YJ, Calado RT, Hathcock KS, Lansdorp PM, Young NS, Hodes RJ. (2010) Telomere length is inherited with resetting of the telomere set-point. *Proc. Natl. Acad. Sci. USA*. **107**(22): 10148–10153.

Chiang YJ, Kim SH, Tessarollo L, Campisi J, Hodes RJ. (2004) Telomere-associated protein TIN2 is essential for early embryonic development through a telomerase-independent pathway. *Mol. Cell. Biol.* **24**(15): 6631–6634.

Chin L, Artandi SE, Shen Q, Tam A, Lee SL, Gottlieb GJ, Greider CW, DePinho RA. (1999) p53 deficiency rescues the adverse effects of telomere loss and cooperates with telomere dysfunction to accelerate carcinogenesis. *Cell.* **97**(4): 527–538.

Choi J, Southworth LK, Sarin KY, Venteicher AS, Ma W, Chang W, Cheung P, Jun S, Artandi MK, Shah N, Kim SK, Artandi SE. (2008) TERT promotes epithelial proliferation through transcriptional control of a Myc- and Wnt-related developmental program. *PLoS Genet.* **4**(1): e10.

Chong L, van Steensel B, Broccoli D, Erdjument-Bromage H, Hanish J, Tempst P, de Lange T. (1995) A human telomeric protein. *Science*. **270**(5242): 1663–1667.

Choudhury AR, Ju Z, Djojosubroto MW, Schienke A, Lechel A, Schaetzlein S, Jiang H, Stepczynska A, Wang C, Buer J, Lee HW, von Zglinicki T, Ganser A, Schirmacher P, Nakauchi H, Rudolph KL. (2007) Cdkn1a deletion improves stem cell function and lifespan of mice with dysfunctional telomeres without accelerating cancer formation. *Nat. Genet.* **39** (1): 99–105.

Clark AJ, Ferrier P, Aslam S, Burl S, Denning C, Wylie D, Ross A, de Sousa P, Wilmut I, Cui W. (2003) Proliferative lifespan is conserved after nuclear transfer. *Nat. Cell Biol.* **5**(6): 535–538.

Cohen SB, Graham ME, Lovrecz GO, Bache N, Robinson PJ, Reddel RR. (2007) Protein composition of catalytically active human telomerase from immortal cells. *Science*. **315**(5820): 1850–1853.

Conrad MN, Wright JH, Wolf AJ, Zakian VA. (1990) RAP1 protein interacts with yeast telomeres *in vivo*: overproduction alters telomere structure and decreases chromosome stability. *Cell.* **63**(4): 739–750.

Coviello-McLaughlin GM, Prowse KR. (1997) Telomere length regulation during postnatal development and ageing in *Mus spretus*. *Nucl. Acids Res.* **25**(15): 3051–3058.

Crabbe L, Jauch A, Naeger CM, Holtgreve-Grez H, Karlseder J. (2007) Telomere dysfunction as a cause of genomic instability in Werner syndrome. *Proc. Natl. Acad. Sci. USA*. **104**(7): 2205–2210.

d'Adda di Fagagna F, Reaper PM, Clay-Farrace L, Fiegler H, Carr P, Von Zglinicki T, Saretzki G, Carter NP, Jackson SP. (2003) A DNA damage checkpoint response in telomere-initiated senescence. *Nature*. **426**(6963): 194–198.

Davoli T, Denchi EL, de Lange T. (2010) Persistent telomere damage induces bypass of mitosis and tetraploidy. *Cell*. **141**(1): 81–93.

de Lange T. (2005) Shelterin: the protein complex that shapes and safeguards human telomeres. *Genes Dev*. **19**(18): 2100–2110.

Denchi EL, de Lange T. (2007) Protection of telomeres through independent control of ATM and ATR by TRF2 and POT1. *Nature*. **448**(7157): 1068–1071.

Dragon F, Pogacic V, Filipowicz W. (2000) *In vitro* assembly of human H/ACA small nucleolar RNPs reveals unique features of U17 and telomerase RNAs. *Mol. Cell. Biol.* **20**(9): 3037–3048.

Du X, Shen J, Kugan N, Furth EE, Lombard DB, Cheung C, Pak S, Luo G, Pignolo RJ, DePinho RA, Guarente L, Johnson FB. (2004) Telomere shortening exposes functions for the mouse Werner and Bloom syndrome genes. *Mol. Cell. Biol.* **24**(19): 8437–8446.

Else T, Theisen BK, Wu Y, Hutz JE, Keegan CE, Hammer GD, Ferguson DO. (2007) Tpp1/Acd maintains genomic stability through a complex role in telomere protection. *Chrom. Res.* **15**(8): 1001–1013.

Erdmann N, Liu Y, Harrington L. (2004) Distinct dosage requirements for the maintenance of long and short telomeres in mTert heterozygous mice. *Proc. Natl. Acad. Sci. USA*. **101**(16): 6080–6085.

Espejel S, Klatt P, Menissier-de Murcia J, Martin-Caballero J, Flores JM, Taccioli G, de Murcia G, Blasco MA. (2004) Impact of telomerase ablation on organismal viability, aging, and tumorigenesis in mice lacking the DNA repair proteins PARP-1, Ku86, or DNA-PKcs. *J. Cell. Biol.* **167**(4): 627–638.

Farazi PA, Glickman J, Horner J, Depinho RA. (2006) Cooperative interactions of p53 mutation, telomere dysfunction, and chronic liver damage in hepatocellular carcinoma progression. *Cancer Res*. **66**(9): 4766–4773.

Feldser D, Strong MA, Greider CW. (2006) Ataxia telangiectasia mutated (Atm) is not required for telomerase-mediated elongation of short telomeres. *Proc. Natl. Acad. Sci. USA*. **103**(7): 2249–2251.

Feldser DM, Greider CW. (2007) Short telomeres limit tumor progression *in vivo* by inducing senescence. *Cancer Cell*. **11**(5): 461–469.

Flores I, Blasco MA. (2009) A p53-dependent response limits epidermal stem cell functionality and organismal size in mice with short telomeres. *PLoS One*. **4**(3): e4934.

Flores I, Canela A, Vera E, Tejera A, Cotsarelis G, Blasco MA. (2008) The longest telomeres: a general signature of adult stem cell compartments. *Genes Dev*. **22**(5): 654–667.

Flores I, Cayuela ML, Blasco MA. (2005) Effects of telomerase and telomere length on epidermal stem cell behavior. *Science*. **309**(5738): 1253–1256.

Flores I, Evan G, Blasco MA. (2006) Genetic analysis of myc and telomerase interactions *in vivo*. *Mol. Cell. Biol.* **26**(16): 6130–6138.

Fu D, Collins K. (2007) Purification of human telomerase complexes identifies factors involved in telomerase biogenesis and telomere length regulation. *Mol. Cell*. **28**(5): 773–785.

Glover L, Alsford S, Beattie C, Horn D. (2007) Deletion of a trypanosome telomere leads to loss of silencing and progressive loss of terminal DNA in the absence of cell cycle arrest. *Nucl. Acids Res.* **35**(3): 872–880.

Gonzalez-Suarez E, Geserick C, Flores JM, Blasco MA. (2005) Antagonistic effects of telomerase on cancer and aging in K5-mTert transgenic mice. *Oncogene*. **24**(13): 2256–2270.

Gonzalez-Suarez E, Samper E, Ramirez A, Flores JM, Martin-Caballero J, Jorcano JL, Blasco MA. (2001) Increased epidermal tumors and increased skin wound healing in transgenic mice overexpressing the catalytic subunit of telomerase, mTERT, in basal keratinocytes. *EMBO J*. **20**(11): 2619–2630.

Goytisolo FA, Samper E, Edmonson S, Taccioli GE, Blasco MA. (2001) The absence of the DNA-dependent protein kinase catalytic subunit in mice results in anaphase bridges and in increased telomeric fusions with normal telomere length and G-strand overhang. *Mol. Cell. Biol.* **21**(11): 3642–3651.

Greenberg RA, Chin L, Femino A, Lee KH, Gottlieb GJ, Singer RH, Greider CW, DePinho RA. (1999) Short dysfunctional telomeres impair tumorigenesis in the INK4a(delta2/3) cancer-prone mouse. *Cell*. **97**(4): 515–525.

Gu BW, Bessler M, Mason PJ. (2008) A pathogenic dyskerin mutation impairs proliferation and activates a DNA damage response independent of telomere length in mice. *Proc. Natl. Acad. Sci. USA*. **105**(29): 10173–10178.

Guo N, Parry EM, Li LS, Kembou F, Lauder N, Hussain MA, Berggren PO, Armanios M. (2011) Short telomeres compromise β-cell signaling and survival. *PLoS One*. **6**(3): e17858.

Hao LY, Armanios M, Strong MA, Karim B, Feldser DM, Huso D, Greider CW. (2005) Short telomeres, even in the presence of telomerase, limit tissue renewal capacity. *Cell*. **123**(6): 1121–1131.

Hao LY, Strong MA, Greider CW. (2004) Phosphorylation of H2AX at short telomeres in T cells and fibroblasts. *J. Biol. Chem.* **279**(43): 45148–45154.

Hathcock KS, Hemann MT, Opperman KK, Strong MA, Greider CW, Hodes RJ. (2002) Haploinsufficiency of mTR results in defects in telomere elongation. *Proc. Natl. Acad. Sci. USA*. **99**(6): 3591–3596.

Hathcock KS, Jeffrey Chiang Y, Hodes RJ. (2005) *In vivo* regulation of telomerase activity and telomere length. *Immunol. Rev.* **205**: 104–113.

He H, Multani AS, Cosme-Blanco W, Tahara H, Ma J, Pathak S, Deng Y, Chang S. (2006) POT1b protects telomeres from end-to-end chromosomal fusions and aberrant homologous recombination. *EMBO J*. **25**(21): 5180–5190.

He H, Wang Y, Guo X, Ramchandani S, Ma J, Shen MF, Garcia DA, Deng Y, Multani AS, You MJ, Chang S. (2009) Pot1b deletion and telomerase haploinsufficiency in mice initiate an ATR-dependent DNA damage response and elicit phenotypes resembling dyskeratosis congenita. *Mol. Cell. Biol.* **29**(1): 229–240.

Hemann MT, Rudolph KL, Strong MA, DePinho RA, Chin L, Greider CW. (2001a) Telomere dysfunction triggers developmentally regulated germ cell apoptosis. *Mol. Biol. Cell*. **12**(7): 2023–2030.

Hemann MT, Strong MA, Hao LY, Greider CW. (2001b) The shortest telomere, not average telomere length, is critical for cell viability and chromosome stability. *Cell*. **107**(1): 67–77.

Herbig U, Jobling WA, Chen BP, Chen DJ, Sedivy JM. (2004) Telomere shortening triggers senescence of human cells through a pathway involving ATM, p53, and p21(CIP1), but not p16(INK4a). *Mol. Cell*. **14**(4): 501–513.

Herrera E, Martinez AC, Blasco MA. (2000) Impaired germinal center reaction in mice with short telomeres. *EMBO J*. **19**(3): 472–481.

Herrera E, Samper E, Martin-Caballero J, Flores JM, Lee HW, Blasco MA. (1999) Disease states associated with telomerase deficiency appear earlier in mice with short telomeres. *EMBO J.* **18**(11): 2950–2960.

Hockemeyer D, Daniels JP, Takai H, de Lange T. (2006) Recent expansion of the telomeric complex in rodents: Two distinct POT1 proteins protect mouse telomeres. *Cell.* **126**(1): 63–77.

Hockemeyer D, Palm W, Else T, Daniels JP, Takai KK, Ye JZ, Keegan CE, de Lange T, Hammer GD. (2007) Telomere protection by mammalian Pot1 requires interaction with Tpp1. *Nat. Struct. Mol. Biol.* **14**(8): 754–761.

Hockemeyer D, Palm W, Wang RC, Couto SS, de Lange T. (2008) Engineered telomere degradation models dyskeratosis congenita. *Genes Dev.* **22**(13): 1773–1785.

Hockemeyer D, Sfeir AJ, Shay JW, Wright WE, de Lange T. (2005) POT1 protects telomeres from a transient DNA damage response and determines how human chromosomes end. *EMBO J.* **24**(14): 2667–2678.

Houghtaling BR, Cuttonaro L, Chang W, Smith S. (2004) A dynamic molecular link between the telomere length regulator TRF1 and the chromosome end protector TRF2. *Curr. Biol.* **14**(18): 1621–1631.

Iwano T, Tachibana M, Reth M, Shinkai Y. (2004) Importance of TRF1 for functional telomere structure. *J. Biol. Chem.* **279**(2): 1442–1448.

Jaskelioff M, Muller FL, Paik JH, Thomas E, Jiang S, Adams AC, Sahin E, Kost-Alimova M, Protopopov A, Cadinanos J, Horner JW, Maratos-Flier E, Depinho RA. (2010) Telomerase reactivation reverses tissue degeneration in aged telomerase-deficient mice. *Nature.* **469**(7328): 102–106.

Ju Z, Jiang H, Jaworski M, Rathinam C, Gompf A, Klein C, Trumpp A, Rudolph KL. (2007) Telomere dysfunction induces environmental alterations limiting hematopoietic stem cell function and engraftment. *Nat. Med.* **13**(6): 742–747.

Karlseder J, Broccoli D, Dai Y, Hardy S, de Lange T. (1999) p53- and ATM-dependent apoptosis induced by telomeres lacking TRF2. *Science.* **283**(5406): 1321–1325.

Karlseder J, Kachatrian L, Takai H, Mercer K, Hingorani S, Jacks T, de Lange T. (2003) Targeted deletion reveals an essential function for the telomere length regulator Trf1. *Mol. Cell. Biol.* **23**(18): 6533–6541.

Kelleher C, Teixeira MT, Forstemann K, Lingner J. (2002) Telomerase: biochemical considerations for enzyme and substrate. *Trends Biochem. Sci.* **27**(11): 572–579.

Khoo CM, Carrasco DR, Bosenberg MW, Paik JH, Depinho RA. (2007) Ink4a/Arf tumor suppressor does not modulate the degenerative conditions or tumor spectrum of the telomerase-deficient mouse. *Proc. Natl. Acad. Sci. USA.* **104**(10): 3931–3936.

Kim H, Lee OH, Xin H, Chen LY, Qin J, Chae HK, Lin SY, Safari A, Liu D, Songyang Z. (2009) TRF2 functions as a protein hub and regulates telomere maintenance by recognizing specific peptide motifs. *Nat. Struct. Mol. Biol.* **16**(4): 372–379.

Kim SH, Beausejour C, Davalos AR, Kaminker P, Heo SJ, Campisi J. (2004) TIN2 mediates functions of TRF2 at human telomeres. *J. Biol. Chem.* **279**(42): 43799–43804.

Kuhlow D, Florian S, von Figura G, Weimer S, Schulz N, Petzke KJ, Zarse K, Pfeiffer AF, Rudolph KL, Ristow M. (2010) Telomerase deficiency impairs glucose metabolism and insulin secretion. *Aging* (Albany NY).

Kurtz S, Shore D. (1991) RAP1 protein activates and silences transcription of mating-type genes in yeast. *Genes Dev.* **5**(4): 616–628.

Lanza RP, Cibelli JB, Blackwell C, Cristofalo VJ, Francis MK, Baerlocher GM, Mak J, Schertzer M, Chavez EA, Sawyer N, Lansdorp PM, West MD. (2000) Extension of cell life-span and telomere length in animals cloned from senescent somatic cells. *Science.* **288**(5466): 665–669.

Latrick CM, Cech TR. (2010) POT1–TPP1 enhances telomerase processivity by slowing primer dissociation and aiding translocation. *EMBO J.* **29**(5): 924–933.

Laud PR, Multani AS, Bailey SM, Wu L, Ma J, Kingsley C, Lebel M, Pathak S, DePinho RA, Chang S. (2005) Elevated telomere-telomere recombination in WRN-deficient, telomere dysfunctional cells promotes escape from senescence and engagement of the ALT pathway. *Genes Dev.* **19**(21): 2560–2570.

Lee HW, Blasco MA, Gottlieb GJ, Horner JW 2nd, Greider CW, DePinho RA. (1998) Essential role of mouse telomerase in highly proliferative organs. *Nature.* **392**(6676): 569–574.

Li B, Oestreich S, de Lange T. (2000) Identification of human Rap1: implications for telomere evolution. *Cell.* **101**(5): 471–483.

Liu D, Safari A, O'Connor MS, Chan DW, Laegeler A, Qin J, Songyang Z. (2004a) PTOP interacts with POT1 and regulates its localization to telomeres. *Nat. Cell. Biol.* **6**(7): 673–680.

Liu L, Blasco M, Trimarchi J, Keefe D. (2002a) An essential role for functional telomeres in mouse germ cells during fertilization and early development. *Dev. Biol.* **249**(1): 74–84.

Liu L, DiGirolamo CM, Navarro PA, Blasco MA, Keefe DL. (2004b) Telomerase deficiency impairs differentiation of mesenchymal stem cells. *Exp. Cell. Res.* **294**(1): 1–8.

Liu L, Franco S, Spyropoulos B, Moens PB, Blasco MA, Keefe DL. (2004c) Irregular telomeres impair meiotic synapsis and recombination in mice. *Proc. Natl. Acad. Sci. USA.* **101**(17): 6496–6501.

Liu Y, Kha H, Ungrin M, Robinson MO, Harrington L. (2002b) Preferential maintenance of critically short telomeres in mammalian cells heterozygous for mTert. *Proc. Natl. Acad. Sci. USA.* **99**(6): 3597–3602.

Liu Y, Snow BE, Hande MP, Yeung D, Erdmann NJ, Wakeham A, Itie A, Siderovski DP, Lansdorp PM, Robinson MO, Harrington L. (2000) The telomerase reverse transcriptase is limiting and necessary for telomerase function *in vivo*. *Curr. Biol.* **10**(22): 1459–1462.

Loayza D, De Lange T. (2003) POT1 as a terminal transducer of TRF1 telomere length control. *Nature.* **423**(6943): 1013–1018.

Longhese MP. (2008) DNA damage response at functional and dysfunctional telomeres. *Genes Dev.* **22**(2): 125–140.

Marion RM, Blasco MA. (2010) Telomere rejuvenation during nuclear reprogramming. *Curr. Opin. Genet. Dev.*

Marion RM, Strati K, Li H, Murga M, Blanco R, Ortega S, Fernandez-Capetillo O, Serrano M, Blasco MA. (2009a) A p53-mediated DNA damage response limits reprogramming to ensure iPS cell genomic integrity. *Nature.* **460**(7259): 1149–1153.

Marion RM, Strati K, Li H, Tejera A, Schoeftner S, Ortega S, Serrano M, Blasco MA. (2009b) Telomeres acquire embryonic stem cell characteristics in induced pluripotent stem cells. *Cell. Stem Cell.* **4**(2): 141–154.

Marrone A, Walne A, Tamary H, Masunari Y, Kirwan M, Beswick R, Vulliamy T, Dokal I. (2007) Telomerase reverse-transcriptase homozygous mutations in autosomal recessive dyskeratosis congenita and Hoyeraal–Hreidarsson syndrome. *Blood.* **110**(13): 4198–4205.

Martinez P, Siegl-Cachedenier I, Flores JM, Blasco MA. (2009a) MSH2 deficiency abolishes the anticancer and pro-aging activity of short telomeres. *Aging Cell.* **8**(1): 2–17.

Martinez P, Thanasoula M, Carlos AR, Gomez-Lopez G, Tejera AM, Schoeftner S, Dominguez O, Pisano DG, Tarsounas M, Blasco MA. (2010) Mammalian Rap1 controls telomere function and gene expression through binding to telomeric and extratelomeric sites. *Nat. Cell Biol.* **12**(8): 768–780.

Martinez P, Thanasoula M, Munoz P, Liao C, Tejera A, McNees C, Flores JM, Fernandez-Capetillo O, Tarsounas M, Blasco MA. (2009b) Increased telomere fragility and fusions resulting from TRF1 deficiency lead to degenerative pathologies and increased cancer in mice. *Genes Dev.* **23**(17): 2060–2075.

Maser RS, Choudhury B, Campbell PJ, Feng B, Wong KK, Protopopov A, O'Neil J, Gutierrez A, Ivanova E, Perna I, Lin E, Mani V, Jiang S, McNamara K, Zaghlul S, Edkins S, Stevens C, Brennan C, Martin ES, Wiedemeyer R, Kabbarah O, Nogueira C, Histen G, Aster J, Mansour M, Duke V, Foroni L, Fielding AK, Goldstone AH, Rowe JM, Wang YA, Look AT, Stratton MR, Chin L, Futreal PA, DePinho RA. (2007) Chromosomally unstable mouse tumours have genomic alterations similar to diverse human cancers. *Nature.* **447**(7147): 966–971.

McNees CJ, Tejera AM, Martinez P, Murga M, Mulero F, Fernandez-Capetillo O, Blasco MA. (2010) ATR suppresses telomere fragility and recombination but is dispensable for elongation of short telomeres by telomerase. *J. Cell. Biol.* **188**(5): 639–652.

Meier UT. (2005) The many facets of H/ACA ribonucleoproteins. *Chromosoma.* **114**(1): 1–14.

Metcalfe JA, Parkhill J, Campbell L, Stacey M, Biggs P, Byrd PJ, Taylor AM. (1996) Accelerated telomere shortening in ataxia telangiectasia. *Nat. Genet.* **13**(3): 350–353.

Meznikova M, Erdmann N, Allsopp R, Harrington LA. (2009) Telomerase reverse transcriptase-dependent telomere equilibration mitigates tissue dysfunction in mTert heterozygotes. *Dis. Model Mech.* **2**(11–12): 620–626.

Miller KM, Rog O, Cooper JP. (2006) Semi-conservative DNA replication through telomeres requires Taz1. *Nature.* **440**(7085): 824–828.

Mochizuki Y, He J, Kulkarni S, Bessler M, Mason PJ. (2004) Mouse dyskerin mutations affect accumulation of telomerase RNA and small nucleolar RNA, telomerase activity, and ribosomal RNA processing. *Proc. Natl. Acad. Sci. USA.* **101**(29): 10756–10761.

Morrish TA, Greider CW. (2009) Short telomeres initiate telomere recombination in primary and tumor cells. *PLoS Genet.* **5**(1): e1000357.

Munoz P, Blanco R, de Carcer G, Schoeftner S, Benetti R, Flores JM, Malumbres M, Blasco MA. (2009) TRF1 controls telomere length and mitotic fidelity in epithelial homeostasis. *Mol. Cell. Biol.* **29**(6): 1608–1625.

Munoz P, Blanco R, Flores JM, Blasco MA. (2005) XPF nuclease-dependent telomere loss and increased DNA damage in mice overexpressing TRF2 result in premature aging and cancer. *Nat. Genet.* **37**(10): 1063–1071.

Murga M, Bunting S, Montana MF, Soria R, Mulero F, Canamero M, Lee Y, McKinnon PJ, Nussenzweig A, Fernandez-Capetillo O. (2009) A mouse model of ATR-Seckel shows embryonic replicative stress and accelerated aging. *Nat. Genet.* **41**(8): 891–898.

O'Connor MS, Safari A, Liu D, Qin J, Songyang Z. (2004) The human Rap1 protein complex and modulation of telomere length. *J. Biol. Chem.* **279**(27): 28585–28591.

Ohki R, Tsurimoto T, Ishikawa F. (2001) *In vitro* reconstitution of the end replication problem. *Mol. Cell. Biol.* **21**(17): 5753–5766.

Okamoto K, Iwano T, Tachibana M, Shinkai Y. (2008) Distinct roles of TRF1 in the regulation of telomere structure and lengthening. *J. Biol. Chem.* **283**(35): 23981–23988.

Palm W, de Lange T. (2008) How shelterin protects mammalian telomeres. *Annu. Rev. Genet.* **42**: 301–334.

Palm W, Hockemeyer D, Kibe T, de Lange T. (2009) Functional dissection of human and mouse POT1 proteins. *Mol. Cell. Biol.* **29**(2): 471–482.

Park JI, Venteicher AS, Hong JY, Choi J, Jun S, Shkreli M, Chang W, Meng Z, Cheung P, Ji H, McLaughlin M, Veenstra TD, Nusse R, McCrea PD, Artandi SE. (2009) Telomerase modulates Wnt signalling by association with target gene chromatin. *Nature.* **460**(7251): 66–72.

Perera SA, Maser RS, Xia H, McNamara K, Protopopov A, Chen L, Hezel AF, Kim CF, Bronson RT, Castrillon DH, Chin L, Bardeesy N, Depinho RA, Wong KK. (2008) Telomere dysfunction promotes genome instability and metastatic potential in a K-ras p53 mouse model of lung cancer. *Carcinogenesis.* **29**(4): 747–753.

Pogacic V, Dragon F, Filipowicz W. (2000) Human H/ACA small nucleolar RNPs and telomerase share evolutionarily conserved proteins NHP2 and NOP10. *Mol. Cell. Biol.* **20**(23): 9028–9040.

Qi L, Strong MA, Karim BO, Armanios M, Huso DL, Greider CW. (2003) Short telomeres and ataxia-telangiectasia mutated deficiency cooperatively increase telomere dysfunction and suppress tumorigenesis. *Cancer Res.* **63**(23): 8188–8196.

Ray A, Runge KW. (1999) The yeast telomere length counting machinery is sensitive to sequences at the telomere–nontelomere junction. *Mol. Cell. Biol.* **19**(1): 31–45.

Rhee DB, Wang Y, Mizesko M, Zhou F, Haneline L, Liu Y. (2010) FANCC suppresses short telomere-initiated telomere sister chromatid exchange. *Hum. Mol. Genet.* **19**(5): 879–887.

Rudolph KL, Chang S, Lee HW, Blasco M, Gottlieb GJ, Greider C, DePinho RA. (1999) Longevity, stress response, and cancer in aging telomerase-deficient mice. *Cell.* **96**(5): 701–712.

Rudolph KL, Millard M, Bosenberg MW, DePinho RA. (2001) Telomere dysfunction and evolution of intestinal carcinoma in mice and humans. *Nat. Genet.* **28**(2): 155–159.

Rufer N, Brummendorf TH, Kolvraa S, Bischoff C, Christensen K, Wadsworth L, Schulzer M, Lansdorp PM. (1999) Telomere fluorescence measurements in granulocytes and T lymphocyte subsets point to a high turnover of hematopoietic stem cells and memory T cells in early childhood. *J. Exp. Med.* **190**(2): 157–167.

Sacco A, Mourkioti F, Tran R, Choi J, Llewellyn M, Kraft P, Shkreli M, Delp S, Pomerantz JH, Artandi SE, Blau HM. (2010) Short telomeres and stem cell exhaustion model Duchenne muscular dystrophy in mdx/mTR mice. *Cell.* **143**(7): 1059–1071.

Sahin E, Depinho RA. . (2010) Linking functional decline of telomeres, mitochondria and stem cells during ageing. *Nature.* **464**(7288): 520–528.

Samper E, Fernandez P, Eguia R, Martin-Rivera L, Bernad A, Blasco MA, Aracil M. (2002) Long-term repopulating ability of telomerase-deficient murine hematopoietic stem cells. *Blood.* **99**(8): 2767–2775.

Samper E, Flores JM, Blasco MA. (2001) Restoration of telomerase activity rescues chromosomal instability and premature aging in Terc$^{-/-}$ mice with short telomeres. *EMBO Rep.* **2**(9): 800–807.

Samper E, Goytisolo FA, Slijepcevic P, van Buul PP, Blasco MA. (2000) Mammalian Ku86 protein prevents telomeric fusions independently of the length of TTAGGG repeats and the G-strand overhang. *EMBO Rep*. **1**(3): 244–252.

Sarin KY, Cheung P, Gilison D, Lee E, Tennen RI, Wang E, Artandi MK, Oro AE, Artandi SE. (2005) Conditional telomerase induction causes proliferation of hair follicle stem cells. *Nature*. **436**(7053): 1048–1052.

Savage SA, Alter BP. (2008) *The role of telomere biology in bone marrow failure and other disorders. Mech. Ageing Dev.* **129**(1–2): 35–47.

Savage SA, Giri N, Baerlocher GM, Orr N, Lansdorp PM, Alter BP. (2008) TINF2, a component of the shelterin telomere protection complex, is mutated in dyskeratosis congenita. *Am. J. Hum. Genet.* **82**(2): 501–509.

Schaetzlein S, Kodandaramireddy NR, Ju Z, Lechel A, Stepczynska A, Lilli DR, Clark AB, Rudolph C, Kuhnel F, Wei K, Schlegelberger B, Schirmacher P, Kunkel TA, Greenberg RA, Edelmann W, Rudolph KL. (2007) Exonuclease-1 deletion impairs DNA damage signaling and prolongs lifespan of telomere-dysfunctional mice. *Cell*. **130**(5): 863–877.

Schaetzlein S, Lucas-Hahn A, Lemme E, Kues WA, Dorsch M, Manns MP, Niemann H, Rudolph KL. (2004) Telomere length is reset during early mammalian embryogenesis. *Proc. Natl. Acad. Sci. USA*. **101**(21): 8034–8038.

Sfeir A, Kabir S, van Overbeek M, Celli GB, de Lange T (2010) Loss of Rap1 induces telomere recombination in the absence of NHEJ or a DNA damage signal. *Science*. **327**(5973): 1657–1661.

Sfeir A, Kosiyatrakul ST, Hockemeyer D, MacRae SL, Karlseder J, Schildkraut CL, de Lange T (2009) Mammalian telomeres resemble fragile sites and require TRF1 for efficient replication. *Cell*. **138**(1): 90–103.

Shay JW, Wright WE. (2005) Senescence and immortalization: role of telomeres and telomerase. *Carcinogenesis*. **26**(5): 867–874.

Siegl-Cachedenier I, Flores I, Klatt P, Blasco MA. (2007a) Telomerase reverses epidermal hair follicle stem cell defects and loss of long-term survival associated with critically short telomeres. *J. Cell. Biol.* **179**(2): 277–290.

Siegl-Cachedenier I, Munoz P, Flores JM, Klatt P, Blasco MA. (2007b) Deficient mismatch repair improves organismal fitness and survival of mice with dysfunctional telomeres. *Genes Dev*. **21**(17): 2234–2247.

Smogorzewska. A, de Lange T. (2002) Different telomere damage signaling pathways in human and mouse cells. *EMBO J*. **21**(16): 4338–4348.

Snow BE, Mateyak M, Paderova J, Wakeham A, Iorio C, Zakian V, Squire J, Harrington L. (2007) Murine Pif1 interacts with telomerase and is dispensable for telomere function *in vivo*. *Mol. Cell. Biol.* **27**(3): 1017–1026.

Stout GJ, Blasco MA. (2009) Genetic dissection of the mechanisms underlying telomere-associated diseases: impact of the TRF2 telomeric protein on mouse epidermal stem cells. *Dis. Model Mech.* **2**(3–4): 139–156.

Strong MA, Vidal-Cardenas SL, Karim B, Yu H, Guo N, Greider CW. (2011) Phenotypes in mTert$^{+/-}$ and mTert$^{-/-}$ mice are due to short telomeres, not telomere-independent functions of telomerase reverse transcriptase. *Mol. Cell. Biol.* **31**: 2369–2379.

Sussel L, Shore D. (1991) Separation of transcriptional activation and silencing functions of the RAP1-encoded repressor/activator protein 1: isolation of viable mutants affecting both silencing and telomere length. *Proc. Natl. Acad. Sci. USA*. **88**(17): 7749–7753.

Takai H, Smogorzewska A, de Lange T. (2003) DNA damage foci at dysfunctional telomeres. *Curr. Biol.* **13**(17): 1549–1556.

Teixeira MT, Arneric M, Sperisen P, Lingner J. (2004) Telomere length homeostasis is achieved via a switch between telomerase- extendible and -nonextendible states. *Cell.* **117**(3): 323–335.

Tejera AM, Stagno d'Alcontres M, Thanasoula M, Marion RM, Martinez P, Liao C, Flores JM, Tarsounas M, Blasco MA. (2010) TPP1 is required for TERT recruitment, telomere elongation during nuclear reprogramming, and normal skin development in mice. *Dev. Cell.* **18**(5): 775–789.

Teo H, Ghosh S, Luesch H, Ghosh A, Wong ET, Malik N, Orth A, de Jesus P, Perry AS, Oliver JD, Tran NL, Speiser LJ, Wong M, Saez E, Schultz P, Chanda SK, Verma IM, Tergaonkar V. (2010) Telomere-independent Rap1 is an IKK adaptor and regulates NF-kappaB-dependent gene expression. *Nat. Cell. Biol.* **12**(8): 758–767.

Tian XC, Xu J, Yang X. (2000) Normal telomere lengths found in cloned cattle. *Nat. Genet.* **26**(3): 272–273.

Tomas-Loba A, Flores I, Fernandez-Marcos PJ, Cayuela ML, Maraver A, Tejera A, Borras C, Matheu A, Klatt P, Flores JM, Vina J, Serrano M, Blasco MA. (2008) Telomerase reverse transcriptase delays aging in cancer-resistant mice. *Cell.* **135**(4): 609–622.

Vaziri et al., 1994 Vaziri H, Dragowska W, Allsopp RC, Thomas TE, Harley CB, Lansdorp PM. (1994) Evidence for a mitotic clock in human hematopoietic stem cells: loss of telomeric DNA with age. *Proc. Natl. Acad. Sci. USA.* **91**(21): 9857–9860.

Venteicher et al., 2009 Venteicher AS, Abreu EB, Meng Z, McCann KE, Terns RM, Veenstra TD, Terns MP, Artandi SE. (2009) A human telomerase holoenzyme protein required for Cajal body localization and telomere synthesis. *Science.* **323**(5914): 644–648.

Vidal-Cardenas and Greider, 2010 Vidal-Cardenas SL, Greider CW. (2010) Comparing effects of mTR and mTERT deletion on gene expression and DNA damage response: a critical examination of telomere length maintenance-independent roles of telomerase. *Nucl. Acids Res.* **38**(1): 60–71.

Vlangos CN, O'Connor BC, Morley MJ, Krause AS, Osawa GA, Keegan CE. (2009) Caudal regression in adrenocortical dysplasia (acd) mice is caused by telomere dysfunction with subsequent p53-dependent apoptosis. *Dev. Biol.* **334**(2): 418–428.

Vulliamy T, Marrone A, Goldman F, Dearlove A, Bessler M, Mason PJ, Dokal I. (2001) The RNA component of telomerase is mutated in autosomal dominant dyskeratosis congenita. *Nature.* **413**(6854): 432–435.

Vulliamy TJ, Dokal I. (2008) Dyskeratosis congenita: the diverse clinical presentation of mutations in the telomerase complex. *Biochimie.* **90**(1): 122–130.

Wakayama T, Shinkai Y, Tamashiro KL, Niida H, Blanchard DC, Blanchard RJ, Ogura A, Tanemura K, Tachibana M, Perry AC, Colgan DF, Mombaerts P, Yanagimachi R. (2000) Cloning of mice to six generations. *Nature.* **407**(6802): 318–319.

Walne AJ, Vulliamy T, Marrone A, Beswick R, Kirwan M, Masunari Y, Al-Qurashi FH, Aljurf M, Dokal I. (2007) Genetic heterogeneity in autosomal recessive dyskeratosis congenita with one subtype due to mutations in the telomerase-associated protein NOP10. *Hum. Mol. Genet.* **16**(13): 1619–1629.

Wang F, Podell ER, Zaug AJ, Yang Y, Baciu P, Cech TR, Lei M. (2007) The POT1–TPP1 telomere complex is a telomerase processivity factor. *Nature.* **445**(7127): 506–510.

Wang Y, Erdmann N, Giannone RJ, Wu J, Gomez M, Liu Y. (2005a) An increase in telomere sister chromatid exchange in murine embryonic stem cells possessing critically shortened telomeres. *Proc. Natl. Acad. Sci. USA*. **102**(29): 10256–10260.

Wang Y, Giannone RJ, Liu Y. (2005b) Telomere sister chromatid exchange in telomerase deficient murine cells. *Cell Cycle*. **4**(10): 1320–1322.

Wong KK, Maser RS, Bachoo RM, Menon J, Carrasco DR, Gu Y, Alt FW, DePinho RA. (2003) Telomere dysfunction and ATM deficiency compromises organ homeostasis and accelerates ageing. *Nature*. **421**(6923): 643–648.

Wong KK, Maser RS, Sahin E, Bailey ST, Xia H, Ji H, McNamara K, Naylor M, Bronson RT, Ghosh S, Welsh R, DePinho RA. (2007) Diminished lifespan and acute stress-induced death in DNA-PKcs-deficient mice with limiting telomeres. *Oncogene*. **26**(20): 2815–2821.

Wright WE, Piatyszek MA, Rainey WE, Byrd W, Shay JW. (1996) Telomerase activity in human germline and embryonic tissues and cells. *Dev. Genet.* **18**(2): 173–179.

Wu L, Multani AS, He H, Cosme-Blanco W, Deng Y, Deng JM, Bachilo O, Pathak S, Tahara H, Bailey SM, Deng Y, Behringer RR, Chang S. (2006) Pot1 deficiency initiates DNA damage checkpoint activation and aberrant homologous recombination at telomeres. *Cell*. **126**(1): 49–62.

Xin H, Liu D, Songyang Z. (2008) The telosome/shelterin complex and its functions. *Genome Biol*. **9**(9): 232.

Xin H, Liu D, Wan M, Safari A, Kim H, Sun W, O'Connor MS, Songyang Z. (2007) TPP1 is a homologue of ciliate TEBP-beta and interacts with POT1 to recruit telomerase. *Nature*. **445**(7127): 559–562.

Yamanaka S, Blau HM. (2010) Nuclear reprogramming to a pluripotent state by three approaches. *Nature*. **465**(7299): 704–712.

Yang Q, Zheng YL, Harris CC. (2005) POT1 and TRF2 cooperate to maintain telomeric integrity. *Mol. Cell. Biol.* **25**(3): 1070–1080.

Ye JZ, Donigian JR, van Overbeek M, Loayza D, Luo Y, Krutchinsky AN, Chait BT, de Lange T. (2004a) TIN2 binds TRF1 and TRF2 simultaneously and stabilizes the TRF2 complex on telomeres. *J. Biol. Chem.* **279**(45): 47264–47271.

Ye JZ, Hockemeyer D, Krutchinsky AN, Loayza D, Hooper SM, Chait BT, de Lange T. (2004b) POT1-interacting protein PIP1: a telomere length regulator that recruits POT1 to the TIN2/TRF1 complex. *Genes Dev*. **18**(14): 1649–1654.

Yuan X, Ishibashi S, Hatakeyama S, Saito M, Nakayama J, Nikaido R, Haruyama T, Watanabe Y, Iwata H, Iida M, Sugimura H, Yamada N, Ishikawa F. (1999) Presence of telomeric G-strand tails in the telomerase catalytic subunit TERT knockout mice. *Genes Cells*. **4**(10): 563–572.

Zalzman M, Falco G, Sharova LV, Nishiyama A, Thomas M, Lee SL, Stagg CA, Hoang HG, Yang HT, Indig FE, Wersto RP, Ko MS. (2010) Zscan4 regulates telomere elongation and genomic stability in ES cells. *Nature*. **464**(7290): 858–863.

Zaug AJ, Podell ER, Nandakumar J, Cech TR. (2010) Functional interaction between telomere protein TPP1 and telomerase. *Genes Dev*. **24**(6): 613–622.

Zhao Y, Sfeir AJ, Zou Y, Buseman CM, Chow TT, Shay JW, Wright WE. (2009) Telomere extension occurs at most chromosome ends and is uncoupled from fill-in in human cancer cells. *Cell*. **138**(3): 463–475.

10

CELLULAR SENESCENCE, TELOMERASE, AND CANCER IN HUMAN CELLS

PHILLIP G. SMIRALDO, JUN TANG, JERRY W. SHAY, AND WOODRING E. WRIGHT

10.1 INTRODUCTION—THE HAYFLICK LIMIT

The field of cellular gerontology, studying aging at the cellular level, was pioneered by Leonard Hayflick. Almost 50 years ago, he discovered that cultured normal human cells have a limited capacity to divide, after which they become senescent. The most convincing evidence that normal human cells have a finite lifespan came from several elegant experiments that he performed while working with Paul Moorhead (Hayflick, 1965; Hayflick and Moorhead, 1961). While culturing primary human fibroblasts, they observed and described three phases of life for a cell population: Phase I—the primary culture, Phase II—months of copious cell growth and division, and Phase III—the period when cell replication diminished and ultimately stopped (Hayflick and Moorhead, 1961). To support the idea that normal human cells have a limited number of divisions, Hayflick and Moorhead cultured three separate populations of human fibroblasts simultaneously: one derived from a male at population doubling (PD) 14, one derived from a female at PD 10, and one that contained an equal number of male (PD 14) and female (PD 10) fibroblasts. When the "older" unmixed male-derived cell population stopped dividing, they investigated the mixed population and discovered that only female cells were present (Hayflick and Moorhead, 1961). Besides providing additional evidence that cells have a limited lifespan, this experiment demonstrated that the older cells

Telomerases: Chemistry, Biology, and Clinical Applications, First Edition.
Edited by Neal F. Lue and Chantal Autexier.

"remembered" that they were old, regardless of being surrounded by younger cells. One additional, and very important, observation that Hayflick reported is that cryogenically preserved cells remembered the number of times that they had divided at the time they were frozen (Hayflick, 1984). Ultimately, these experiments demonstrated that a counting mechanism, which Hayflick termed the "replicometer," was somehow programmed into each cell and once this biological clock (as opposed to a chronological clock) had expired, the cell would stop dividing (Hayflick, 1998). In honor of Hayflick's discoveries that human cells have a finite lifespan, the term "Hayflick limit" is sometimes used to describe the maximum number of times a given cell type can divide. Today, this withdrawal from the cell cycle after a certain number of cellular divisions is termed senescence, which can be triggered as a result of shortened telomeres. Studies on replicative senescence have begun to provide valuable information towards our understanding of organismal aging and, additionally, have created new opportunities in the area of regenerative medicine. Equally important, cancer cells have evolved the ability to overcome senescence by using mechanisms capable of maintaining telomere lengths (such as expressing telomerase), which enables cancer cells to divide indefinitely.

10.2 TELOMERES AND SENESCENCE

The term "senescence" is used to describe a signal-transduction pathway leading to irreversible growth arrest of cells in culture, accompanied by a distinct set of phenotypic changes (Reddel, 2000). Cellular senescence can be triggered by a number of endogenous or exogenous stresses that may not be mutually exclusive, including oxidative damage, overexpression of oncogenes, chromatin changes, and DNA damage (Reddel, 2010; Shay and Wright, 2007). The focus of this chapter is what we term "replicative" senescence, which is initiated by critically shortened telomeres.

Telomeres are nucleoprotein structures found at the ends of all linear chromosomes that function to stabilize chromosome termini and allow cells to distinguish natural chromosome ends from DNA double-stranded breaks (Blackburn, 1991). Each telomere consists of many kilobases of a repeated hexamer sequence followed by a single-stranded 3′ overhang at the extreme terminus (Henderson and Blackburn, 1989); the overhangs range from ~12 to several hundred nucleotides in length (Cimino-Reale et al., 2001; Makarov et al., 1997; Wright et al., 1997; Zhao et al., 2008). The overhangs not only serve as the location to which telomerase binds for telomere elongation, but also as a site for telomere-specific binding proteins. These overhang-binding proteins form a complex with other factors that bind to double-stranded telomeric DNA that collectively function to control telomerase activity and protect chromosome ends, potentially by folding each telomere into a Holliday junction-like structure, called a t-loop (telomeric loop) (Griffith et al., 1999).

In most human somatic cells, telomeres progressively shorten each time a cell divides (Harley et al., 1990). There have been several mechanisms described that contribute to telomere shortening in the cell. First, during lagging-strand

DNA synthesis, DNA polymerases are unable to fill in the gap between the final RNA priming event and the end of the chromosome (referred to as the end-replication problem) (Olovnikov, 1973; Watson, 1972). Once the final RNA primer is removed, this gap could represent a 10–12 nucleotide region (if the RNA primer is positioned at the very end of the chromosome) or a larger gap (if the final priming event occurs several hundred base pairs from the chromosome end) (Wright et al., 1997). Second, processing of telomeres by an ill-defined nuclease to increase the length of the single-stranded telomeric overhangs will decrease the length of the DNA template for the next round of DNA synthesis (Shay and Wright, 2006). Finally, other processes, such as oxidative damage can contribute to telomere shortening (von Zglinicki, 2002).

When telomere lengths are shortened to a certain threshold, cells undergo replicative senescence (mortality stage 1, M1), which encompasses a complex set of changes within the cell that include cytoskeletal alterations, changes in gene expression and secretory proteins, and permanent withdrawal from the cell cycle (Coppe et al., 2010; Wright et al., 1989). Senescent cells are metabolically active (even though they are growth arrested), resistant to apoptosis, and can remain alive for many years (Reddel, 2010). It is important to note that replicative senescence does not require that all telomeres within the cell be short; it can be initiated by a subset of short telomeres (Hemann et al., 2001; Zou et al., 2004). Interestingly, senescent cells with shortened telomeres contain nuclear DNA damage repair foci. These foci localize to dysfunctional telomeric ends and contain a high concentration of proteins involved in DNA damage checkpoint control and double-stranded break repair (d'Adda di Fagagna et al., 2003; Satyanarayana et al., 2004a; Sedelnikova et al., 2004; Takai et al., 2003). It is not clear, however, what the length of a telomere must be for recognition by the DNA damage-sensing mechanisms. A telomere that still contains telomeric repeats, but is too short to form the protective t-loop structure, could trigger the initial damage signal (as opposed to a telomere that has completely exhausted its repeats). Regardless, it is clear that short, dysfunctional telomeres are not able to perform one of their critical functions, which is to allow the cell to distinguish natural chromosome ends from DNA double-stranded breaks. It is likely that this persistent DNA damage response to "unrepairable" shortened telomeres initiates cell-cycle arrest (Shay and Wright, 2006).

In cell culture models, cellular senescence can be circumvented when important cell-cycle checkpoint genes, such as p53, are inactivated (Hara et al., 1991; Shay et al., 1991b). If M1 is bypassed, cells enter an extended proliferation status and continue division. As these cells continue to divide, their telomeres become so short that they eventually fail to protect the ends of many chromosomes from the competition of DNA repair. Sensing these uncapped chromosome ends as DNA damage, DNA repair mechanisms (specifically, nonhomologous end joining, NHEJ) ligate chromosomes together to form dicentric or multicentric chromosomes (Maser and DePinho, 2002). These gross chromosomal abnormalities initiate a breakage–fusion–bridge cycle during mitosis and eventually cause mitotic catastrophe. This state of rampant chromosomal instability and widespread apoptosis is termed crisis (mortality stage 2, M2) (Maser and DePinho, 2002).

To address the problem of how cells deal with their shortened telomeres between M1 and M2, cytogenetic analysis was performed on human fibroblasts that had bypassed M1 (by ectopic expression HPV16 E6/E7) (Zou et al., 2009). Post M1 in E6/E7-expressing cells, an increased number of telomere associations (TAs; where the ends of chromosomes touch without any evidence of cytogenetic abnormalities) was observed, yet this did not affect growth rate or cell death. Knockdown of ligase IV had no effect on the frequency of TAs, suggesting that TAs might represent DNA repair intermediates where the final NHEJ step (ligation) remains blocked. Once TAs were detected in NHEJ-competent cells, the fibroblasts went through an additional seven to eight doublings before dicentric chromosomes with ligated ends were observed. M2 occurs once cells have accumulated a sufficient number of dicentric chromosomes, complex chromosomal rearrangements, and multicentric chromosomes to cause mitotic catastrophe. These data suggest that short telomeres trigger a DNA damage response and form NHEJ intermediates (noncovalent structures, TAs) many cell doublings prior to the onset of ligation events, which eventually lead to cell death/M2 (Fig. 10.1; Zou et al., 2009).

It is possible for cells to overcome the replicative-induced M1 and M2 stages by activating a mechanism that elongates or stabilizes telomere lengths. Telomerase compensates for the continued erosion of telomeres that occurs with each cellular division, thus a balance is formed between processes that lengthen and shorten telomeres (Collins and Mitchell, 2002; Lingner and Cech, 1998; Nugent and Lundblad, 1998). The core enzyme of telomerase consists of an RNA component (hTERC or hTR) (Feng et al., 1995), which provides the template for the *de novo* synthesis of telomeric DNA, and a catalytic subunit (hTERT) (Nakamura et al., 1997), which has reverse transcriptase-like activity. Proof that telomere shortening is a cause of cellular senescence was provided by introducing exogenous *hTERT* into normal telomerase-silent cells and showing that it was sufficient to activate telomerase activity and maintain telomere lengths, resulting in the bypass of M1 and cell immortalization (Bodnar et al., 1998). Therefore, telomeres are the molecular clocks that "count" the number of times a cell has divided and determine when cellular senescence and crisis occur.

Collectively, cellular senescence and crisis are thought of as potent barriers to prevent uncontrolled cell division (Shay et al., 1991a,b; Wright et al., 1989; Wright and Shay, 2001). However, there may be consequences associated with the activation of these mechanisms. Recent studies have demonstrated that senescent cells (induced by replicative aging, nontelomeric DNA damage, oncogene expression, or strong mitogenic/stress signals) secrete a number of factors, including interleukins, chemokines, growth factors, proteases, insoluble proteins, extracellular matrix components, and reactive oxygen species. This common set of factors is referred to as the senescence-associated secretory phenotype (SASP) (for a complete review of SASP, the reader is directed to Coppe et al., 2010). Additionally, telomere length regulates the expression of interferon stimulated gene 15 (ISG15), which may in turn contribute to chronic inflammatory states associated with human aging; ISG15 RNA and protein levels increase in human cells with short telomeres (Lou et al., 2009). Therefore, cells with short telomeres or that are senescent may have the ability to alter the tissue

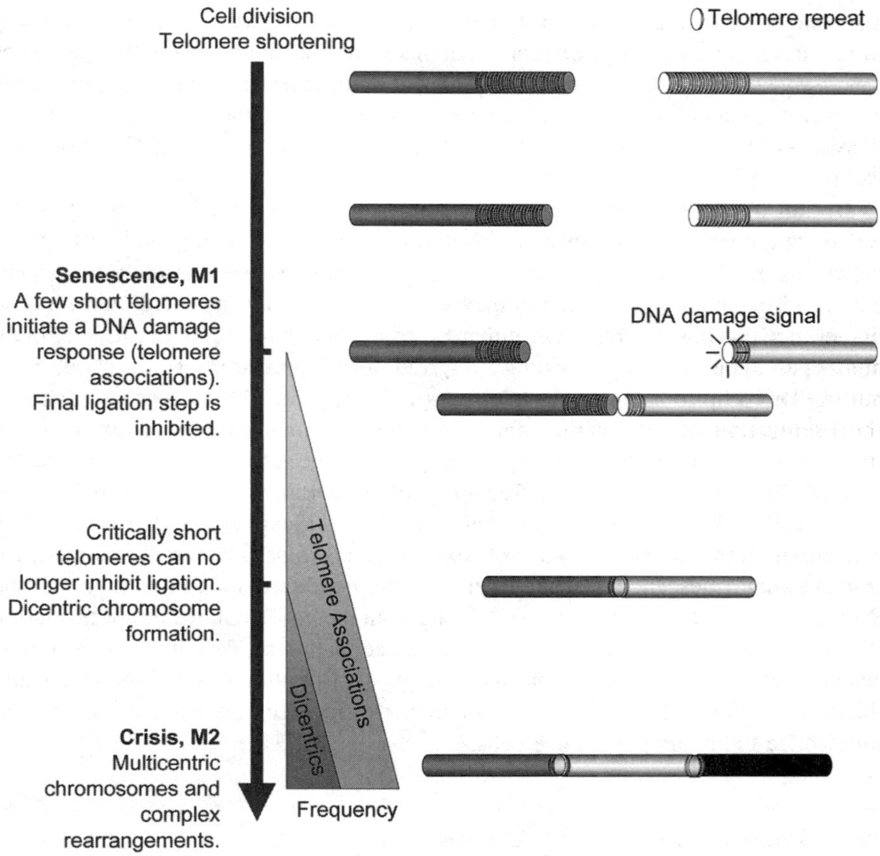

FIGURE 10.1 Telomere states between M1 and M2. As normal human cells divide, telomeres shorten with each round of cellular division. Once a subset of telomeres has shortened to a certain threshold, a DNA damage response is initiated that confers the formation of telomere associations (TAs) and the activation of replicative senescence (M1). TAs occur when the ends of chromosomes touch without any evidence of cytogenetic abnormalities and when a constriction in the diameter of the chromosome is apparent. TAs appear to be noncovalent DNA repair intermediates that are blocked from becoming actual end-fusion events. If M1 is bypassed, the cells continue to divide, the telomeres continue to erode, and an increased number of TAs is observed. Dicentric chromosomes are formed when some telomeres become sufficiently short and that they no longer inhibit ligation. As the cells continue to divide, a further increase in the frequency TAs and an increased number of dicentric chromosomes is observed. Crisis (M2) occurs once cells have accumulated enough dicentric chromosomes, complex chromosomal rearrangements, and multicentric chromosomes to cause mitotic catastrophe (Zou et al., 2009).

microenvironment in deleterious ways, including the promotion of tumorigenesis (Coppe et al., 2010).

At face value, the strategy to use genome instability induced by critically short telomeres to eliminate cells in replicative-based crisis is a sensible one. However, this mechanism, as we understand it, may have a serious potential flaw. The type of genome instability conferred by dysfunctional telomeres and breakage–fusion–bridge cycles can easily contribute to chromosomal ploidy abnormalities or deletion, amplification, and rearrangements of genes, all of which could contribute to the progression of cancer (Davoli et al., 2010; Feldser et al., 2003; Goytisolo and Blasco, 2002; Maser and DePinho, 2002).

Not all mammals seem to use replicative aging. For example, regarding senescence, telomeres, and telomerase regulation, there are fundamental differences between mice and humans. Laboratory mouse telomeres are extremely long compared to human telomeres and telomerase activity is found in several adult mouse tissues in which it is suppressed in humans (Wright and Shay, 2000). In culture, mouse fibroblasts typically stop dividing (referred to as premature senescence) after 10–15 PDs (despite having long telomeres), yet they readily undergo spontaneous immortalization without manipulation. In contrast, human fibroblasts almost never spontaneously immortalize and, even after Rb and p53 activities have been blocked with SV40 large T antigen, the frequency of immortalization is only 10^{-7} (Shay et al., 1993). When mouse embryonic fibroblasts are grown in low oxygen, a concentration that is similar to normal oxygen levels in tissues, there is no evidence of premature senescence (Parrinello et al., 2003). Additionally, mice that lack the RNA component of telomerase ($mTR^{-/-}$) have an increased risk for tumor formation (Blasco et al., 1997). This is in stark contrast to human cancers, which almost always require telomerase activity for ongoing proliferation (see below) (Shay and Bacchetti, 1997). These data point to major differences between the causes of senescence in human and mouse cells.

10.3 TELOMERASE AND CANCER

The biggest difference between tumor and normal cells is their proliferative capacity (Phatak and Burger, 2007). While normal cells can only divide a certain number of times, cancer cells have a limitless replicative potential, one of the original six hallmarks of cancer (Hanahan and Weinberg, 2000). At birth, human telomeres are 15–20 kb in length. As we age, telomeres progressively shorten in most cells and in human tissues from middle-aged individuals, telomere lengths are about half that of newborns. The average telomere length in cancer cells is only ~5 kb, which is generally much shorter compared to its surrounding tissue. However, unlike normal cells, tumor cells show no decrease in average telomere lengths with successive cell divisions (Harley et al., 1990; Hastie et al., 1990; Holt et al., 1996). This observation suggests that telomerase is required to maintain telomere lengths in order for cells to escape replicative senescence and proliferate indefinitely. Consistent with this idea, ectopic expression of *hTERT* was a key step in creating human tumor cells from normal cells, *in vitro*, with "defined" genetic elements (Hahn et al., 1999a).

Because cancer cells require multiple mutations within a single cell to become malignant (Hahn et al., 1999a), they must have the ability to divide enough times to accumulate a correct set of genetic alterations without undergoing replicative senescence. In most cases, at least six mutations are required to form a cancer cell (Vogelstein and Kinzler, 1993). Consider that a mutation occurs in a single cell. This cell would have to expand to at least a million cells (approximately 20 divisions) before there is reasonable probability for an additional cancer-promoting mutation to occur. Since most cancer mutations are recessive (e.g., tumor suppressor genes), this cell would have to expand again to a million cells to eliminate the remaining wild-type copy of the gene. To acquire the next mutation, this cell would have to expand to a million cells yet again, and so on. Therefore, it is likely that all malignant tumors need to become immortal to sustain their growth and telomerase activity could be a rate-limiting step required for the continuing proliferation of advanced cancers (Shay and Wright, 2006). Support for this concept came following the development of a critically important assay, the telomeric repeat amplification protocol, also known as TRAP. Using this sensitive PCR-based assay, telomerase activity was detected in approximately 90% of primary human tumors (Kim et al., 1994). Conversely, if telomerase activity is disrupted in a telomerase-positive cancer cell line, telomere shortening, chromosomal instability, and cell death occurs, demonstrating that telomere stability and telomerase activity are two defining hallmarks of cells that are able to divide indefinitely (Hahn et al., 1999b). Although the ALT mechanism, which involves DNA recombination at the telomeres, is able to maintain telomere lengths (Bryan et al., 1995; Dunham et al., 2000; Kass-Eisler and Greider, 2000), the mechanism for telomere maintenance in human tumors is almost always through activation or upregulation of telomerase.

Because telomerase activity is necessary for unobstructed cell division in cancer cells, telomerase has been proposed to be a biomarker for early cancer detection, prognosis, and monitoring for residual disease (Shay and Bacchetti, 1997; Shay and Gazdar, 1997). The TRAP assay can detect as few as 1–10 telomerase-positive cells in a mixed population and, therefore, may permit the detection of early cancer lesions prior to the onset of tissue invasion. In general, a good correlation has been found between telomerase activity and the histological grade of the tumor (Shay and Wright, 2007). The development of clinical telomerase diagnostics for screening or monitoring patients, however, is still undergoing validation and standardization studies.

10.3.1 Telomerase Chemotherapy Approaches

Therapy for patients with advanced cancer generally includes surgical tumor resection, intensive multimodal chemotherapy, radiation, or a combination of these regimens. The ideal cancer treatment would specifically target cancerous cells and have little or no effect on normal cells. Because of the compelling correlation that most normal cells are telomerase silent while telomerase activity is detected in nearly all cancers, telomerase has emerged as an almost universal target for cancer therapeutics. Since telomerase activity is absent from most human somatic cells, telomerase-based therapies should possess greater specificity, lower toxicity, and reduced side-effects compared to conventional chemotherapeutic approaches.

Here, we describe some of the general schemes for targeting telomerase and the advantages and challenges associated with each strategy (Table 10.1).

10.3.2 Immunotherapy

Using immunotherapy approaches to cause tumor regression is based upon the hypothesis that most, if not all, tumors express antigens that T-lymphocytes can recognize and target for lysis (Vonderheide, 2002). The benefit of immunotherapy strategies is that cancer cell death would be immediate. One major hurdle to overcome, however, is to identify antigens that are cancer-specific (absent from normal cells) and applicable to a broad range of tumor types. Because telomerase is required for long-term growth of tumor cells, hTERT is a very promising target for cancer immunotherapy. It has been shown that peptides derived from hTERT are naturally processed by tumor cells, presented as epitopes by the major histocompatibility complex (MHC), and serve as a target for antigen-specific cytotoxic T-lymphocytes (Minev et al., 2000; Vonderheide et al., 1999). There have been two promising approaches that have advanced to clinical trials. First, an *ex vivo* process is used where immature dendritic cells (the most effective antigen-presenting cells) are isolated from the patient's blood, pulsed with *hTERT* RNA, and then returned to the patient's body where they activate cytotoxic T-lymphocytes to kill the tumor cells that express telomerase and present telomerase epitopes (Su et al., 2005). The second strategy, which is similar to classic protein subunit-based vaccines, employs an injectable formulation of promiscuous peptides derived from the active site of hTERT. After immunization, the immune system recognizes the hTERT peptides as a foreign antigen and will destroy the telomerase-positive cancer cells that present it as an epitope (Bernhardt et al., 2006; Brunsvig et al., 2006). In either strategy, cancer cells could attempt to escape immune surveillance by shutting down *hTERT* gene expression. This, to the benefit of the patient, would cause telomere shortening in cancer cells and consequently have deleterious effects on tumor growth.

One major concern of activating the immune system to attack hTERT antigen-presenting cells is that autoimmunity may develop against normal cells in which telomerase has been detected (such as hematopoietic stem cells, activated lymphocytes, basal keratinocytes, gonadal cells, and certain proliferative epithelial cells). However, the results of immunotherapy Phase I and II clinical trials using the hTERT vaccine have not revealed any serious adverse effects. There was no evidence of depression of stem cells in bone marrow and no evidence of autoimmune disease in long-term survivors who received the monthly booster vaccines (Bernhardt et al., 2006; Brunsvig et al., 2006). The development of these promising approaches for a telomerase-based universal cancer vaccine is encouraging and might be even more effective when used to treat patients with less advanced disease. Interestingly, if telomerase immunotherapy proves to be an effective strategy for cancer treatments, it would open the doors for preventative immunotherapy. Since it is predicted that approximately 90% of cancers will express the *hTERT* gene, a telomerase vaccine could be used for healthy patients with a high risk for developing cancer based upon genetic factors or medical history (Shay and Keith, 2008).

TABLE 10.1 Advantages and Disadvantages of Different Telomerase-Based Anticancer Therapies

Approach	Possible Advantages	Possible Disadvantages
Immunotherapy		
• *ex vivo* priming of APC cells with *hTERT* RNA • hTERT peptide vaccine	• Telomerase is a broadly expressed tumor-associated antigen • Cell death not subject to phenotypic lag • Long-term effects with periodic boosting • Could be used for cancer-preventative strategies	• Significant manufacturing and regulatory challenges with ex vivo priming approach • Potential autoimmune response • Immune system may be compromised in some patients
Suicide gene therapy		
• Pro-drug conversion to toxic substrates • Engineered adenoviruses	• Specific against *hTR*- or *h*TERT-expressing cells. • Cell death not subject to phenotypic lag • Adenoviruses can spread to adjacent cells after cell lysis	• Significant manufacturing and regulatory challenges • Immune response to vector delivery system
Telomerase activity inhibition		
• Small molecule inhibitors • Small interfering RNA against *hTR* or *hTERT* • Mutant *hTERT* overexpression • Alternative splicing of *hTERT* • Telomerase assembly • G-quadruplex stabilizing compounds • Inhibiting C-strand fill-in following telomerase extension	• Specific against tumor cells requiring telomerase activity for unlimited cell division • Standard clinical development route	• Delayed cell killing/phenotypic lag • Long-term treatments could affect normal cells that express telomerase
Agents that target the telomere		
• Mutant hTR expression • Telomere capping proteins • G-quadruplex stabilizing compounds	• Cell death may not subject to phenotypic lag	• Possible toxicity to normal cells and tissues

10.3.3 Gene Therapy

Cancer gene therapy approaches attempt to exploit specific factors that are differentially expressed in cancer compared to normal cells. Because cancer cells express telomerase, a potential strategy is to hijack the molecular mechanisms responsible for regulating telomerase expression and use them to drive a suicide gene or replication-competent adenovirus, thus rapidly killing telomerase-expressing cells (Shay and Keith, 2008).

There have been multiple strategies described for the suicide gene approach in cancer cells. For example, Plumb and colleagues placed the bacterial nitroreductase gene under the control of the regulatory sequences of the *hTR* or *hTERT* gene promoters (Plumb et al., 2001). Only cells that express telomerase will express the nitroreductase gene, which converts the pro-drug CB1954 to a cytotoxic form, thus killing the cancer cell. The use of this system is comprised of many events, each with challenges to overcome: adenoviral delivery of the suicide gene construct (cancerous cells must be infected without initiating an immune response to the delivery system), infecting sufficient cells in all regions of the tumor (so that bystander killing of most uninfected cells can still occur), expression of the nitroreductase gene by the telomerase promoter (some cancer cells may weakly express telomerase and, therefore, not express nitroreductase at sufficiently high levels to convert the pro-drug to the cytotoxic form), and the addition of the pro-drug (effective distribution of the drug). This strategy showed promise in the effective killing of tumor cell lines, *in vitro*, derived from cervical, ovarian, lung, and colon cancers (Bilsland et al., 2003; Plumb et al., 2001). Moreover, sensitization to the pro-drug was retained, *in vivo*, in xenograft studies using cervical and small-cell lung cancer cells. It is true however, that toxicity and cell death was restricted to cell lines expressing high levels of telomerase (Bilsland et al., 2003; Plumb et al., 2001). Therefore, for this strategy to be used in a clinical setting, only cancers that express high levels of telomerase (as determined by screening tumor biopsies) will be sufficiently sensitized for treatment.

An alternative suicide gene strategy is to use adenoviruses that have been manipulated to selectively target and destroy telomerase-positive cancer cells. Cancer-cell specificity is accomplished by using the promoter region of the *hTERT* gene to regulate the replication of the adenovirus. Telomerase-expressing cancer cells infected with this manipulated adenovirus will replicate the virus, lyse, and release new virus to adjacent cells. When the same engineered viruses infect normal somatic cells that do not express telomerase, there should be no replication of the virus or killing effects (Keith et al., 2004, 2007). One of the newest oncolytic viruses, CG5757, has demonstrated promise for cancer treatments by having a high degree of specificity and effectiveness in xenograft models (Li et al., 2005). In order to further increase tumor selectivity, this virus was generated by replacing the nonselective *E1a* and *E1b* endogenous viral promoters with the promoter regions of the human *E2F1* and *hTERT* genes, respectively. The *E2F1* promoter is activated in retinoblastoma- (Rb-) defective tumor cells (the Rb pathway is disrupted in approximately 85% of cancers) while approximately 90% of cancers express telomerase. *In vitro*, expression of the *E1a* and *E1b* genes was only detected in Rb-defective and hTERT-positive cancer cells while the virus did not replicate in normal cells. *In vivo*, a good safety profile and a strong antitumor activity were demonstrated in xenograft experiments using bladder, lung, and prostate cancer models (Li et al., 2005). Therefore, the use of oncolytic viruses is showing promise for effective treatments of a broad range of human cancers.

10.4 SMALL MOLECULE AND OLIGONUCLEOTIDE TELOMERASE INHIBITORS

The ultimate goal of using small molecules for telomerase-based cancer therapy is to disrupt the ability of telomerase to elongate and maintain telomere lengths. One can imagine targeting any of the steps involved in telomerase action including telomerase gene transcription, RNA maturation, holoenzyme assembly, or interactions with telomeres (Harley, 2008). As such, multiple approaches have been attempted with varying levels of success. Regardless of strategy, if telomerase activity is inhibited, the telomeres in cancer cells will shorten with each cell division, ultimately leading to cancer cell crisis and death. One concern is that telomerase-inhibition therapy might affect normal telomerase-positive reproductive cells and other proliferative cells of renewal tissues. However, since normal cells from such tissues generally divide only transiently and have much longer telomeres than most tumor cell populations, antitelomerase treatment for tumors could be designed to end before any significant telomere depletion occurs in these cell types (Gellert et al., 2006; Shay and Wright, 2005, 2006, 2007). One obstacle for telomerase inhibition-based strategies is that there will be a phenotypic lag or delayed cell death; the cancer cells must undergo enough cell divisions for the telomeres to become sufficiently short to initiate cellular crisis. Due to this limitation, telomerase inhibitors would likely be used in combination with chemotherapy, radiation, or surgical resection (all of which would decrease the initial tumor burden) to target chemoresistant or stem-like residual cancer cells (Shay and Wright, 2005).

The only telomerase inhibitor currently in clinical trials is the oligonucleotide-based molecule, Imetelstat (GRN163L). Because of the unique structure of this compound, it is able to overcome two major challenges of oligonucleotide-based therapies—how to get oligonucleotides into the cell and how to get them to the target without being degraded by nucleases. GRN163L is a lipidated 13-mer thiophosphoramidate that is complementary to the 13-nucleotide-long template region of hTR and its primary mechanism of action is to bind the hTR component of telomerase. *In vitro*, GRN163L binds to the active site and directly blocks the ability of telomerase to elongate telomeric substrates. *In vivo*, GRN163L might also prevent hTR from forming a complex with hTERT (Dikmen et al., 2005; Djojosubroto et al., 2005; Gellert et al., 2006). Efficacy studies with GRN163L were conducted in mouse xenograft models representing a range of human tumor types including lung (Dikmen et al., 2005; Jackson et al., 2007), breast (Gellert et al., 2006; Gomez-Millan et al., 2007; Hochreiter et al., 2006), prostate (Asai et al., 2003), liver (Djojosubroto et al., 2005), brain (Marian et al., 2010; Ozawa et al., 2004), bladder

(Dikmen et al., 2008), and hematological cancers (Akiyama et al., 2003; Wang et al., 2004). GRN163L has since entered into clinical trials for chronic lymphocytic leukemia, multiple myeloma, and a variety of solid tumors such as nonsmall cell lung cancer and breast cancer (Harley, 2008).

10.4.1 Telomere Directed Therapeutics

Mammalian telomeres are bound by a complex of proteins to form protective "caps," which function to stabilize chromosome termini and allow cells to distinguish natural chromosome ends from DNA double-stranded breaks. This protein complex, termed shelterin, consists of six interdependent core proteins that associate with or bind to telomeres in a DNA sequence-specific manner (de Lange, 2005). If the shelterin complex is disrupted, the telomeric cap would not function and DNA damage responses would be initiated at the telomeres. For example, in human fibrosarcoma cells, expression of a mutant version of *TRF2* (a gene encoding one of the shelterin components), which lacks an N-terminal domain and the DNA binding domain, caused a high number of telomere–telomere fusions and cell death (Karlseder et al., 1999, 2002; van Steensel et al., 1998). Therefore, an alternative strategy for targeting telomerase-positive cancer cells might be to disrupt the protective telomeric cap, which would initiate DNA damage responses at telomeres and result in senescence or apoptosis. A major drawback of this approach is specificity. Because normal cells require shelterin complexes for telomere protection, the ability to target cancer cells without affecting normal cells could limit the development of this strategy.

Kim et al. (2001) ectopically expressed a mutant version of the *hTR* gene in telomerase-positive prostate or breast cancer cell lines. Because this version of hTR contained base changes in the sequence that hTERT uses as a template to elongate telomeres, telomerase adds a different sequence to telomeric ends (rather than the normal TTAGGG sequence). Following expression (even at low levels) of this mutant *hTR* gene, decreased cellular viability and increased apoptosis was observed, presumably because the shelterin or end-binding complexes could not recognize and bind to the mutant telomeres. In addition, the mutant sequences may have prevented the establishment of stable t-loops. Importantly, this strategy targets cancer cells for death very rapidly before either telomere shortening or inhibition of endogenous telomerase activity was detected (Kim et al., 2001). Although telomerase-silent cells should be resistant to this approach, normal proliferative stem cells that express telomerase may be severely affected since they possess normal cell-cycle checkpoints. It could be possible, however, to generate constructs to express the mutant *hTR* gene that are inducible or under the control of tumor-enhanced promoters, such as *E2F1*, which would allow normal cells to remain unaffected.

Telomeric DNA is unique, not just because of its position on chromosomes, but also for its sequence. It has been demonstrated that guanine-rich telomeric oligonucleotides, *in vitro*, can fold into various conformations and higher-order structures, termed G-quadruplexes, which are based on the quartet structure of Hoogsteen-hydrogen bonding (Henderson et al., 1987; Sundquist and Klug, 1989;

Williamson et al., 1989). Cations present in the cellular environment, such as K^+ and Na^+, play a critical role in stabilizing the quadruplex by occupying the central cavity and neutralizing the electrostatic repulsion (Simonsson, 2001). With the effect of molecular crowding, which mimics the physiological intracellular environment, the stability of G-quadruplexes is increased dramatically and can actually compete with duplex formation (Miyoshi et al., 2004). Additionally, G-quadruplex formation preferentially occurs at the 3′ ends of telomeres, as opposed to internal positions, providing a molecular basis for telomerase inhibition (Tang et al., 2008). Although evidence of G-quadruplexes, *in vivo*, is still accumulating (Lipps and Rhodes, 2009) and their physiological role is ambiguous, it has been shown that induction of G-quadruplexes results in inhibition of telomerase activity (Oganesian et al., 2006; Zahler et al., 1991; Zaug et al., 2005). Therefore, ligands (either natural molecules or synthetic compounds) that can induce the formation or stabilization of G-quadruplexes have the potential to block the access of telomerase binding to telomeres, thus disrupting telomere length maintenance in cancer cells.

To date, many G-quadruplex stabilizing ligands have been developed that inhibit telomerase activity, including: pentacyclic acridine RHPS4, perylene diimide PIPER, porphyrin TMPyP4, trisubstituted acridine BRACO19, and telomestatin from the bacteria *Streptomyces anulatus* (Bryan and Baumann, 2010). However, further investigation of basic G-quadruplex biology is needed before the clinical application of such drugs is possible. For example, the conformation of G-quadruplexes might vary under different conditions, which could affect the binding and stabilization efficiency of the compounds. Additionally, genome-wide sequence analysis has revealed that hundreds of thousands of regions within the human genome have the potential of forming G-quadruplexes; the manner in which this could affect transcribed RNA is unknown (Huppert and Balasubramanian, 2005; Todd et al., 2005). Therefore, the specificity of these compounds to target telomeres rather than other DNA or RNA molecules is one of the current challenges in this field.

10.4.2 Stem Cells

Although there are many cell types in which telomere shortening might contribute to the physiological declines associated with aging, increasing focus has centered on the role of stem-cell senescence. Part of this shift is due to the identification of stem-cell failure in diseases resulting from mutations in *hTERT* or *hTR*, such as DKC (dyskeratosis congenita), sporadic aplastic anemias, IPF (idiopathic pulmonary fibrosis), and liver disorders (Calado et al., 2009; Garcia et al., 2007). In addition, there is a significant overlap between the symptoms in these diseases and in premature aging syndromes such as Hutchinson–Gilford Progeria. These symptoms not only include anemia and early cancer lesions such as leukoplakia, but also dermatological changes in skin pigmentation and nail dystrophy, which overlap with DKC (Hofer et al., 2005). The molecular defect in progeria is found in the lamin A/C gene, which has important roles in nuclear structure and regulation. Although this mutation is unlikely to directly impact telomeres, the working hypothesis is that the telomere shortening that has been observed in progeria fibroblasts and stem cells results from increased cell turnover due to lamin A/C dysfunction; this can eventually lead to cell death and premature stem-cell depletion.

There is good evidence that rodents do not use telomere shortening/replicative aging as a tumor-suppression mechanism (Sherr and DePinho, 2000; Wright and Shay, 2000). An important difference between human and mouse biology is reflected in their different rates of stem-cell turnover. DNA replication is the major source for the accumulation of mutations, which is why the germ-cell lineage develops very early during embryogenesis and undergoes the minimum number of divisions between generations. In humans, generally quiescent stem cells divide infrequently and, when they do, they spin-off transient amplifying cells and return to their resting state. It is the transient amplifying cells that are responsible for most of the cell division to produce the mass of functioning cells for their respective tissues. In mice, more than 50% of hematopoietic stem cells are labeled within 1 week of continuous BrdU exposure, indicating a rapid turnover (Cheshier et al., 1999). However, less than 40% of hematopoietic stem cells became labeled after 20 weeks of continuous BrdU treatment in baboons (Mahmud et al., 2001). The rate of stem-cell turnover in these two species is very roughly proportional to their lifespan, so one might speculate that the pattern of cell division between stem and transient amplifying cells is adjusted so that approximately the same number of stem-cell divisions occurs regardless of lifespan. Although we do not believe that telomere shortening/replication contributes to the physiological declines of aging in the mouse, the telomere shortening that accompanies stem-cell turnover in humans is likely to produce effects in tissues subject to increased turnover due to disease or in exceptionally elderly individuals.

10.5 CONCLUSIONS

There have been many significant advances in the fields of senescence, telomerase, and telomere stability since Hayflick's first publications describing the limited proliferative potential of human cells. However, many gaps in our understanding of these processes remain. What we do know is that maintenance of functional telomeres is essential for all proliferating cells. In order to overcome replicative-based senescence and to divide enough times to accumulate mutations in proto-oncogenes and tumor-suppressor genes, most tumor cells have escaped replicative aging by expressing telomerase. Telomerase, therefore, is an attractive target for cancer diagnosis and therapy. However, telomerase expression and activity can vary greatly among cancers (Shay and Bacchetti, 1997). Therefore, one key area to pursue is determining the types of cancers that exhibit clinically useful correlations between telomerase activity and either diagnostic or prognostic outcomes (Shay and Wright, 2007). In spite of significant efforts by a variety of pharmaceutical companies, telomerase proved refractory as a target for classic cell-permeable small molecule inhibitor discovery. One major challenge is to discover additional targets in the telomerase pathway that might prove more amenable to inhibition by small molecules. A second challenge to progress this field more rapidly is to develop better tissue culture and animal models for preclinical testing. Compared to humans, current inbred rodent strains are not the best models for studying telomerase therapeutics because of the marked differences in telomere biology and telomerase regulation (so one cannot use mouse models of endogenous tumor formation) (Keith et al., 2002). The current preclinical testing uses human xenograft models, which are done largely in the absence of immune and inflammatory responses that are part of the normal tumor microenvironment. Regardless, it will be extremely exciting to see how well the current telomere- and telomerase-based cancer therapies function in advanced-stage clinical trials and how the next generation of reagents evolves to increase performance.

This chapter has primarily focused on the ability of cancer cells to exploit telomerase in order to divide indefinitely. However, telomerase, itself, is not the enemy. It is, simply put, "the right weapon in the wrong hands" (Satyanarayana et al., 2004b). Transient expression of telomerase has been used to extend the lifespan of normal healthy cells, without the threat of cellular transformation (Steinert et al., 2000). Therefore, in addition to being a target for anticancer therapies, telomerase could be a useful tool for the treatment of age-related diseases and for regenerative medicine.

REFERENCES

Akiyama M, Hideshima T, Shammas MA, Hayashi T, Hamasaki M, Tai YT, Richardson P, Gryaznov S, Munshi NC, Anderson KC. (2003) Effects of oligonucleotide N3′ → P5′ thio-phosphoramidate (GRN163) targeting telomerase RNA in human multiple myeloma cells. *Cancer Res.* **63**: 6187–6194.

Asai A, Oshima Y, Yamamoto Y, Uochi TA, Kusaka H, Akinaga S, Yamashita Y, Pongracz K, Pruzan R, Wunder E, Piatyszek M, Li S, Chin AC, Harley CB, Gryaznov S. (2003) A novel telomerase template antagonist (GRN163) as a potential anticancer agent. *Cancer Res.* **63**: 3931–3939.

Bernhardt SL, Gjertsen MK, Trachsel S, Moller M, Eriksen JA, Meo M, Buanes T, Gaudernack G. (2006) Telomerase peptide vaccination of patients with non-resectable pancreatic cancer: a dose escalating phase I/II study. *Br. J. Cancer* **95**: 1474–1482.

Bilsland AE, Anderson CJ, Fletcher-Monaghan AJ, McGregor F, Evans TR, Ganly I, Knox RJ, Plumb JA, Keith WN. (2003) Selective ablation of human cancer cells by telomerase-specific adenoviral suicide gene therapy vectors expressing bacterial nitroreductase. *Oncogene* **22**: 370–380.

Blackburn EH. (1991) Structure and function of telomeres. *Nature* **350**: 569–573.

Blasco MA, Lee HW, Hande MP, Samper E, Lansdorp PM, DePinho RA, Greider CW. (1997) Telomere shortening and tumor formation by mouse cells lacking telomerase RNA. *Cell* **91**: 25–34.

Bodnar AG, Ouellette M, Frolkis M, Holt SE, Chiu CP, Morin GB, Harley CB, Shay JW, Lichtsteiner S, Wright WE. (1998) Extension of life-span by introduction of telomerase into normal human cells. *Science* **279**: 349–352.

Brunsvig PF, Aamdal S, Gjertsen MK, Kvalheim G, Markowski-Grimsrud CJ, Sve I, Dyrhaug M, Trachsel S, Moller M, Eriksen JA, Gaudernack G. (2006) Telomerase peptide vaccination: a phase I/II study in patients with non-small cell lung cancer. *Cancer Immunol. Immunother.* **55**: 1553–1564.

Bryan TM, Baumann P. (2010) G-quadruplexes: from guanine gels to chemotherapeutics. *Methods Mol. Biol.* **608**: 1–16.

Bryan TM, Englezou A, Gupta J, Bacchetti S, Reddel RR. (1995) Telomere elongation in immortal human cells without detectable telomerase activity. *EMBO J* **14**: 4240–4248.

Calado RT, Regal JA, Kleiner DE, Schrump DS, Peterson NR, Pons V, Chanock SJ, Lansdorp PM, Young NS. (2009) A spectrum of severe familial liver disorders associate with telomerase mutations. *PLoS One* **4**: e7926.

Cheshier SH, Morrison SJ, Liao X, Weissman IL. (1999) *In vivo* proliferation and cell cycle kinetics of long-term self-renewing hematopoietic stem cells. *Proc. Natl. Acad. Sci. USA* **96**: 3120–3125.

Cimino-Reale G, Pascale E, Battiloro E, Starace G, Verna R, D'Ambrosio E. (2001) The length of telomeric G-rich strand 3′-overhang measured by oligonucleotide ligation assay. *Nucl. Acids Res.* **29**: E35.

Collins K, Mitchell JR. (2002) Telomerase in the human organism. *Oncogene* **21**: 564–579.

Coppe JP, Desprez PY, Krtolica A, Campisi J. (2010) The senescence-associated secretory phenotype: the dark side of tumor suppression. *Annu. Rev. Pathol.* **5**: 99–118.

d'Adda di Fagagna F, Reaper PM, Clay-Farrace L, Fiegler H, Carr P, Von Zglinicki T, Saretzki G, Carter NP, Jackson SP. (2003) A DNA damage checkpoint response in telomere-initiated senescence. *Nature* **426**: 194–198.

Davoli T, Denchi EL, de Lange T. (2010) Persistent telomere damage induces bypass of mitosis and tetraploidy. *Cell* **141**: 81–93.

de Lange T. (2005) Shelterin: the protein complex that shapes and safeguards human telomeres. *Genes Dev.* **19**: 2100–2110.

Dikmen ZG, Gellert GC, Jackson S, Gryaznov S, Tressler R, Dogan P, Wright WE, Shay JW. (2005) *In vivo* inhibition of lung cancer by GRN163L: a novel human telomerase inhibitor. *Cancer Res.* **65**: 7866–7873.

Dikmen ZG, Wright WE, Shay JW, Gryaznov SM. (2008) Telomerase targeted oligonucleotide thio-phosphoramidates in T24-luc bladder cancer cells. *J. Cell. Biochem.* **104**: 444–452.

Djojosubroto MW, Chin AC, Go N, Schaetzlein S, Manns MP, Gryaznov S, Harley CB, Rudolph KL. (2005) Telomerase antagonists GRN163 and GRN163L inhibit tumor growth and increase chemosensitivity of human hepatoma. *Hepatology* **42**: 1127–1136.

Dunham MA, Neumann AA, Fasching CL, Reddel RR. (2000) Telomere maintenance by recombination in human cells. *Nat. Genet.* **26**: 447–450.

Feldser DM, Hackett JA, Greider CW. (2003) Telomere dysfunction and the initiation of genome instability. *Nat. Rev. Cancer* **3**: 623–627.

Feng J, Funk WD, Wang SS, Weinrich SL, Avilion AA, Chiu CP, Adams RR, Chang E, Allsopp RC, Yu J, et al. (1995) The RNA component of human telomerase. *Science* **269**: 1236–1241.

Garcia CK, Wright WE, Shay JW. (2007) Human diseases of telomerase dysfunction: insights into tissue aging. *Nucl. Acids Res.* **35**: 7406–7416.

Gellert GC, Dikmen ZG, Wright WE, Gryaznov S, Shay JW. (2006) Effects of a novel telomerase inhibitor, GRN163L, in human breast cancer. *Breast Cancer Res. Treat.* **96**: 73–81.

Gomez-Millan J, Goldblatt EM, Gryaznov SM, Mendonca MS, Herbert BS. (2007) Specific telomere dysfunction induced by GRN163L increases radiation sensitivity in breast cancer cells. *Int. J. Radiat. Oncol. Biol. Phys.* **67**: 897–905.

Goytisolo FA, Blasco MA. (2002) Many ways to telomere dysfunction: *in vivo* studies using mouse models. *Oncogene* **21**: 584–591.

Griffith JD, Comeau L, Rosenfield S, Stansel RM, Bianchi A, Moss H, de Lange T. (1999) Mammalian telomeres end in a large duplex loop. *Cell* **97**: 503–514.

Hahn WC, Counter CM, Lundberg AS, Beijersbergen RL, Brooks MW, Weinberg RA. (1999a) Creation of human tumour cells with defined genetic elements. *Nature* **400**: 464–468.

Hahn WC, Stewart SA, Brooks MW, York SG, Eaton E, Kurachi A, Beijersbergen RL, Knoll JH, Meyerson M, Weinberg RA. (1999b) Inhibition of telomerase limits the growth of human cancer cells. *Nat. Med.* **5**: 1164–1170.

Hanahan D, Weinberg RA. (2000) The hallmarks of cancer. *Cell* **100**: 57–70.

Hara E, Tsurui H, Shinozaki A, Nakada S, Oda K. (1991) Cooperative effect of antisense-Rb and antisense-p53 oligomers on the extension of life span in human diploid fibroblasts, TIG-1. *Biochem. Biophys. Res. Commun.* **179**: 528–534.

Harley CB. (2008) Telomerase and cancer therapeutics. *Nat. Rev. Cancer* **8**: 167–179.

Harley CB, Futcher AB, Greider CW. (1990) Telomeres shorten during ageing of human fibroblasts. *Nature* **345**: 458–460.

Hastie ND, Dempster M, Dunlop MG, Thompson AM, Green DK, Allshire RC. (1990) Telomere reduction in human colorectal carcinoma and with ageing. *Nature* **346**: 866–868.

Hayflick L. (1965) The limited *in vitro* lifetime of human diploid cell strains. *Exp. Cell. Res.* **37**: 614–636.

Hayflick L. (1984) The coming of age of WI-38. *Adv. Cell. Cult.* **3**: 303–316.

Hayflick L. (1998) How and why we age. *Exp. Gerontol.* **33**: 639–653.

Hayflick L, Moorhead PS. (1961) The serial cultivation of human diploid cell strains. *Exp. Cell. Res.* **25**: 585–621.

Hemann MT, Strong MA, Hao LY, Greider CW. (2001) The shortest telomere, not average telomere length, is critical for cell viability and chromosome stability. *Cell* **107**: 67–77.

Henderson E, Hardin CC, Walk SK, Tinoco I, Jr., Blackburn EH. (1987) Telomeric DNA oligonucleotides form novel intramolecular structures containing guanine–guanine base pairs. *Cell* **51**: 899–908.

Henderson ER, Blackburn EH. (1989) An overhanging 3′ terminus is a conserved feature of telomeres. *Mol. Cell. Biol.* **9**: 345–348.

Hochreiter AE, Xiao H, Goldblatt EM, Gryaznov SM, Miller KD, Badve S, Sledge GW, Herbert BS. (2006) Telomerase template antagonist GRN163L disrupts telomere maintenance, tumor growth, and metastasis of breast cancer. *Clin. Cancer Res.* **12**: 3184–3192.

Hofer AC, Tran RT, Aziz OZ, Wright W, Novelli G, Shay J, Lewis M. (2005) Shared phenotypes among segmental progeroid syndromes suggest underlying pathways of aging. *J. Gerontol. A Biol. Sci. Med. Sci.* **60**: 10–20.

Holt SE, Shay JW, Wright WE. (1996) Refining the telomere–telomerase hypothesis of aging and cancer. *Nat. Biotechnol.* **14**: 836–839.

Huppert JL, Balasubramanian S. (2005) Prevalence of quadruplexes in the human genome. *Nucl. Acids Res.* **33**: 2908–2916.

Jackson SR, Zhu CH, Paulson V, Watkins L, Dikmen ZG, Gryaznov SM, Wright WE, Shay JW. (2007) Antiadhesive effects of GRN163L—an oligonucleotide N3′ → P5′ thio-phosphoramidate targeting telomerase. *Cancer Res.* **67**: 1121–1129.

Karlseder J, Broccoli D, Dai Y, Hardy S, de Lange T. (1999) p53- and ATM-dependent apoptosis induced by telomeres lacking TRF2. *Science* **283**: 1321–1325.

Karlseder J, Smogorzewska A, de Lange T. (2002) Senescence induced by altered telomere state, not telomere loss. *Science* **295**: 2446–2449.

Kass-Eisler A, Greider CW. (2000) Recombination in telomere-length maintenance. *Trends Biochem. Sci.* **25**: 200–204.

Keith WN, Bilsland A, Evans TR, Glasspool RM. (2002) Telomerase-directed molecular therapeutics. *Expert Rev. Mol. Med.* **4**: 1–25.

Keith WN, Bilsland A, Hardie M, Evans TR. (2004) Drug insight: cancer cell immortality-telomerase as a target for novel cancer gene therapies. *Nat. Clin. Pract. Oncol.* **1**: 88–96.

Keith WN, Thomson CM, Howcroft J, Maitland NJ, Shay JW. (2007) Seeding drug discovery: integrating telomerase cancer biology and cellular senescence to uncover new therapeutic opportunities in targeting cancer stem cells. *Drug Discov. Today.* **12**: 611–621.

Kim MM, Rivera MA, Botchkina IL, Shalaby R, Thor AD, Blackburn EH. (2001) A low threshold level of expression of mutant-template telomerase RNA inhibits human tumor cell proliferation. *Proc. Natl. Acad. Sci. USA.* **98**: 7982–7987.

Kim NW, Piatyszek MA, Prowse KR, Harley CB, West MD, Ho PL, Coviello GM, Wright WE, Weinrich SL, Shay JW. (1994) Specific association of human telomerase activity with immortal cells and cancer. *Science* **266**: 2011–2015.

Li Y, Idamakanti N, Arroyo T, Thorne S, Reid T, Nichols S, VanRoey M, Colbern G, Nguyen N, Tam O, Working P, Yu DC. (2005) Dual promoter-controlled oncolytic adenovirus CG5757 has strong tumor selectivity and significant antitumor efficacy in preclinical models. *Clin. Cancer Res* **11**: 8845–8855.

Lingner J, Cech TR. (1998) Telomerase and chromosome end maintenance. *Curr. Opin. Genet. Dev.* **8**: 226–232.

Lipps HJ, Rhodes D. (2009) G-quadruplex structures: *in vivo* evidence and function. *Trends Cell. Biol.* **19**: 414–422.

Lou Z, Wei J, Riethman H, Baur JA, Voglauer R, Shay JW, Wright WE. (2009) Telomere length regulates ISG15 expression in human cells. *Aging (Albany NY)* **1**: 608–621.

Mahmud N, Devine SM, Weller KP, Parmar S, Sturgeon C, Nelson MC, Hewett T, Hoffman R. (2001) The relative quiescence of hematopoietic stem cells in nonhuman primates. *Blood* **97**: 3061–3068.

Makarov VL, Hirose Y, Langmore JP. (1997) Long G tails at both ends of human chromosomes suggest a C strand degradation mechanism for telomere shortening. *Cell* **88**: 657–666.

Marian CO, Cho SK, McEllin BM, Maher EA, Hatanpaa KJ, Madden CJ, Mickey BE, Wright WE, Shay JW, Bachoo RM. (2010) The telomerase antagonist, imetelstat, efficiently targets glioblastoma tumor-initiating cells leading to decreased proliferation and tumor growth. *Clin Cancer Res.* **16**: 154–163.

Maser RS, DePinho RA. (2002) Connecting chromosomes, crisis, and cancer. *Science* **297**: 565–569.

Minev B, Hipp J, Firat H, Schmidt JD, Langlade-Demoyen P, Zanetti M. (2000) Cytotoxic T cell immunity against telomerase reverse transcriptase in humans. *Proc. Natl. Acad. Sci. USA* **97**: 4796–4801.

Miyoshi D, Matsumura S, Nakano S, Sugimoto N. (2004) Duplex dissociation of telomere DNAs induced by molecular crowding. *J. Am. Chem. Soc.* **126**: 165–169.

Nakamura TM, Morin GB, Chapman KB, Weinrich SL, Andrews WH, Lingner J, Harley CB, Cech TR. (1997) Telomerase catalytic subunit homologs from fission yeast and human. *Science* **277**: 955–959.

Nugent CI, Lundblad V. (1998) The telomerase reverse transcriptase: components and regulation. *Genes Dev.* **12**: 1073–1085.

Oganesian L, Moon IK, Bryan TM, Jarstfer MB. (2006) Extension of G-quadruplex DNA by ciliate telomerase. *EMBO J.* **25**: 1148–1159.

Olovnikov AM. (1973) A theory of marginotomy. The incomplete copying of template margin in enzymic synthesis of polynucleotides and biological significance of the phenomenon. *J. Theor. Biol.* **41**: 181–190.

Ozawa T, Gryaznov SM, Hu LJ, Pongracz K, Santos RA, Bollen AW, Lamborn KR, Deen DF. (2004) Antitumor effects of specific telomerase inhibitor GRN163 in human glioblastoma xenografts. *Neuro. Oncol.* **6**: 218–226.

Parrinello S, Samper E, Krtolica A, Goldstein J, Melov S, Campisi J. (2003) Oxygen sensitivity severely limits the replicative lifespan of murine fibroblasts. *Nat. Cell. Biol.* **5**: 741–747.

Phatak P, Burger AM. (2007) Telomerase and its potential for therapeutic intervention. *Br. J. Pharmacol.* **152**: 1003–1011.

Plumb JA, Bilsland A, Kakani R, Zhao J, Glasspool RM, Knox RJ, Evans TR, Keith WN. (2001) Telomerase-specific suicide gene therapy vectors expressing bacterial nitroreductase sensitize human cancer cells to the pro-drug CB1954. *Oncogene* **20**: 7797–7803.

Reddel RR. (2000) The role of senescence and immortalization in carcinogenesis. *Carcinogenesis* **21**: 477–484.

Reddel RR. (2010) Senescence: an antiviral defense that is tumor suppressive? *Carcinogenesis* **31**: 19–26.

Satyanarayana A, Greenberg RA, Schaetzlein S, Buer J, Masutomi K, Hahn WC, Zimmermann S, Martens U, Manns MP, Rudolph KL. (2004a) Mitogen stimulation cooperates with telomere shortening to activate DNA damage responses and senescence signaling. *Mol. Cell. Biol.* **24**: 5459–5474.

Satyanarayana A, Manns MP, Rudolph KL. (2004b) Telomeres, telomerase and cancer: an endless search to target the ends. *Cell Cycle* **3**: 1138–1150.

Sedelnikova OA, Horikawa I, Zimonjic DB, Popescu NC, Bonner WM, Barrett JC. (2004) Senescing human cells and ageing mice accumulate DNA lesions with unrepairable double-strand breaks. *Nat. Cell. Biol.* **6**: 168–170.

Shay JW, Bacchetti S. (1997) A survey of telomerase activity in human cancer. *Eur. J. Cancer* **33**: 787–791.

Shay JW, Gazdar AF. (1997) Telomerase in the early detection of cancer. *J. Clin. Pathol.* **50**: 106–109.

Shay JW, Keith WN. (2008) Targeting telomerase for cancer therapeutics. *Br. J. Cancer* **98**: 677–683.

Shay JW, Pereira-Smith OM, Wright WE. (1991a) A role for both RB and p53 in the regulation of human cellular senescence. *Exp. Cell. Res.* **196**: 33–39.

Shay JW, Van Der Haegen BA, Ying Y, Wright WE. (1993) The frequency of immortalization of human fibroblasts and mammary epithelial cells transfected with SV40 large T-antigen. *Exp. Cell. Res.* **209**: 45–52.

Shay JW, Wright WE. (2005) Mechanism-based combination telomerase inhibition therapy. *Cancer Cell* **7**: 1–2.

Shay JW, Wright WE. (2006) Telomerase therapeutics for cancer: challenges and new directions. *Nat. Rev. Drug Discov.* **5**: 577–584.

Shay JW, Wright WE. (2007) Hallmarks of telomeres in ageing research. *J. Pathol.* **211**: 114–123.

Shay JW, Wright WE, Werbin H. (1991b) Defining the molecular mechanisms of human cell immortalization. *Biochim. Biophys. Acta.* **1072**: 1–7.

Sherr CJ, DePinho RA. (2000) Cellular senescence: mitotic clock or culture shock? *Cell* **102**: 407–410.

Simonsson T. (2001) G-quadruplex DNA structures—variations on a theme. *Biol. Chem.* **382**: 621–628.

Steinert S, Shay JW, Wright WE. (2000) Transient expression of human telomerase extends the life span of normal human fibroblasts. *Biochem. Biophys. Res. Commun.* **273**: 1095–1098.

Su Z, Dannull J, Yang BK, Dahm P, Coleman D, Yancey D, Sichi S, Niedzwiecki D, Boczkowski D, Gilboa E, Vieweg J. (2005) Telomerase mRNA-transfected dendritic cells stimulate antigen-specific CD8+ and CD4+ T cell responses in patients with metastatic prostate cancer. *J. Immunol.* **174**: 3798–3807.

Sundquist WI, Klug A. (1989) Telomeric DNA dimerizes by formation of guanine tetrads between hairpin loops. *Nature* **342**: 825–829.

Takai H, Smogorzewska A, de Lange T. (2003) DNA damage foci at dysfunctional telomeres. *Curr. Biol.* **13**: 1549–1556.

Tang J, Kan ZY, Yao Y, Wang Q, Hao YH, Tan Z. (2008) G-quadruplex preferentially forms at the very 3′ end of vertebrate telomeric DNA. *Nucl. Acids Res.* **36**: 1200–1208.

Todd AK, Johnston M, Neidle S. (2005) Highly prevalent putative quadruplex sequence motifs in human DNA. *Nucleic Acids Res* **33**: 2901–2907.

van Steensel B, Smogorzewska A, de Lange T. (1998) TRF2 protects human telomeres from end-to-end fusions. *Cell* **92**: 401–413.

Vogelstein B, Kinzler KW. (1993) The multistep nature of cancer. *Trends Genet.* **9**: 138–141.

von Zglinicki T. (2002) Oxidative stress shortens telomeres. *Trends Biochem. Sci.* **27**: 339–344.

Vonderheide RH. (2002) Telomerase as a universal tumor-associated antigen for cancer immunotherapy. *Oncogene* **21**: 674–679.

Vonderheide RH, Hahn WC, Schultze JL, Nadler LM. (1999) The telomerase catalytic subunit is a widely expressed tumor-associated antigen recognized by cytotoxic T lymphocytes. *Immunity* **10**: 673–679.

Wang ES, Wu K, Chin AC, Chen-Kiang S, Pongracz K, Gryaznov S, Moore MA. (2004) Telomerase inhibition with an oligonucleotide telomerase template antagonist: *in vitro* and *in vivo* studies in multiple myeloma and lymphoma. *Blood* **103**: 258–266.

Watson JD. (1972) Origin of concatemeric T7 DNA. *Nat. New Biol.* **239**: 197–201.

Williamson JR, Raghuraman MK, Cech TR. (1989) Monovalent cation-induced structure of telomeric DNA: the G-quartet model. *Cell* **59**: 871–880.

Wright WE, Pereira-Smith OM, Shay JW. (1989) Reversible cellular senescence: implications for immortalization of normal human diploid fibroblasts. *Mol. Cell Biol.* **9**: 3088–3092.

Wright WE, Shay JW. (2000) Telomere dynamics in cancer progression and prevention: fundamental differences in human and mouse telomere biology. *Nat. Med.* **6**: 849–851.

Wright WE, Shay JW. (2001) Cellular senescence as a tumor-protection mechanism: the essential role of counting. *Curr. Opin. Genet Dev.* **11**: 98–103.

Wright WE, Tesmer VM, Huffman KE, Levene SD, Shay JW. (1997) Normal human chromosomes have long G-rich telomeric overhangs at one end. *Genes Dev.* **11**: 2801–2809.

Zahler AM, Williamson JR, Cech TR, Prescott DM. (1991) Inhibition of telomerase by G-quartet DNA structures. *Nature* **350**: 718–720.

Zaug AJ, Podell ER, Cech TR. (2005) Human POT1 disrupts telomeric G-quadruplexes allowing telomerase extension *in vitro*. *Proc. Natl. Acad. Sci. USA* **102**: 10864–10869.

Zhao Y, Hoshiyama H, Shay JW, Wright WE. (2008) Quantitative telomeric overhang determination using a double-strand specific nuclease. *Nucl. Acids Res.* **36**: e14.

Zou Y, Misri S, Shay JW, Pandita TK, Wright WE. (2009) Altered states of telomere deprotection and the two-stage mechanism of replicative aging. *Mol. Cell. Biol.* **29**: 2390–2397.

Zou Y, Sfeir A, Gryaznov SM, Shay JW, Wright WE. (2004) Does a sentinel or a subset of short telomeres determine replicative senescence? *Mol. Biol. Cell.* **15**: 3709–3718.

11

TELOMERASE, RETROTRANSPOSONS, AND EVOLUTION

Irina R. Arkhipova

11.1 INTRODUCTION

Telomerase reverse transcriptase (TERT) is so far the only known reverse transcriptase (RT) harboring the features of a *bona fide* nuclear gene, which has an important function in the eukaryotic cell. Traditionally, it has been called a "specialized reverse transcriptase," because it is highly adapted to a specific G-rich template RNA and can be regarded as a ribonucleoprotein enzyme in which both the protein and RNA moieties act together to add a large number of short tandem G-rich repeat units to the exposed chromosome termini. The evolutionary origins of telomerase have been debated since the discovery of the characteristic RT motifs in the genes coding for ciliate, yeast, and human telomerases by Cech and his colleagues (Lingner et al., 1997; Nakamura et al., 1997). This seminal finding challenged the previously established view that all RTs originate from various kinds of cellular or viral selfish genetic elements, and immediately triggered the search for a presumptive ancestral RT form which existed during early eukaryotic evolution and gave rise to telomerase (Eickbush, 1997; Nakamura and Cech, 1998). While the origin of telomerase is almost unambiguously placed at the point of divergence of the earliest eukaryotic organisms, the exact sequence of events leading from prokaryotic RTs to eukaryotic telomerases remains a subject of debate, and will likely remain obscured by the very ancient nature of the prokaryote-to-eukaryote transition. However, attempts to penetrate the depth of time at the dawn of eukaryotic evolution and to reconstruct the events that could have led to such a major saltatory transition as formation of linear

Telomerases: Chemistry, Biology, and Clinical Applications, First Edition.
Edited by Neal F. Lue and Chantal Autexier.

chromosomes are certainly fascinating. It would hardly be an overstatement to say that early eukaryotic evolution was inextricably linked to the emergence of linear chromosomes and the concomitant appearance of the mechanism capable of maintaining linear chromosome ends.

The principal lines of reasoning behind the attempts to uncover these early events are based on the widespread occurrence of several different types of RT-related sequences in prokaryotic genomes. If we disregard a highly unlikely possibility that all of these RTs were introduced to prokaryotes from eukaryotes *via* horizontal transfer, then the most likely sequence of events involved recruitment of an ancestral prokaryotic RT for maintenance of linear chromosome ends. The point of contention in this scenario is usually the nature of this ancestral RT. Two possible candidates have been considered: prokaryotic group II introns, and non-LTR retrotransposons which could have emerged in earliest eukaryotes. Since both of these types represent transposable elements, the evolutionary transition to telomerase would necessitate domestication of an initially mobile genetic unit, which was converted from multicopy to single-copy and underwent disassociation from its template, so that the latter was moved to an unlinked genomic location. However, it is also possible that a single-copy RT gene, initially present in a prokaryotic genome, became specialized for extending 3′-OH termini of linear DNA *via* acquisition of additional domains and association with an unlinked G-rich template. In the following sections of this chapter, I will consider these possibilities in detail, taking into account the properties of known RT-containing elements and their evolutionary relationships, and will evaluate different scenarios in an attempt to trace the chain of evolutionary transitions connecting RTs of retrotransposons, telomerases, and their possible evolutionary intermediates.

In essence, the telomerase-based solution to chromosome-end maintenance consists of two principal components: the added terminal sequences *per se*, and the enzymatic machinery that generates them, that is, TERT with the corresponding accessory factors. Terminal sequences are known to consist of short tandem G-rich repeats, and telomerase is known to synthesize those repeats using the existing 3′-OH end of the chromosome as a primer and the specialized telomerase RNA (TER or TR) as a template (Fig. 11.1) (reviewed in Autexier and Lue, 2006).

The chapter is subdivided into four major parts. The first part offers a very brief overview of the relatively small number of known prokaryotic linear chromosomes. It is aimed at understanding the problems that could arise upon linearization of circular replicons, and of general strategies that may be used to generate and maintain such linear replicons. The second part reviews the ways in which organisms can cope with total loss of telomerase-based chromosome-end maintenance, and re-emphasizes profound similarities between telomerases and retrotransposons in utilizing RNA-dependent DNA synthesis to preserve linear chromosome ends (Fig. 11.1). The third part examines the best-known cases in which telomeric repeats and retrotransposons occur together at the chromosome termini, and evaluates the possible significance of such co-occurrence for evolutionary transitions between retroelements and telomerases. The final part compares the properties of the major types of RTs in extant prokaryotes, as well as in their eukaryotic counterparts, in an attempt to understand which of the RT types were more likely to participate in a transition to a telomerase-like

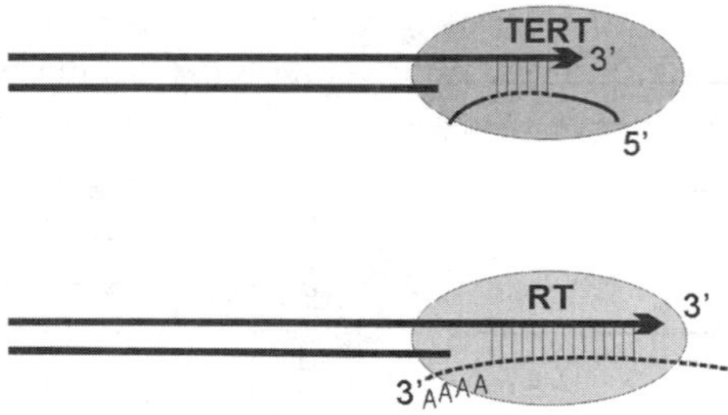

FIGURE 11.1 Similarity between telomerase-mediated and retrotransposon-mediated telomere elongation. The enzyme (TERT or RT) performs RNA-dependent DNA synthesis, using either the short segment of telomerase RNA (repeatedly) or the full-length retrotransposon RNA (continuously) as a template. DNA, thick lines; RNA, thin lines; RNA template region, dashed lines; enzyme, shaded ovals; thin vertical lines, complementary base pairs.

ancestral RT in a primordial eukaryote. While the details of this transition, which occurred well over a billion years ago, will forever remain mysterious, the eventual solution of the end-replication problem, that is telomerase-based chromosome-end maintenance, is well understood and nearly universal in extant eukaryotes. In figuring out the details of this transition, which may sometimes amount to a guessing game rather than rigorous scientific investigation (Nakamura and Cech, 1998), the best we can do is to make an educated guess, which would offer the most likely evolutionary scenario while keeping inevitable contradictions to a minimum.

11.2 CIRCULAR VERSUS LINEAR CHROMOSOMES IN BACTERIA

An overwhelming majority of bacterial chromosomes exist in the circular form and are not faced with problems associated with incomplete DNA replication at the ends, or with the propensity of linear DNA ends to be degraded by exonucleases. Moreover, no linear chromosomes have yet been reported in archaea. It is certainly reasonable to assume that ancestral eukaryotes underwent transition from circular to linear chromosomes, and that in the course of this transition they had to develop novel evolutionary solutions to the newly arising problems of chromosome-end restoration and protection. Thus, a brief description of linear chromosomes in the prokaryotic world may help in understanding general strategies that can be used by bacteria to solve the problems arising from linearization of an originally circular chromosome.

11.2.1 Major Types of Bacterial "Telomeres"

Linear chromosomes in prokaryotes, in contrast to eukaryotes, represent an exception rather than the rule. The best-studied cases occur in highly dissimilar bacterial groups, and underscore the major types of organization of prokaryotic linear replicons (Fig. 11.2a). The free-living soil bacteria *Streptomyces* (Actinobacteria), with very large (6–9 Mb) GC-rich genomes, have relatively long terminal inverted repeats

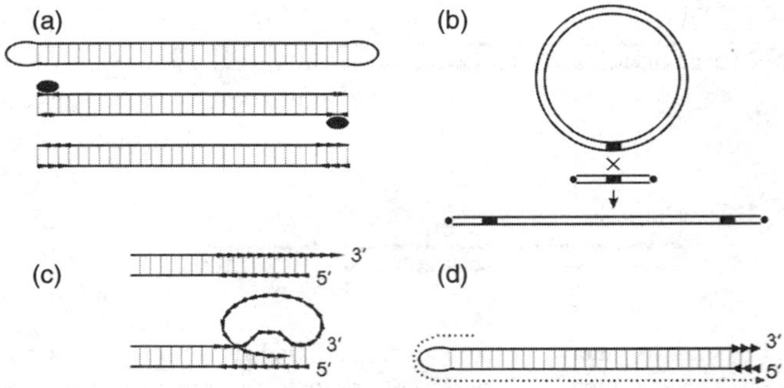

FIGURE 11.2 The diversity of prokaryotic "telomeres." (a) Types of terminal structures exemplified by *Borrelia* (Chaconas, 2005), *Streptomyces* (Chen et al., 2002), and *Tetrahymena* (Morin and Cech, 1986, 1988). (b) Recombination between a circular chromosome and a linear plasmid with protected termini (black circles). (c) Protection of linear DNA ends *via* formation of a t-loop, which forms an end-structure analogous to that of a closed hairpin, but requires terminal repeats (de Lange, 2004). (d) "Clothespin" structure of linear pFOXC retroplasmids in *Fusarium oxysporum* combines two different types of linear DNA end-structures (closed hairpin and terminal repeats) (Simpson et al., 2004; Walther and Kennell, 1999). Thin vertical lines, Watson–Crick base pairing; thin dotted line, RNA transcript. Not to scale.

(TIRs) at the chromosome ends, which are capped by covalently bound terminal proteins (TP) (reviewed in Chen et al., 2002). The highly pathogenic obligate parasites of the genus *Borrelia* (Spirochaetes), with compact (~0.9 Mb) AT-rich genomes, protect their ends by covalently closed terminal hairpins, which are generated after chromosome replication with the aid of specialized telomere resolvases encoded on circular plasmids (reviewed in Chaconas and Chen, 2005).

In certain bacteriophages, the recombinase-type enzymes that generate the hairpin ends have been dubbed "protelomerases" (Aihara et al., 2007; Deneke et al., 2000, 2002). In this case, however, the analogy with telomerases is rather superficial, and it may be argued that the relatively recent linearization of these chromosomes makes them of limited use when it comes to tracing the origin of linear chromosomes during early eukaryogenesis. While considering linear replicons in prokaryotes, it should be kept in mind that the solutions adopted by prokaryotes relatively recently in evolution would not necessarily parallel those solutions that ancestral eukaryotes came to adopt. Nevertheless, the diversity of underlying scenarios makes it possible to outline some of the general principles that may be used to solve similar problems by taking advantage of diverse available resources.

In both *Borrelia* and *Streptomyces*, it has been proposed that the most likely scenario for linearization of initially circular chromosomes was *via* homologous recombination with a linear plasmid encoding the enzymatic function required to regenerate its ends (Fig. 11.2b) (Casjens, 1999; Chen et al., 2002; Volff and Altenbuchner, 2000). Although in this case the enzymatic machinery (tyrosine recombinase, or YR) bears no resemblance to telomerase, the principle of recombining a linear plasmid with a circular chromosome certainly deserves attention, and will be discussed in more detail below.

To prevent exonucleolytic degradation, eukaryotes typically avoid exposure of DNA ends to exonucleases by protein binding and formation of the so-called t-loops, instead of using covalently closed hairpins, thereby achieving essentially the same "hairpin effect" in a t-loop by burying the exposed 3′ G-rich overhang within the duplex part of the telomere *via* strand displacement and D-loop formation (Fig. 11.2c; reviewed in de Lange, 2004). The t-loops, however, require a certain degree of sequence redundancy, so that a single-stranded end would be able to invade the DNA duplex and to base-pair with the strand of the opposite polarity in that duplex, which may have favored acquisition of telomeric repeats (see de Lange, 2004).

Terminal redundancy is also used to counteract terminal loss in linear replicons. Mitochondrial genomes, which descend from endosymbiotic proteobacteria, exist in a linear form in many fungal and plant species, and exhibit a variety of terminal structures (reviewed in Nosek et al., 1998). In particular, the ends of mtDNA in certain ciliated protozoans carry variable numbers of tandem repeats, which bear no resemblance to nuclear telomeric repeats, do not exhibit any sequence conservation between species, and sometimes even between telomeres, and are thought to be maintained by homologous recombination (Morin and Cech, 1986, 1988).

Thus, despite the lack of telomerase-based mechanisms for maintenance and protection of linear chromosome ends, prokaryotes can illustrate the general strategies which can be used to perform these functions: either the same protein binds at multiple chromosome ends, or the same sequence is present at multiple ends. Would a combination of these two strategies synergistically enhance the advantages of having the same sequences and the same set of proteins at each of the chromosome termini? Apparently, such a combination has become possible for specialized RTs, aka telomerases, which have the potential to recognize their template sequences in *trans*, and to disperse these sequences in tandemly repeated arrays at multiple chromosome ends. If these sequences are also capable of binding proteins, one could argue that telomerase-based end maintenance may turn out to be the most efficient solution for maintaining and protecting chromosome ends.

11.2.2 A Special Type of Linear Replicon: Mitochondrial Retroplasmids

An overview of prokaryotic linear replicons would be incomplete without describing linear retroplasmids. While the best-studied retroplasmids inhabiting mitochondria of certain *Neurospora* strains (Mauriceville and Varkud) represent circular replicons, their counterparts in *Fusarium oxysporum* (pFOXC1, pFOXC2, pFOXC3) are linear rather than circular (Kistler et al., 1997; Kuiper and Lambowitz, 1988). The *Fusaruim* mitochondrial plasmids are thought to be propagated by reverse transcription, as they encode a single open-reading frame with RT activity (Walther and Kennell, 1999). Their ends are organized in a peculiar way, representing an interesting mix of two different types of termini. While the 5′ end represents a covalently closed hairpin, as in certain bacterial linear chromosome ends described above, the 3′ end carries a 5-bp sequence which is tandemly repeated 3–5 times, seemingly resembling the redundant termini of eukaryotic chromosomes. The mechanism of formation of a terminal hairpin, however, is fundamentally different from that in *Borrelia*: there is no resolvase involved in closing the hairpin, and instead, both complementary strands forming the hairpin can be transcribed through the hairpin region (Fig. 11.2d). This leads to formation of a transcript with a terminal hairpin at the 5′ end. While it is not known exactly how the 3′ terminal tandem repeats are formed, the analogy with telomeric repeats has been pointed out, although in this case the terminal repeat sequences are not G-rich (ATCTA/TAGAT).

The pFOX retroplasmid RT has a high degree of specificity for its transcript, and apparently initiates cDNA synthesis at or near the 3′ end of the transcript containing terminal repeats. The priming mechanism is hypothesized to be a "snap-back" of the 3′-end on itself using a short region of microcomplementarity within the telomeric-like repeats (Simpson et al., 2004). In contrast, the RTs of circular retroplasmids have the ability to initiate cDNA synthesis *de novo* after recognizing a tRNA-like structure ending in $(CCA)_2$ at the 3′ end of the transcript (Wang and Lambowitz, 1993; Chiang and Lambowitz, 1997). No integration events have yet been observed for any of the linear pFOXC plasmids, but occasional integration of *Neurospora* mitochondrial retroplasmids is thought to involve template jumps onto mt rRNA and subsequent integration of hybrid cDNA by homologous recombination with rDNA (Chiang et al., 1994). The RTL (RT-like) gene from mitochondrial DNA of the green alga *Chlamydomonas reinhardtii* (Boer and Gray, 1988) was also hypothesized to be a product of a homologous recombination event involving a chimeric template-switched cDNA (Chiang et al., 1994).

It has been argued that retroplasmids, especially the linear ones, could represent "molecular fossils" related to the progenitors of the telomerase complex, as they form RNP complexes and can initiate reverse transcription on an RNA template which ends in a short simple repeat (Walther and Kennell, 1999; Wang and Lambowitz, 1993). If this is the case, the principal requirements for transition to the telomerase-based mechanism could be disassociation between the RT-encoding ORF and its template, acquisition of *trans*-recognition between these two components, and incorporation of the ORF and the template-coding sequence into chromosomal DNA, possibly by recombination. While plasmids are generally regarded as selfish genetic elements, a plasmid-derived RT might fulfill these requirements upon integration into the chromosome, so that it would become a single-copy gene. If retroplasmids similar to pFOXC, which forms an RNP complex with its own template in *Fusarium* mitochondria, were to be considered as evolutionary precursors to telomerases, the transition step would also necessarily include acquisition of an RNA-binding domain which can specifically recognize the RNA template in *trans*.

11.3 WHAT HAPPENS WHEN TELOMERASE GETS LOST?

11.3.1 Takeover of Chromosome Ends by Retrotransposons in *Drosophila melanogaster*

Even before it was discovered that telomerase is a RT, one exception to the general rule of telomerase-based chromosome-end maintenance was already evident: *Drosophila* chromosomes were found to end in moderately repetitive telomeric

TABLE 11.1 Major Classes of Eukaryotic Retroelements

Class	Group (Clade)	Endonuclease
LTR	Hepadnavirus	N/A
	Caulimovirus	N/A
	Gypsy	Integrase (IN), C-terminal to RT
	Retrovirus	Integrase (IN), C-terminal to RT
	Copia	Integrase (IN), N-terminal to RT
Non-LTR (LINE)	CRE/Gil/R2	Restriction enzyme-like (REL)
	L1/RTE/I/CR1/Jockey	Apurinic-apyrimidinic (APE)
DIRS		Tyrosine recombinase (YR)
Penelope		GIY-YIG (or none)
TERT		N/A

elements, HeT-A, which did not resemble G-rich telomeric repeats present in the majority of other eukaryotes (Biessmann et al., 1990). Following the discovery of another telomeric transposable element, TART, which contained an ORF coding for RT similar to that of non-LTR retrotransposons (Levis et al., 1993), the HeT-A elements were also classified as non-LTR retrotransposons, although they entirely lack an RT domain (Table 11.1; Fig. 11.3a). A common feature shared between both families, which allows their classification as non-LTR retrotransposons, is the presence of a *gag*-like ORF with three Zn knuckle motifs (Biessmann et al., 1992; Danilevskaya et al., 1992, 1994; Levis et al., 1993). HeT-A and TART retrotransposons form very long, head-to-tail, interspersed, and strictly polar arrays at each of the *Drosophila* telomeres, whereby the oligo(A) stretch at the 3′ end is invariably joined to the rest of the chromosome, so that the 5′ end of each element becomes the new chromosomal end after each retrotransposition event.

More recently, a third telomeric RT-containing retrotransposon named TAHRE was identified in the complete *Drosophila* genome (Abad et al., 2004). It is highly homologous to HeT-A in its nucleotide sequence, but in addition it contains an 1103-amino acid ORF coding for an AP-like endonuclease (EN) and RT, which all of the HeT-A elements completely lack. One full-length and three 3′-truncated copies of the TAHRE element were found in the *D. melanogaster* genome, and only two of these copies, one full-length and one truncated in the middle of the *gag* gene, have the potential to give rise to a full-length RT, making TAHRE the closest approximation to a single-copy RT gene responsible for movement of HeT-A, and possibly illustrating the process of retrotransposon domestication. The TAHRE RT protein is only 65% identical to TART RT, and the copy number of full-length TART elements was estimated to vary between 7 and 11 in different *D. melanogaster* stocks (George et al., 2006).

It is not my intention to review in this chapter all of the properties of telomere-associated retrotransposons in *Drosophila*, which have been the subject of numerous review articles (for recent reviews, see Mason et al., 2008; Pardue and DeBaryshe, 2008). Instead, I will focus on the peculiar features of telomeric retrotransposons (TRs) which may provide clues to the overwhelming dominance of RT-based

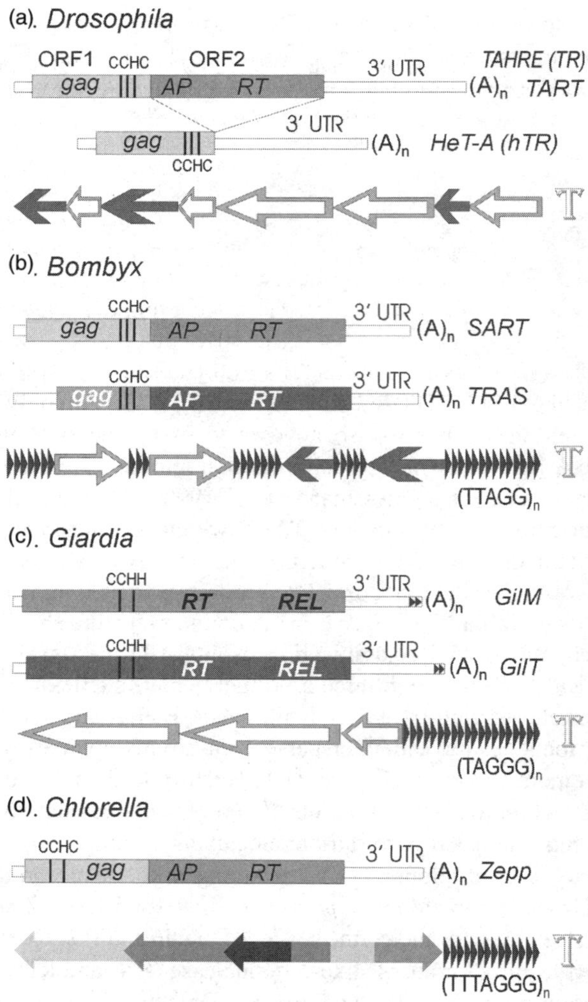

FIGURE 11.3 Non-LTR retrotransposons at telomeres. Each panel (a through d) shows the ORF structure of the corresponding retrotransposable elements at the top, and their arrangement at telomeres at the bottom. Small triangles represent telomeric repeats. T, telomere; RT, reverse transcriptase domain; AP, apurinic–apyrimidinic endonuclease domain; REL, restriction enzyme-like endonuclease (EN) domain; $(A)_n$, poly(A) tail. (a) Telomeric retrotransposons (TR and half-TR) in *Drosophila* (after Pardue and DeBaryshe, 2008; Villasante et al., 2008). The TR structure fits both TAHRE and TART elements. (b) Telomere-specific retrotransposons in *Bombyx mori* (after Fujiwara et al., 2005). (c) Subtelomeric retrotransposons in *Giardia lamblia* (after Arkhipova and Morrison, 2001). (d) Telomere-associated retrotransposons in *Chlorella vulgaris* (after Higashiyama et al., 1997).

chromosome end maintenance in eukaryotes. One of the most intriguing features of the retrotransposon-based system in *Drosophila* is the apparent inability of each individual retroelement to provide all of the functions necessary for its terminal retrotransposition. Instead, there is a large degree of interdependence between elements with respect to the functions that each element is unable to provide on its own. Based on the current knowledge of non-LTR retrotransposon replication mechanisms (Eickbush and Jamburuthugoda, 2008), HeT-A retrotransposition may be inferred to depend on a *trans*-acting RT, such as that of TAHRE or TART. The former would be more likely to recognize the HeT-A template because of extensive nucleotide sequence similarity between HeT-A and TAHRE (up to 80%). However, the overall picture appears a lot more complex than just the dependence on a *trans*-acting RT. An additional level of interdependence is introduced by telomeric targeting, which is presumably accomplished by the *gag*-like ORF of the nonautonomous HeT-A retrotransposon. Studies of fluorescent protein-tagged ORF1 showed that the Gag proteins of either TART or TAHRE fail to localize to *Drosophila* chromosome ends in S3 tissue culture cells in the absence of HeT-A Gag (Fuller et al., 2010; Rashkova et al., 2002a,b). Thus, the HeT-A Gag not only has the ability to localize to telomeres, but it also appears to assist the two other telomere-associated retrotransposons in telomeric localization, so that in this interesting ménage à trois each of the elements cannot rely solely on itself for terminal retrotransposition, and in this sense stops being selfish in terms of promoting its own proliferation.

Another aspect of "selflessness" of telomere-associated retrotransposons in *D. melanogaster* was illustrated by the promoter architecture of HeT-A (Danilevskaya et al., 1997). Unlike the majority of non-LTR retrotransposons, which possess a completely internal RNA polymerase II promoter allowing them to preserve all of the essential promoter elements after retrotransposition, the HeT-A promoter is located within the 3′ end of the element. This 3′ promoter drives transcription of another HeT-A element located immediately downstream, instead of driving transcription of the copy in which the promoter resides. Indeed, when HeT-A was attached to a terminally deleted telomere containing an exposed promoterless *yellow* gene, the 3′ terminal HeT-A promoter was able to drive *yellow* transcription (Kahn et al., 2000). This promoter architecture of HeT-A, however, is not preserved in *D. virilis* (Traverse et al., 2010).

Different evolutionary scenarios were put forward to explain the dominance of retrotransposon arrays at *Drosophila* telomeres. The most plausible chain of events involves recruitment of non-LTR retrotransposons from the jockey clade (Table 11.1 and Section 4.1 below) to the chromosome ends (Abad et al., 2004). In this scenario, an ancestral TAHRE-like retrotransposon in the lineage leading to Dipteran insects was co-opted as the sole source of RT for nonautonomous derivatives with an internal deletion spanning the RT domain. These nonautonomous elements subsequently evolved into HeT-A, while preserving sufficient homology to TAHRE, so as to be able to take advantage of the TAHRE RT. Similar diversification could have occurred in primitive eukaryotes during transition to telomerase-based chromosome end maintenance (Eickbush, 1997; Zimmerly et al., 1995). An alternative hypothesis was put forward by Pardue and colleagues on the basis of their findings that, out of the three telomere-associated *Drosophila* retrotransposons, only the HeT-A *gag*-like ORF has the capacity to provide telomeric targeting (Fuller et al., 2010; Rashkova et al., 2002a,b). These authors hypothesized that TRs in *Drosophila* evolved from the telomerase-based system *via* fusion of a gene required for telomeric targeting to the telomerase RNA template, leading to formation of HeT-A (Pardue et al., 1997; Pardue and DeBaryshe, 2008). While this scenario is difficult to reconcile with molecular phylogenetic evidence, it nevertheless re-emphasizes the "non-selfish" nature of HeT-A, which cannot be fully responsible for its own propagation, but can assist other retroelements in telomeric targeting, and, in turn, depends on an unlinked source of RT for its transposition. To summarize, the lessons learned from *D. melanogaster* telomeres have greatly enriched our knowledge about the requirements necessary for retroelement adaptation to perform the role of telomere extenders, and emphasized the dependence on an unlinked template in a multicomponent system, increasing the potential for host control over telomere lengthening.

11.3.2 Telomeric Retrotransposons in *D. yakuba, D. virilis*, and Other Drosophilids

The exceptional nature of *D. melanogaster* telomeres, which received the most thorough experimental attention, has eventually proved to be the rule rather than exception in other Drosophilid insects. Beginning with the studies of the closely related *D. yakuba* from the melanogaster subgroup, where a targeted search for telomere-associated retrotransposons revealed both HeT-A-like and TART-like elements (Casacuberta and Pardue, 2002; Danilevskaya et al., 1998), the same pattern of TR occurrence was subsequently extended to *D. virilis*, a distantly related species which is separated from *D. melanogaster* by more than 60 million years (Casacuberta and Pardue, 2003a,b). Thus, it was convincingly established that retrotransposon-mediated telomere maintenance represents an evolutionarily robust mechanism of protecting and extending chromosome ends, which has been in use for tens of millions of years, predating diversification of the genus *Drosophila* (reviewed in Pardue and DeBaryshe, 2003, 2008; Pardue et al., 2005). The features of TRs thought to be of most significance for their role at telomeres are the long and intrinsically repetitive 3′ UTRs, which could participate in establishment of telomeric heterochromatin, and the telomeric targeting capabilities provided by the HeT-A Gag not only for itself, but for the other TRs as well (Fuller et al., 2010; Rashkova et al., 2002a,b).

With the 12 sequenced *Drosophila* genomes now in hand (Clark et al., 2007) and the telomerase gene lacking in all of them, it became possible to investigate the patterns and driving forces behind evolution of TRs throughout the entire genus *Drosophila* (Villasante et al., 2007, 2008). All of the 12 sequenced species contain between two and ten different families of TRs. The study of Villasante et al. (2007) established the recurring loss of RT-coding ORF2 from TRs, which leads to formation of the so-called half-TRs, coding only for Gag proteins. It also validated the congruence between the phylogeny of TRs and that of their hosts, pointing to a single event in the early evolutionary history of drosophilids which recruited existing retrotransposons to perform the function of chromosome end maintenance.

Perhaps the most interesting feature of TRs detected in the study of Villasante et al. (2007) is the surprising propensity of TRs to lose their RT-containing ORFs and to form half-TRs coding only for Gag (Fig. 11.3a). Their evolutionary predecessors, non-LTR retrotransposons from the jockey clade, do not exhibit this propensity, and are typically composed of both *gag* and *pol* domains. It is reasonable to suggest that this unusual propensity for RT loss is connected with the acquisition of cellular function by the TRs. While most autonomous parasitic retroelements with the potential to propagate themselves usually carry all of the functions required for retrotransposition on the same copy in *cis*, any domesticated retroelement with a host function is expected to be subject to host control over the rate of retrotransposition. This could best be achieved by separation of the RT-encoding ORF onto a single or only a few master elements, and by subsequent proliferation of nonautonomous elements with telomere-targeting properties by a *trans*-acting RT. A plausible scenario for ORF2 loss would be the formation of a *gag*-encoding nonautonomous element *via* splicing of ORF2, with subsequent *trans*-retrotransposition of nonautonomous copies to telomeres. Such an arrangement could become selectively advantageous, as it would permit host control over nonautonomous copies while endowing them with telomere-targeting capabilities.

11.3.3 Telomere Maintenance in Non-Drosophilid Insects and Other Arthropods

The phylum Arthropoda in general appears to be not too much different from all other eukaryotes with respect to telomerase-mediated chromosome end maintenance. The TERT genes are present in many insect and other arthropod species, including ticks, body lice, pea aphids, wasps, honeybees, flour beetles, silkworms, and water fleas (Table 11.2). The prevalent form of telomeric repeats in arthropods is the pentanucleotide TTAGG, with a minor variation of TCAGG in *Tribolium castaneum* (Osanai et al., 2006; Robertson and Gordon, 2006). However, a tendency for sporadic loss of TTAGG repeats (or their conversion to other sequence variants) can be observed among several insect orders, as indicated by lack of hybridization and of PCR products using the $(TTAGG)_n$ probe (Frydrychová et al., 2004; Vítková et al., 2005). No telomeric repeats of any kind could be detected in genomic sequences of dipteran insects, in agreement with the absence of a TERT gene from completed genome projects of *Anopheles gambiae, Aedes aegypti*, and *Drosophila* spp. (Clark et al., 2007; Holt et al., 2002; Nene et al., 2007). Previously, *Chironomus* and *Anopheles* chromosomes were shown to end in long complex tandem repeats, which were proposed to undergo recombination-based elongation (reviewed in Biessmann and Mason, 1997, 2003; Kamnert et al., 1997). Overall, the hypothesis that telomerase-based mechanism of chromosome end maintenance was lost in an ancestor of dipteran insects long before having been replaced by retrotransposons (Biessmann and Mason, 1997; Pardue et al., 1997) has now gained wide recognition and is well-supported by molecular and phylogenetic evidence. Recombination-based backup mechanisms could provide a substitute for loss of telomerase-based chromosome end maintenance, and minimize terminal DNA loss in the absence of telomeric repeats at the chromosome ends prior to recruitment of retrotransposons to telomeres. Thus, multiple telomere-maintenance mechanisms constitute a potent way of buffering against terminal gene loss, and in general it is not surprising that vital cellular functions are typically being supported by partially redundant pathways.

TABLE 11.2 TERT Genes and Telomeric Repeats in Sequenced Arthropod Genomes

Arthropod spp.	Class; Order	TERT Genbank Contig	Tel. Repeats
Ixodes scapularis	Arachnida; Ixodida	ABJB010618527.1	TTAGG
Daphnia pulex	Crustacea; Diplostraca	jgi060905: scaffold_47	TTAGG
Pediculus humanus corporis	Insecta; Phthiraptera	EEB15290	TTAGG
Acyrthosiphon pisum	Insecta; Hemiptera	XP_001946970	TTAGG
Rhodnius prolixus	Insecta; Hemiptera	–	TTAGG
Apis mellifera	Insecta; Hymenoptera	NP_001035771	TTAGG
Nasonia vitripennis	Insecta; Hymenoptera	NW_001816348.1	+/–
Nasonia longicornis	Insecta; Hymenoptera	ADAP01066960.1	+/–
Nasonia giraulti	Insecta; Hymenoptera	ADAO01117422.1	+/–
Tribolium castaneum	Insecta; Coleoptera	ABD17350	TCAGG
Bombyx mori	Insecta; Lepidoptera	ABF56516	TTAGG
Bombyx mandarina	Insecta; Lepidoptera	ABF56517	TTAGG
Anopheles gambiae	Insecta; Diptera	–	–
Culex quinquefasciatus	Insecta; Diptera	–	–
Aedes aegypti	Insecta; Diptera	–	–
Drosophila melanogaster	Insecta; Diptera	–	–
Drosophila simulans	Insecta; Diptera	–	–
Drosophila erecta	Insecta; Diptera	–	–
Drosophila sechellia	Insecta; Diptera	–	–
Drosophila yakuba	Insecta; Diptera	–	–
Drosophila ananassae	Insecta; Diptera	–	–
Drosophila pseudoobscura	Insecta; Diptera	–	–
Drosophila persimilis	Insecta; Diptera	–	–
Drosophila willistoni	Insecta; Diptera	–	–
Drosophila virilis	Insecta; Diptera	–	–
Drosophila mojavensis	Insecta; Diptera	–	–
Drosophila grimshawi	Insecta; Diptera	–	–

11.3.4 Yeast Telomeres

The baker's yeast, *Saccharomyces cerevisiae*, arguably the best-studied eukaryotic model organism, can also provide a glimpse into the sequence of events which could occur upon loss of telomerase activity. Like most eukaryotes, yeast chromosomes are capped by telomeric repeats, which are added by telomerase (reviewed in McEachern et al., 2000; Zakian, 1995). In mutant yeast strains deficient for one of the telomerase subunits, most of the cells suffer from chromosome deprotection, resulting in end-to-end fusions and chromosomal instability likely caused by breakage–fusion–bridge cycles. Some cells can survive the crisis by taking advantage of telomerase-independent alternative telomere maintenance mechanisms (reviewed in Lundblad, 2003; Biessmann and Mason, 2003). Type I survivors amplify the Y′ subtelomeric tandem repeats, while Type II survivors use recombination-based mechanisms to extend their existing telomeric repeat tracts. Interestingly, yeast LTR retrotransposons Ty1 are mobilized in *est2* mutants lacking the catalytic subunit of telomerase (Scholes et al., 2003). Unprotected telomeres induce

DNA damage checkpoint signaling, and the stress signal activates Ty1 retrotransposition. This, in turn, can lead to mobilization of the Y′ repeats, which get incorporated into subtelomeric regions as part of chimeric cDNAs, generated by the Ty1 RT and primed by Ty1 sequences (Maxwell et al., 2004). However, the presence of additional copies of Ty1 sequences at telomeres induces ectopic recombination with Ty1 copies which are normally dispersed throughout the chromosomes, and leads to a high frequency of chromosomal rearrangements (Maxwell and Curcio 2008). Thus, while retrotransposon sequences may be driven to telomeres by various means, the true telomere specialization can be achieved only with acquisition of telomeric targeting, which would essentially exclude a retrotransposon family from nontelomeric regions and confine it to telomeres.

Importantly, yeast also served to illustrate this important principle of retrotransposon telomeric targeting not in direct connection with telomere maintenance. The yeast LTR retrotransposon Ty5 is found exclusively at telomeres and mating-type loci, and was shown to be targeted to silent chromatin (Zou et al., 1995, 1996; Zou and Voytas, 1997). Targeting preferences were provided by the Ty5-encoded integrase (IN), the C-terminal domain of which has a six-amino-acid motif interacting specifically with Sir4, a structural component of silent chromatin (Xie et al., 2001). Interestingly, the specificity of this interaction depends on phosphorylation of the targeting motif, and is reduced or abolished in the absence of phosphorylation under stress conditions (Dai et al., 2007). The targeting motif was apparently "borrowed" by Ty5 from the cellular protein Esc1, which tethers telomeres to the nuclear periphery, and interacts with the same domain of Sir4 as does the Ty5 integrase (Brady et al., 2007). Thus, a relatively small amino acid motif may be sufficient to determine chromosomal preferences of retrotransposon integration. Either convergent evolution or domain swapping could have played a role in acquisition of targeting properties by the Ty5 IN protein, and similar events may also confer different targeting properties to other retrotransposons.

11.4 TELOMERIC REPEATS AND RETROTRANSPOSONS: TOGETHER AT THE CHROMOSOME ENDS

In this section, we will attempt to understand how organisms can quickly adapt to the sudden loss of the telomerase-based chromosome end maintenance pathway, by describing telomere structures that may be regarded as canonical telomeres with built-in backup mechanisms of restoring terminal sequences. Such backup mechanisms could potentially take over the role of chromosome end maintenance in case the telomerase function is impaired.

11.4.1 Non-LTR Retrotransposons

11.4.1.1 SART/TRAS in Bombyx mori One of the best-known examples of co-occurrence of telomeric repeats and retrotransposons at the chromosome ends is the silkworm, *B. mori* (reviewed in Fujiwara et al., 2005). In a survey of seven insect

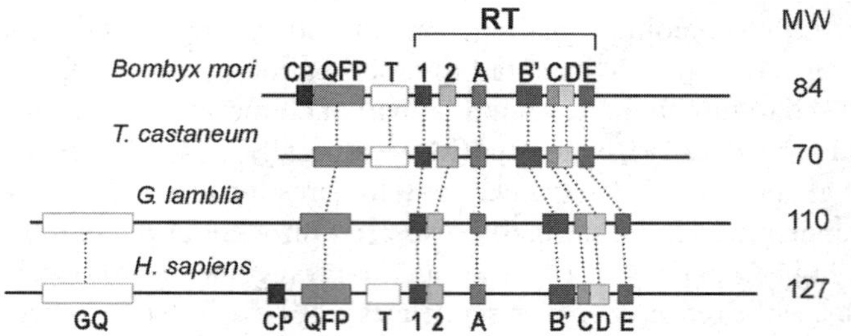

FIGURE 11.4 Structural organization of TERTs from *Bombyx mori*, *Tribolium castaneum*, *Giardia lamblia*, and *Homo sapiens* (after Osanai et al., 2006 and Malik et al., 2000a,b). Conserved motifs are shown as boxes. Motifs 1 through E comprise the core RT domain and are present in most RTs; motifs GQ, CP, QFP, and T are characteristic of TERTs only.

species using the standard TRAP (telomeric repeat amplification protocol) assay for telomerase activity, such activity was detected in cockroaches, crickets, and two lepidopteran species, but not in dipterans (*Drosophila, Sarcophaga*) or in the lepidopteran *B. mori* (Sasaki and Fujiwara, 2000). However, the *B. mori* genome harbors the TERT gene as well as TTAGG telomeric repeats, while the genome of *Tribolium castaneum* contains telomerase and a variant telomeric repeat sequence TCAGG (Okazaki et al., 1993; Osanai et al., 2006). The *Bombyx* and *Tribolium* TERT genes are unusual in that they do not contain the TEN (telomerase essential N-terminal) domain with the GQ motif characteristic of other TERT genes (Fig. 11.4), which is proposed to interact with both telomerase RNA and telomere DNA (Jacobs et al., 2006). In addition, the *B. mori* TERT gene is intronless, which is rather unusual for TERT genes (Osanai et al., 2006). These findings led Fujiwara and colleagues to hypothesize that the loss of the N-terminal TERT domain, possibly *via* retrotransposition and 5′ truncation of the TERT gene, could have led to significant reduction of TERT activity in *Bombyx*, and perhaps to the eventual loss of the telomerase gene in the ancestor of dipteran insects. Indeed, no remnants of the TERT gene or G-rich telomeric repeats could be identified in any of the sequenced dipteran genomes (Table 11.2).

In the context of very low TERT activity, what could possibly prevent silkworm telomeres from drastic shortening? The answer may lie in the capacity of several *B. mori* non-LTR retrotransposon families to integrate site-specifically into telomeric repeat sequences. The two best-studied elements, TRAS1 and SART1, which belong to the R1 clade of non-LTR retrotransposons, insert site-specifically into the $(CCTAA)_n$ and $(TTAGG)_n$ telomeric strands, respectively, and are always oriented opposite to each other (Fig. 11.3b; Fig. 11.5). TRAS and SART insertion occurs in subtelomeric regions of the chromosome, leaving at least 6–8 kb of telomeric repeats at the distal end. The *B. mori* genome harbors about 1000 copies of TRs, and they have the capacity to occupy about 10 Mb of telomeric DNA, adding, on average, about 180 kb to each *B. mori* telomere (Fujiwara et al., 2005).

While these elements may still fit the definition of parasitic DNA, which serves the only purpose of self-proliferation, it is equally possible that their presence at the

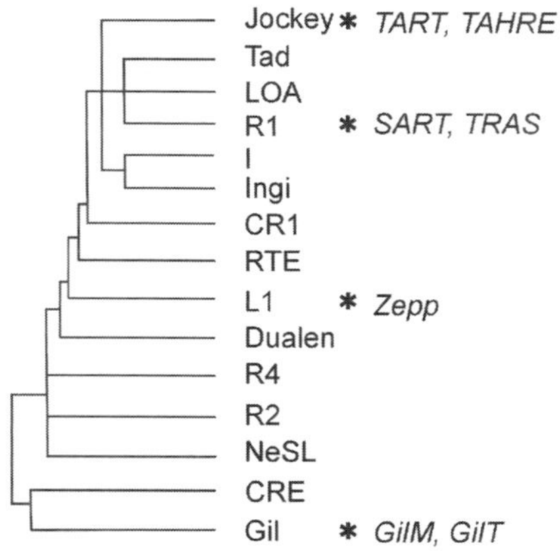

FIGURE 11.5 Independent acquisition of telomere specificity by non-LTR retrotransposons. Clades containing telomere-associated retrotransposons are marked by asterisks on the generalized cladogram of non-LTR retrotransposons compiled after Eickbush and Malik (2002), Arkhipova and Morrison (2001), and Kojima and Fujiwara (2005).

telomeres may benefit the host by adding extra DNA at the termini, which would otherwise undergo terminal DNA loss. These elements are exceptionally well adapted to retrotransposition into telomeric regions: not only the apurinic–apyrimidinic (AP)-like EN domains of both SART and TRAS were shown to specifically recognize and cleave the telomeric repeat pentanucleotide, but the SART1 ORF1 was also demonstrated to play an essential role in telomeric targeting, similar to the role ORF1 plays in targeting the HeT-A element to *Drosophila* telomeres (Matsumoto et al., 2004). Thus, it is quite possible that actively ongoing retrotransposition of these elements into subtelomeric regions may help to compensate for insufficient activity of the *B. mori* telomerase.

11.4.1.2 Gil(=Genie) in Giardia lamblia Another example of coexistence of telomere-associated non-LTR retrotransposons and telomeric repeats is the parasitic protozoan, *G. lamblia (aka G. intestinalis)* (Arkhipova and Morrison, 2001; Burke et al., 2002). Its telomerase also exhibits noncanonical sequence features, such as the lack of the highly conserved T motif (Fig. 11.4) (Malik et al., 2000a,b), and telomeric repeats are represented by the pentanucleotide TAGGG (Adam et al., 1991; Hou et al., 1995). Two different non-LTR retrotransposon families, GilM and GilT, are found at a subset of *Giardia* telomeres, where they are often subject to 5′ truncation and subsequent addition of telomeric repeats at the 5′ ends (Fig. 11.3c). Unlike SART/TRAS, but similar to HeT-A, insertions of GilM and GilT are strictly polar, so that the 3′ end of the element is oriented proximally, and the distal 5′ end is preceded by $(TACCC)_n$, a reverse-complement of *Giardia* telomeric repeats. The elements typically occur in head-to-tail tandem arrays, and are characterized by relatively long 3′ UTR regions. Although they encode an EN domain of the site-specific REL (restriction enzyme-like) type, the EN domain is apparently not responsible for site-specific insertion into telomeric repeats, as is the case with SART/TRAS, since telomeric repeats are not found in the 3′ flanking regions of GilM or GilT. One may entertain the possibility that the EN domain exhibits specificity for insertion into the very end/beginning of the element itself, leading to formation of tandem arrays. It is not known whether the nucleic acid binding domain in the GilM/GilT ORF confers any telomere-targeting properties. Nevertheless, telomere-associated retrotransposons of *Giardia* represent an interesting case, displaying features which are somewhat intermediate between SART/TRAS and HeT-A/TAHRE/TART. The capacity to form head-to-tail tandem arrays at chromosome termini may signify an important step in transition from a selfish transposable element to a *bona fide* TR.

11.4.1.3 Zepp in Chlorella vulgaris Telomeric insertions were also reported for the *Chlorella* non-LTR retrotransposon *Zepp*, copies of which frequently insert into each other, forming nested clusters that are often joined to TTTAGGG telomeric repeats (Higashiyama et al., 1997; Noutoshi et al., 1998; Yamamoto et al., 2000, 2003) (Fig. 11.3d). Self-insertions occur in the same orientation, but without any preferential site of integration within the element, and cause target site duplications of variable length, as expected for non-LTR retrotransposons. Occasionally, copies are found in an inverted arrangement, reminiscent of "twin priming" (Ostertag and Kazazian, 2001). Thus, while the 3′ end with the poly(A) tract is typically oriented towards the centromere, as in *Drosophila*, in inverted copies the 3′ end may become oriented towards the telomere, undergo terminal erosion, and become capped by telomeric repeats. Several chromosome ends were found to carry *Zepp* elements but no telomeric repeats, indicating that deprotected chromosome ends can initially acquire retrotransposons at the termini. Moreover, *Zepp* elements were added to the ends of newly formed minichromosomes upon irradiation, so that one end of the minichromosome would have only telomeric repeats, and the other only *Zepp* elements (Yamamoto et al., 2003). Thus, while these elements are not present at every chromosome end and cannot be regarded as obligatory components of *Chlorella* telomeres, they apparently have the capacity to aid in restoration of broken and deprotected chromosome ends, including cases where chromosome breakage was induced by ionizing radiation.

Overall, the above examples of telomere-associated non-LTR retrotransposons demonstrate that the ability to associate with telomeres can arise independently multiple times in evolution, in retrotransposons belonging to very different non-LTR clades (jockey, R1, Gil/Genie, and L1 clades; Fig. 11.5) (Eickbush and Malik, 2002). In agreement with their patchy phylogenetic distribution, these elements have acquired different means for accumulating at telomeres, which include specialization of the element-encoded EN for insertion into telomeric repeats, targeting to chromosome ends by the element-encoded proteins *via* possible interactions with components of the telomere-capping complex, or even the propensity for self-insertion. All of the non-LTR elements described in this section, however, appear to use their EN domain for insertion into telomeres, and may differ mechanistically from the *Drosophila* telomere-associated retrotransposons, which are widely believed to transpose *via* attachment of the poly(A) tail in the RNA template directly to the chromosome end and subsequent reverse transcription primed by the terminal 3′-OH in the chromosomal DNA. Nevertheless, it is worth mentioning that both TART and TAHRE elements contain an intact AP-like EN domain, the function of which was never tested experimentally and remains a mystery.

11.4.2 Endonuclease-Deficient Penelope-Like Retroelements

A different class of retroelements, *Penelope*-like elements (PLEs), is associated with chromosome ends in rotifers of the class Bdelloidea (Gladyshev and Arkhipova, 2007). PLEs comprise a distinct class of retrotransposons which differs from LTR and non-LTR retrotransposons by the presence of the GIY–YIG EN domain and by separate phylogenetic placement in the RT phylogeny (Arkhipova et al., 2003; Lyozin et al., 2001). Cloning and sequencing of chromosome ends from two species of bdelloid rotifers, *Adineta vaga* and *Philodina roseola*, revealed that bdelloid telomeres harbor a specialized subset of PLEs characterized by complete absence of an EN domain. Most surprisingly, similar EN-deficient PLEs were found at the chromosome ends in sequenced genomes of several fungi, protists, and plants (the inky cap mushroom *Coprinus cinereus*, the white rot fungus *Phanerochaete chrysosporium*, the pennate diatom *Phaeodactylum tricornutum*, and the spike moss *Selaginella moellendorffii*).

All of these retroelements have a surprisingly similar organization. They occur in head-to-tail tandem arrays, in which members of different families may be interspersed, and the polarity of these arrays is such that the 5′ end of the element is distal, and often becomes truncated by addition of species-specific telomeric repeats. In addition, each 3′ UTR also contains a short stretch of telomeric repeats, in the same orientation. Thus, the coding strand of the retroelement corresponds to the C-rich strand of the telomere. In addition to the RT-containing ORF2, these elements also contain an upstream ORF1, which does not exhibit any characteristic features other than a conserved coiled-coil motif (Fig. 11.6a). Interestingly, EN-containing PLEs, none of which are known to exhibit telomere associations, do not have an ORF1. Therefore, it may be hypothesized that the ORF1 product could be involved in telomeric targeting, as is the case with HeT-A and SART elements (Matsumoto et al., 2004; Rashkova et al., 2002a,b). Target-primed reverse transcription (TPRT) at the chromosome ends could be facilitated by base-pairing interactions between the short stretches of reverse-complement telomeric repeats at the 3′ end of the RNA template and the exposed G-rich overhangs at the chromosome termini. Thus, telomerase-mediated and retrotransposon-mediated addition of DNA to chromosome ends exhibit the same polarity, whereby both telomerase RNA template (TER) and retrotransposon RNA template could be reverse-transcribed to add cDNA to the G-rich strand at the telomere.

11.4.3 Mammalian L1: Driven to Telomeres Upon Loss of Endonuclease Activity

It turns out that not only those retrotransposons which naturally lack the EN domain can undergo terminal transposition, but even those that do contain an EN could end up

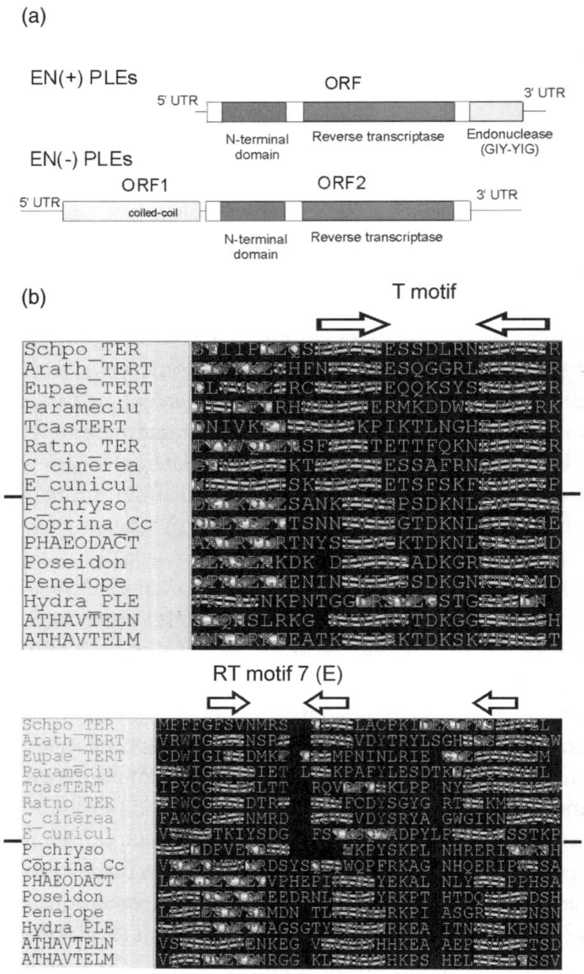

FIGURE 11.6 Structural organization of *Penelope*-like elements (a) and their similarity to telomerases (b). The EN(−) (endonuclease-deficient) PLEs exhibit specificity for telomeres in diverse eukaryotes. In panel (b), secondary structure predictions for representative TERT (top 8) and PLE (bottom 8) sequences are compared in selected portions of the RT amino acid alignment, showing the N-terminal T-motif region and the C-terminal motif 7 of the core RT domain. Arrows designate characteristic beta-hairpins in the secondary structure. Sequences were viewed with the aid of a structure-based sequence alignment program (STRAP) (http://www.bioinformatics.org/strap/). (See the color version of this figure in Color Plates section.)

at telomeres when their EN domain is artificially disabled. This phenomenon was observed for the human L1 element, the most prominent non-LTR retrotransposon in our genome, when its active copy with the mutagenized EN domain was introduced into a Chinese hamster ovary (CHO) tissue culture cell line. This line was characterized by dysfunctional telomeres, and was deficient in activity of one of the components responsible for non-homologous end joining (NHEJ), namely the DNA-dependent protein kinase catalytic subunit (DNA-PKcs) (Morrish et al., 2007). In these cells, on average, one-third of transposition events registered by the standard retrotransposition assay had the 3′ end of the L1 element joined to reverse-complement telomeric repeats, $(CCCTAA)_n$. Although the eventual localization of these insertions was not terminal because of subsequent rearrangements, these findings do demonstrate that EN-containing non-LTR retrotransposons, which normally insert into internal chromosomal locations, can occasionally use deprotected chromosome ends for priming of reverse transcription in the absence of EN activity, in a way similar to telomerases (Curcio and Belfort, 2007).

The above findings underscored the affinity of non-LTR retroelements for double-stranded DNA breaks (DSB), or deprotected chromosome ends which may also be recognized as DSBs. Thus, Morrish et al. (2007) proposed that EN-independent retrotransposition is an ancestral mechanism of RNA-mediated DNA repair associated with non-LTR retrotransposons, which may have been used for addition of retroelements to chromosome ends prior to the acquisition of an EN domain.

11.5 TRANSITION TO LINEAR CHROMOSOMES AND THE EMERGENCE OF TELOMERASE-BASED CHROMOSOME END MAINTENANCE

As mentioned above, two principal scenarios for telomerase evolution have been entertained. In the most popular evolutionary scenario of telomerase origin from a domesticated retrotransposon, the transposon-encoded RT, once present in multiple copies, is postulated to have lost all of its copies but one, which then became fully responsible for generation of telomeric repeats, and acquired the status of a single-copy gene (Eickbush, 1997; Zimmerly et al., 1995). If a bacterial-type retrotransposon was domesticated in an ancestral protoeukaryote upon linearization of its chromosome(s), such domestication presumably entailed loss of EN domain, loss of self-recognition, disassociation of RT from its template, and loss of multicopy status. Another scenario is that of an ancestral RT gene evolving into a telomerase, which should be reconciled with the fact that retrotransposons (group II introns) already existed in prokaryotes before the emergence of telomerase. So, has a primordial retrotransposon lost its intragenomic proliferating capacity to become a telomerase, or was telomerase initially an immobile RT gene? To address this question, we need to review the properties of the known RT types, in order to find out which of these could be the most likely candidate for a telomerase ancestor.

11.5.1 Major Types of Prokaryotic RTs

Three major groups of prokaryotic retroelements have been generally recognized: retrons, group II introns, and, more recently, diversity-generating retroelements (DGRs). Retron RTs were the first RTs to be discovered in prokaryotes (Lampson et al., 1989; Lim and Maas, 1989). They occur in a wide variety of bacteria, but their role still remains enigmatic. The retron RT generates extrachromosomal copies of the so-called msDNA (multicopy single-stranded DNA) covalently linked to the template RNA *via* a 2′–5′ phosphodiester linkage, and these copies do not integrate into chromosomes. The most numerous are group II introns, which consist of an RT gene embedded into a catalytic self-splicing structure (reviewed in Belfort et al., 2002; Lambowitz and Zimmerly, 2004, 2010; Toro et al., 2007). These elements specifically insert into a single genomic location by a process termed retrohoming, but are also capable of occasional insertion into noncognate sites, the behavior expected of a retrotransposon. Finally, DGRs are a relatively recent addition to the major prokaryotic RT groupings (reviewed in Medhekar and Miller, 2007). Although DGRs are more closely related to group II intron RTs than to retron RTs, they are not mobile elements. The best-studied phage DGR confers variability to the phage tail protein in the region contacting the bacterium during phage infection, and it is thought that other DGRs also generate diversity at nearby genomic loci coding for proteins with highly variable regions (VRs). Both retrons and DGRs can be found on bacterial chromosomes, plasmids, or phages.

The recent increase in availability of prokaryotic genome sequences helped to reveal that bacterial genomes also contain many “uncharacterized” RT sequences which do not fit into any of the three groups described above (Kojima and Kanehisa, 2008; Simon and Zimmerly, 2008). These sequences can be subdivided into several groups, members of which, however, are not nearly as numerous as the three well-known classes—in fact, most of these groups contain only a few members and are found in very few species. Moreover, there is no convincing evidence of retromobility for any of these “unknown” groups, with only one suspected case identified so far (Simon and Zimmerly, 2008). Finally, most prokaryotic RTs are of very limited complexity in terms of domain architecture, which typically does not include any extra domains beyond the seven core palm/finger RT motifs and thumb. According to Simon and Zimmerly (2008), there are 17 RT groups classified as “unknown,” which contain 3–8 members from diverse bacterial species, and 36 unclassified singleton RT ORFs that cannot be assigned to any group. Of the “unknown” groups, only four (UG1, UG5, UG6, and UG8, consisting of 8, 8, 4, and 5 members, respectively) contain additional C-terminal extensions fused to the RT, which are mostly represented by various “domains of unknown function” (DUF). Thus, an overwhelming majority of prokaryotic RTs possess the simplest possible domain architecture, consisting only of the RT moiety and no extra domains.

On balance, it may be concluded that, while prokaryotes contain single-copy as well as multicopy RTs, convincing evidence of retromobility is displayed only by a single class of prokaryotic retroelements, namely group II introns. These mobile elements, despite distinctive preference for their cognate insertion sites (“homing sites”), can also be found dispersed at various genomic locations, in some cases reaching copy numbers as high as 28 group II introns per genome ($\sim$1% of the genome) in the cyanobacterium *Thermosynechococcus elongatus* (Simon and Zimmerly, 2008). It may be thought that the retromobility success of group II introns is associated with the presence of a C-terminal EN domain (typically containing the conserved HNH motif), and the efficiency of retrotransposition, as opposed to retrohoming, may depend on relaxed recognition specificity of the EN for its homing site. Some of the group II introns, however, lack EN domains altogether, and their dispersal is therefore replication-dependent, as they cannot rely on EN-generated nicks in chromosomal DNA to prime reverse transcription, and need to use the available 3′ OH groups, which are most frequently generated from the Okazaki fragments during replication (Ichiyanagi et al., 2002, 2003). This limitation would make EN(−) intron dispersal less efficient than that of the EN-containing introns, and EN(−) introns are thought to be ancestral to EN(+) introns (Lambowitz and Zimmerly, 2004; Toro et al., 2007).

Is it possible to tell which of the known prokaryotic retroelements, if any, are more likely to have evolved into telomerases? Group II introns are similar to TERTs in that they also use TPRT (Zimmerly et al., 1995), but they exhibit more similarity to non-LTR retrotransposons in their sequence, as they share the so-called motif 2a in the core RT domain (Malik et al., 1999). Retrons are somewhat mechanistically similar to TERTs in that they represent single-copy genes making multiple copies of a template which eventually might have the potential to be dispersed. However, their atypical priming mode, highly diverged sequence motifs in the RT domain, and the inability of extrachromosomal msDNA to attach to any other DNA make retrons unlikely candidates for telomerase ancestors. DGRs are highly specialized and, moreover, highly mutagenic, which would interfere with sequence conservation necessary for protein binding. Thus, it is now appropriate to review domain architectures of the known eukaryotic retroelements and to compare their properties with telomerases in search of the most closely related groups.

11.5.2 The Diversity of Eukaryotic Retroelements

Eukaryotic retroelements, to which telomerase belongs, are subdivided into several major classes (Table 11.1). It may be asked which of these classes, if any, could have shared a common ancestor with telomerases more recently than other classes of eukaryotic RTs. Several approaches can be combined in order to determine ancestral relationships. Traditional methods of phylogenetic analysis, while useful, cannot provide robust clade support values to the deepest-branching clades due to limitations on the number of informative characters in a single RT gene, and to ancient timing of clade separation and diversification. Phylogenetic network representations show that there is a lot of uncertainty in RT phylogenies due to conflicting phylogenetic signals, although the major RT groups are easily definable (Fig. 11.7). The core RT domain encompassing the seven highly conserved motifs spans less than 300 amino acids in length, many of which are located in interdomain regions and do not exhibit any conservation at all. Thus, in addition to conventional phylogenetic analysis, it may be useful to take into account domain architectures, as well as the most conserved secondary structure features, which are not necessarily manifested at the level of amino acid sequence identity/similarity.

So far, TERTs represent the only known single-copy genes among eukaryotic RTs, while the vast majority of eukaryotic RTs belong to different types of mobile elements and viruses. The most abundant classes are LTR retrotransposons and non-LTR

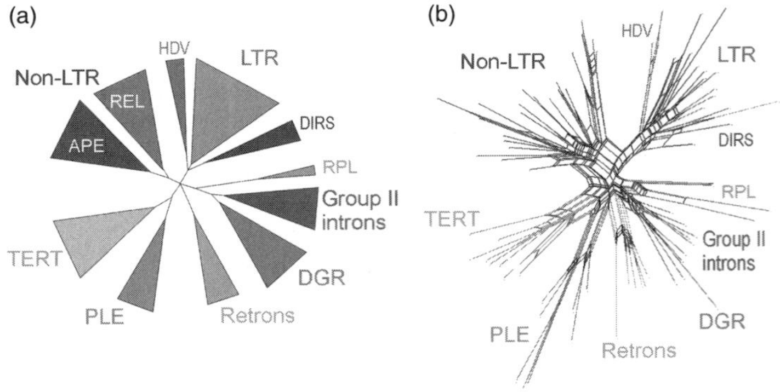

FIGURE 11.7 Phylogenetic relationships between different RT classes. (a) Diagrammatic representation of an unrooted phylogram showing each RT class as a triangle with the size approximately reflecting the diversity within the group, as in Eickbush and Malik (2002). The LTR group includes retroviruses, LTR-copia, LTR-gypsy, and caulimoviruses. RPL, mitochondrial retroplasmids, and RTL elements; HDV, hepadnaviruses; for other abbreviations, see text. (b) Phylogenetic network showing relative positions of each RT group and visualizing conflicting signals and areas of reticulate events in the overall tree-like phylogeny (maximum likelihood distance for an alignment of *ca* 600 RT amino acids) (SplitsTree4.1; Huson and Bryant 2006). Adapted after Gladyshev and Arkhipova (2011). (See the color version of this figure in Color Plates section.)

retrotransposons, which were traditionally distinguished on the basis of the presence/absence of terminally redundant sequences called long terminal repeats (LTRs). Additional, less numerous and less studied classes are PLEs and the DIRS elements (Evgen'ev and Arkhipova, 2005; Poulter and Goodwin, 2005). Table 11.1 lists these classes, as well as the major groups (or clades) comprising these classes. As is evident from this table, each group of retrotransposons (excluding pararetroviruses which normally do not integrate into chromosomes) may be easily defined by the type of the EN domain which is typically associated with it.

Historically, the LTR/non-LTR distinction emerged because the first-discovered RTs were those of retroviruses (Baltimore, 1970; Temin and Mizutani, 1970), which, in their integrated form, are framed by LTRs. This retrovirus-like LTR structure was also characteristic of the first-studied eukaryotic mobile elements from *Drosophila* and yeast (Bayev et al., 1980; Gafner and Philippsen, 1980; Levis et al., 1980), which were subsequently shown to go through the same stages of replication cycle as retroviruses (Arkhipova et al., 1986). Thus, initially LTRs were thought to be typical of all RT-containing elements, and the lack of LTRs in several independently described retrotransposable elements, such as I-element, F-element, and R1/R2 ribosomal insertions in *Drosophila*, L1 from mice and humans, and Ingi from trypanosomes, prompted their designation as non-LTR retrotransposons by Xiong and Eickbush (1988). These authors argued that non-LTR elements represented a distinct group, rather than the result of independent losses of LTRs. Not having something, however, does not constitute a good distinguishing feature, and attempts have recently been made to classify retroelements based on their priming mechanism (TP/EP, target-primed/extrachromosomally primed) (Beauregard et al., 2008).

This classification, which puts together such groups as LINEs, PLEs, and TERTs, is useful but somewhat too inclusive, as TPRT appears to be the ancestral priming mechanism which is used by most of the early-branching RTs. Virus-like elements with extrachromosomal priming appeared in eukaryotes as soon as cDNA synthesis could be moved to the cytoplasm with the acquisition of RNase H (RH), which led to loss of dependence on nuclear host factors for replication (Malik and Eickbush 2001).

11.5.3 Phylogenetic Relationships Within and Between RT Classes

Ultimately, RT classes can be defined as monophyletic clades identified in conventional phylogenetic analyses and supported by other distinctive synapomorphies (shared derived characters), such as domain architecture and conserved secondary structure elements (Figs. 11.6,11.7). Identification of putative evolutionary intermediates which could have emerged during transitional periods, may also greatly assist in defining otherwise obscure relationships.

As an example, consider two major groups of non-LTR retrotransposons, namely the “early-branching” ones containing the C-terminal restriction EN-like (REL) domain, and the “late-branching” ones containing the N-terminal AP EN domain. This branching order was initially determined in a neighbor-joining phylogenetic tree rooted with group II introns (Malik et al., 1999), and it was argued that the more ancient REL domain (which is more typical of prokaryotes) was displaced by the AP domain (which is present mostly in eukaryotes) in the course of evolution of the non-LTR lineage. This scenario received the most convincing support with detection of a non-LTR retrotransposon *Dualen* in the unicellular green alga *Chlamydomonas reinhardtii,* which possesses not one, but both of these domains simultaneously, with AP at the N-terminus and REL at the C-terminus (Kojima and Fujiwara, 2005). Moreover, phylogenetic placement of the *Dualen* RT is intermediate between the REL branch and the AP branch. It is therefore likely that such a structure served as an intermediate during the evolutionary transition from early-branching to late-branching non-LTR retrotransposons. Note that overall the early-branching elements are more divergent than the late-branching ones, in agreement with their more ancient origin, so that the entire non-LTR clade does not always receive statistically robust phylogenetic support, although several shared characteristic residues in the core catalytic RT motifs are highly indicative of common descent.

The other large class of eukaryotic retroelements, LTR retrotransposons, also represent a highly divergent assemblage of genetic elements (including those of viral and retrotransposon origin), which also rarely forms a phylogenetically robust grouping but is widely believed to have shared a common ancestor, mostly due to the presence of the RH domain. In addition to retroviruses and LTR retrotransposons coding for an integrase-type EN, the class also includes two viral groups collectively named pararetroviruses (Temin, 1985), which include caulimoviruses and hepadnaviruses. Pararetroviruses replicate *via* reverse transcription, but do not encode any integration machinery, and are propagated extrachromosomally by viral infection. Since pararetroviruses lack integrated forms, they have no LTRs, but because of internal priming with tRNA there could be a potential for LTR formation in a hypothetical proviral form, if one existed. The RT of plant caulimoviruses is phylogenetically very close to the Gypsy group of retrotransposons, and it is thought that these viruses arose when the gypsy-like RT-RH module was captured by a plant DNA virus (Malik et al., 2000a,b). Hepadnaviruses occupy a more basal position, but, like all other members of this large class of retroelements, their RT contains a C-terminal RH domain, which is essential for completion of the viral replication cycle. The increased level of complexity of retrovirus-like genetic elements, which could have been formed as a result of fusion of a retroelement with a DNA transposon coding for a DDE-type transposase/integrase (Capy et al., 1996, 1997; Eickbush and Malik, 2002), and contain gag and protease in addition to RT/RH/IN, also agrees with their origin well within eukaryotes.

The DIRS elements, while also harboring the RT–RH domain fusion, contain a tyrosine recombinase (YR) rather than integrase, and are often grouped into a separate class (Poulter and Goodwin, 2005). Finally, the least numerous but the most relevant for this review class of eukaryotic retroelements, the PLEs, is characterized by yet another type of EN, GIY–YIG, located on the C-terminal side of the RT (Lyozin et al., 2001). Its prokaryotic counterparts are found in group I introns (reviewed in Van Roey and Derbyshire, 2005), which move as DNA rather than RNA, again pointing at the possibility of retrotransposon formation *via* fusion between RT and a DNA mobile element. As described above, PLEs exhibit a remarkable property of transposing to telomeres in the absence of the GIY–YIG domain. Most importantly, PLE RTs form a sister clade to TERT RTs in independent phylogenetic analyses of the region spanning the core RT domain with the seven most conserved RT motifs (Arkhipova et al., 2003; Chang et al., 2007; Doulatov et al., 2004). Moreover, even beyond the core RT motifs, there are highly conserved secondary structure elements which are present in both PLEs and TERTs, but not in other types of RTs, favoring the hypothesis that these additional N- and C-terminal extensions were acquired in the common ancestor of PLEs and TERTs (Fig. 11.6B). It remains to be seen whether the characteristic beta-hairpin, which spans the T motif of telomerases (Lingner et al., 1997), is also involved in interactions with the RNA template in PLEs. Overall, several independent lines of evidence strongly suggest that the RTs of PLEs and TERTs shared a common ancestor with each other more recently than with other classes of retroelements.

Finally, there is a distinct group of polymerases, viral RNA-dependent RNA polymerases (RdRPs), which are not RTs, but nevertheless share sufficient sequence similarity with RTs, so that one is often tempted to include them into RT phylogenies. This temptation is driven by the assumption that RdRP could be the primordial enzyme that existed in the RNA world (Maizels and Weiner, 1993; Weiner and Maizels, 1987), and therefore must be ancestral to all RTs. However, the error-prone nature of RNA replication makes the rate of sequence evolution for RNA viruses so high that they exist as "quasi-species" always balancing at the brink of mutational meltdown (Domingo and Holland, 1997). Consequently, inclusion of highly divergent RdRPs into RT phylogenies typically causes serious problems due to substantial differences in evolutionary rates for RNA- and DNA-based evolution. Koonin et al. (2008) argued that the major classes of RdRPs of picorna-like viruses emerged at the early stage of eukaryogenesis and originated from bacterial RTs similar to widespread group II introns, rather than from RdRPs of bacteriophages, questioning the validity of inclusion of phage RdRPs into RT phylogenies for tree-rooting purposes. Overall, viral RdRPs should not necessarily be regarded as ancestral to eukaryotic RTs, and there is no need to force the incorporation of members of this highly divergent group into RT phylogenies for the purpose of determining the branching order of eukaryotic RTs.

11.5.4 Target-Primed Reverse Transcription (TPRT) and the Evolution of Template/Primer Requirements

RTs generally cannot initiate DNA synthesis *de novo,* with the only known exception of fungal retroplasmids (Wang and Lambowitz, 1993). Most eukaryotic and prokaryotic RTs prime DNA synthesis using the 3′ hydroxyl, while bacterial retrons utilize the 2′-OH group. The priming hydroxyl in LTR elements is provided extrachromosomally: by a host tRNA (in retroviruses, LTR retrotransposons, and caulimoviruses); by cleaved genomic RNA (in self-priming LTR retrotransposons); or by a priming protein (hepadnaviruses). Non-LTR retrotransposons, and presumably PLEs, undergo TPRT, a process during which cleaved chromosomal DNA at the target site is used to prime cDNA synthesis (Luan et al., 1993; reviewed in Eickbush and Jamburuthugoda, 2008). TPRT is also utilized by group II introns, which are the most abundant group of bacterial retroelements, and presumably by DGRs (Guo et al., 2008; Zimmerly et al., 1995). Thus, TPRT may be regarded as an ancient and widespread priming mechanism, and extrachromosomal priming likely evolved later, upon relocation of cDNA synthesis into the cytoplasm with the emergence of the eukaryotic nucleus. In this regard, it should be emphasized that telomerases are also taking advantage of the TPRT mechanism, using the 3′-OH of the G-rich overhang in chromosomal DNA to repeatedly initiate reverse transcription of the short template segment of the TER RNA (Fig. 11.1). The TPRT mechanism is ideally suited to assist in genome protection, whereby the 3′OH at the end of a DNA nick, a double-stranded DNA break (DSB), or simply the unprotected end of a linear chromosomal DNA molecule can be extended by polymerization to provide extra DNA for further repair and healing of broken ends. Retrotransposons were reported to use both DSBs and chromosome ends for priming of DNA synthesis in cells deficient for non-homologous end-joining (NHEJ), and this ancestral ability of RT to utilize exposed 3′OH groups in DNA to prime RNA-templated DNA synthesis could have been advantageous during transition to linear chromosomes (Morrish et al., 2002, 2007).

One of the most intriguing aspects of telomere biology is the highly conserved sequence composition of the telomeric repeat unit. From early-branching unicellular eukaryotes to fungi, plants, and animals, these repeats usually represent some variation of a 5–8 nt G-rich sequence (although more diverse G-rich variants, such as in yeasts, also exist), with an overall GT-bias on the telomerase-synthesized strand which forms the G-rich overhang (Cohn et al., 1998; McEachern et al., 2000). Not surprisingly, such unique sequence composition creates an attractive target for sequence-specific single-stranded and double-stranded DNA-binding proteins, which play important roles in forming protective caps at the chromosome ends.

Moreover, the propensity of such sequences to form quadruplex structures ("G-quartets") may be biologically relevant for end recognition and for regulation of telomerase access to the termini (Fang and Cech, 1993; Neidle and Parkinson, 2003; Oganesian and Bryan, 2007; Sen and Gilbert, 1992; Zhang et al., 2010). Thus, it is likely that, in early eukaryotes, the association between telomerase RT and the G-rich template was strongly favored by selection, and this favorable combination was retained throughout the course of eukaryotic evolution.

Since the gene for telomerase RNA template, TER, is not linked to the gene coding for the TERT catalytic subunit, it may be asked whether the G-rich template was originally linked to the telomerase precursor, that is, whether both TERT and TER were represented in an ancestral retroelement composed of the TERT gene and a linked template, which subsequently lost the ability to propagate itself due to the loss of linkage between TER and TERT. Alternatively, the TERT gene could have acquired TER specificity *via* N-terminal fusion with an RNA-binding protein that had the affinity for TER-encoded RNA that could be used as a template. Whether originally mobile or immobile, the telomerase precursor would have had to confer selective advantage to the cells carrying it by becoming capable of adding extra nucleotides to the ends of linear DNA molecules in *trans*.

It is therefore of interest to examine prokaryotic RTs for linkage between RTs and their templates. Interestingly, for every well-studied prokaryotic RT group, the template is usually linked to the RT gene. For group II introns, the template is the intron sequence itself, part of which also codes for RT. For retrons, the RT reverse-transcribes the short segment of RNA upstream of the RT coding sequence, which is framed by inverted repeats and provides the "branching G" residue with the 2′-OH group for priming (reviewed in Lampson et al., 2005). Finally, DGRs use as a template the short region upstream of the RT gene, called TR (template region), which gets mutagenized during reverse transcription, and replaces a homologous region called VR in the course of "mutagenic homing" (reviewed in Medhekar and Miller, 2007). Nothing, however, is known about templates and/or priming mechanisms for numerous RT-related genes, such as Abi RTs, CRISPR-associated RTs, "unknown" groups 1–9, or unclassified singleton RTs (Simon and Zimmerly, 2008).

In the prokaryotic cell, where transcription and translation are coupled, a newly translated RT would be in close proximity to its template RNA, which would also become a preferential template for reverse transcription. Thus, given our current knowledge about prokaryotic RTs, the logical course of events during telomerase evolution would be disassociation between the RT gene and its originally linked template. To reinforce this argument, one might even hypothesize that, initially, a short repetitive G-rich sequence could have been generated by the ancestral RT itself, given the ability of many RTs to generate short tandem microsatellite-like repeats, especially in the initial steps of engagement near the 3′ end of the RNA template (Kajikawa and Okada, 2002; Ricchetti and Buc, 1996). In certain groups of non-LTR retrotransposons, such as CR1, these short microsatellite-like repeats are usually element-specific and their formation depends on RT properties (Kajikawa and Okada, 2002; Kapitonov and Jurka, 2003). An ancestral RT, which "stuttered" on a G-rich microsatellite-like sequence and developed the capacity to disperse it *in trans*, could have been selected for at the time of transition from circular to linear chromosomes. On the other hand, we cannot disregard the possibility that a primordial RT gene, which existed in an ancestral eukaryote, was fused to an N-terminal domain that would ensure its specific interactions with an unlinked G-rich template, a combination which then became selectively advantageous.

11.6 CONCLUDING REMARKS

Forty years after the discovery of RT, a unique enzyme capable of performing RNA-templated DNA synthesis (Baltimore, 1970; Temin and Mizutani, 1970), it becomes increasingly clear that the role of reverse transcription is not limited to proliferation of viruses and selfish genetic elements. The initial interest in telomerase as a RT in 1997 originally led to the question of whichever came first, telomerases or retrotransposons, and this question has also evolved over time. Perhaps it is no longer the most relevant question to be asked, as it is equally possible that a primordial RT could have given rise to both retrotransposons and telomerases, so that *cis*-retromobility and *trans*-dispersal of short repeats by an RT never had to replace one another. Given that RT is simply another type of DNA polymerase belonging to a much larger family of "right-hand" polymerases, including the Klenow fragment of DNA polymerase I, this may be a good time to reconsider our approach to thinking about RT evolution and to acknowledge that ancestral RTs were not necessarily associated with mobile elements.

Among prokaryotic RTs, it is difficult to pinpoint the RT type that would most resemble the ancestor of telomerase, and it does not necessarily exist in extant prokaryotic organisms. It is likely, however, that such an ancestral form already possessed the N- and C-terminal extensions which can now be found in both TERTs and PLEs and confer the important features permitting specific interactions with the RNA template. Future progress in defining the ancestral forms will probably depend on identification of additional evolutionary intermediates, whose structural and functional properties could help to fill in the existing gaps in retroelement phylogeny. Given the current speed of advances in genome sequencing technologies, the number of odd life forms with partially or fully sequenced genomes is likely to serve as an ever-expanding source of novel RT types in a quest for understanding the nature of major evolutionary transitions which occurred billions of years ago.

ACKNOWLEDGMENTS

The author thanks Joan Curcio, Marlene Belfort, and the editors, Neal F. Lue and Chantal Autexier, for critical reading of the manuscript and helpful comments. The work in her laboratory was supported by the grant MCB-0821956 from the National Science Foundation.

REFERENCES

Abad JP, De Pablos B, Osoegawa K, De Jong PJ, Martín-Gallardo A, Villasante A. (2004) TAHRE, a novel telomeric retrotransposon from *Drosophila melanogaster*, reveals the origin of *Drosophila* telomeres. *Mol. Biol. Evol.* **21**: 1620–1624.

Adam RD, Nash TE, Wellems TE. (1991) Telomeric location of *Giardia* rDNA genes. *Mol. Cell. Biol.* **11**: 3326–3330.

Aihara H, Huang WM, Ellenberger T. (2007) An interlocked dimer of the protelomerase TelK distorts DNA structure for the formation of hairpin telomeres. *Mol. Cell.* **27**: 901–913.

Arkhipova IR, Mazo AM, Cherkasova VA, Gorelova TV, Schuppe NG, Ilyin YV (1986) The steps of reverse transcription of Drosophila mobile dispersed genetic elements and U3-R-U5 structure of their LTRs. *Cell* **44**: 555–563.

Arkhipova IR, Morrison HG. (2001) Three retrotransposon families in the genome of *Giardia lamblia*: two telomeric, one dead. *Proc. Natl. Acad. Sci. USA*. **98**: 14497–14502.

Arkhipova IR, Pyatkov KI, Meselson M, Evgen'ev MB. (2003) Retroelements containing introns in diverse invertebrate taxa. *Nat. Genet.* **33**: 123–124.

Autexier C, Lue NF. (2006) The structure and function of telomerase reverse transcriptase. *Annu. Rev. Biochem.* **75**: 493–517.

Baltimore D. (1970) RNA-dependent DNA polymerase in virions of RNA tumor viruses. *Nature*. **226**: 1209–1211.

Bayev AA Jr, Krayev AS, Lyubomirskaya NV, Ilyin YV, Skryabin KG, Georgiev GP. (1980) The transposable element Mdg3 in *Drosophila melanogaster* is flanked with the perfect direct and mismatched inverted repeats. *Nucl. Acids Res.* **8**: 3263–3273.

Beauregard A, Curcio MJ, Belfort M. (2008) The take and give between retrotransposable elements and their hosts. *Annu. Rev. Genet.* **42**: 587–617.

Belfort M, Derbyshire V, Parker MM, Cousineau B, Lambowitz AM. (2002) Mobile introns: pathways and proteins. In: Craig NL, Craigie R, Gellert M, Lambowitz A,eds. *Mobile DNA II*. Washington, DC: ASM Press, pp. 761–783.

Biessmann H, Mason JM. (1997) Telomere maintenance without telomerase. *Chromosoma*. **106**: 63–69.

Biessmann H, Mason JM, Ferry K, d'Hulst M, Valgeirsdottir K, Traverse KL, Pardue ML. (1990) Addition of telomere-associated HeT DNA sequences "heals" broken chromosome ends in *Drosophila*. *Cell.* **61**: 663–673.

Biessmann H, Champion LE, C'Hair M, Ikenaga K, Kasravi B, Mason JM. (1992) Frequent transpositions of *Drosophila melanogaster* HeT-A transposable elements to receding chromosome ends. *EMBO J.* **11**: 4459–4469.

Biessmann H, Mason JM. (1997) Telomere maintenance without telomerase. *Chromosoma*. **106**: 63–69.

Biessmann H, Mason JM. (2003) Telomerase-independent mechanisms of telomere elongation. *Cell. Mol. Life Sci.* **60**: 2325–2333.

Boer PH, Gray MW. (1988) Genes encoding a subunit of respiratory NADH dehydrogenase (ND1) and a reverse transcriptase-like protein (RTL) are linked to ribosomal RNA gene pieces in *Chlamydomonas reinhardtii* mitochondrial DNA. *EMBO J.* **11**: 3501–3508.

Brady TL, Fuerst PG, Dick RA, Schmidt C, Voytas DF. (2008) Retrotransposon target site selection by imitation of a cellular protein. *Mol. Cell. Biol.* **28**: 1230–1239.

Burke WD, Malik HS, Rich SM, Eickbush TH. (2002) Ancient lineages of non-LTR retrotransposons in the primitive eukaryote, *Giardia lamblia*. *Mol. Biol. Evol.* **19**: 619–630.

Capy P, Vitalis R, Langin T, Higuet D, Bazin C. (1996) Relationships between transposable elements based upon the integrase-transposase domains: is there a common ancestor? *J. Mol. Evol.* **42**: 359–368.

Capy P, Langin T, Higuet D, Maurer P, Bazin C. (1997) Do the integrases of LTR-retrotransposons and class II element transposases have a common ancestor? *Genetica*. **100**: 63–72.

Casacuberta E, Pardue ML. (2002) Coevolution of the telomeric retrotransposons across *Drosophila* species. *Genetics*. **161**: 1113–1124.

Casacuberta E, Pardue ML. (2003a) Transposon telomeres are widely distributed in the *Drosophila* genus: TART elements in the virilis group. *Proc. Natl. Acad. Sci. USA* **100**: 3363–3368.

Casacuberta E, Pardue ML. (2003b) HeT-A elements in *Drosophila virilis*: retrotransposon telomeres are conserved across the *Drosophila* genus. *Proc. Natl. Acad. Sci. USA*. **100**: 14091–14096.

Casjens S. (1999) Evolution of the linear DNA replicons of the *Borrelia spirochetes*. *Curr. Opin. Microbiol.* **2**: 529–534.

Chaconas G, Chen CW. (2005) Linear chromosomes in bacteria: no longer going around in circles. In *The Bacterial Chromosome*. Higgins, N.P. (ed.). Washington, DC: American Society for Microbiology Press pp. 525–539.

Chaconas G. (2005) Hairpin telomeres and genome plasticity in Borrelia: all mixed up in the end. *Mol. Microbiol.* **58**: 625–635.

Chang GS, Hong Y, Ko KD, Bhardwaj G, Holmes EC, Patterson RL, van Rossum DB (2008) Phylogenetic profiles reveal evolutionary relationships within the "twilight zone" of sequence similarity. *Proc Natl Acad Sci USA* **105**: 13474–13479.

Chen CW, Huang CH, Lee HH, Tsai HH, Kirby R. (2002) Once the circle has been broken: dynamics and evolution of *Streptomyces* chromosomes. *Trends Genet*. **18**: 522–529.

Chiang CC, Kennell JC, Wanner LA, Lambowitz AM. (1994) A mitochondrial retroplasmid integrates into mitochondrial DNA by a novel mechanism involving the synthesis of a hybrid cDNA and homologous recombination. *Mol. Cell. Biol.* **14**: 6419–6432.

Chiang CC, Lambowitz AM. (1997) The Mauriceville retroplasmid reverse transcriptase initiates cDNA synthesis *de novo* at the 3′ end of tRNAs. *Mol. Cell. Biol.* **17**: 4526–4535.

Clark AG et al: *Drosophila* 12 Genomes Consortium (2007) Evolution of genes and genomes on the *Drosophila* phylogeny. *Nature*. **450**: 203–218.

Cohn M, McEachern MJ, Blackburn EH. (1998) Telomeric sequence diversity within the genus *Saccharomyces*. *Curr. Genet.* **33**: 83–91.

Curcio MJ, Belfort M. (2007) The beginning of the end: links between ancient retroelements and modern telomerases. *Proc. Natl. Acad. Sci. USA*. **104**: 9107–9108.

Dai J, Xie W, Brady TL, Gao J, Voytas DF. (2007) Phosphorylation regulates integration of the yeast Ty5 retrotransposon into heterochromatin. *Mol. Cell.* **27**: 289–299.

Danilevskaya ON, Petrov DA, Pavlova MN, Koga A, Kurenova EV, Hartl DL. (1992) A repetitive DNA element, associated with telomeric sequences in *Drosophila melanogaster*, contains open reading frames. *Chromosoma*. **102**: 32–40.

Danilevskaya ON, Slot F, Pavlova M, Pardue ML. (1994) Structure of the *Drosophila* HeT-A transposon: a retrotransposon-like element forming telomeres. *Chromosoma*. **103**: 215–224.

Danilevskaya ON, Arkhipova IR, Traverse KL, Pardue ML. (1997) Promoting in tandem: the promoter for telomere transposon HeT-A and implications for the evolution of retroviral LTRs. *Cell*. **88**: 647–655.

Danilevskaya ON, Tan C, Wong J, Alibhai M, Pardue ML. (1998) Unusual features of the *Drosophila melanogaster* telomere transposable element HeT-A are conserved in *Drosophila yakuba* telomere elements. *Proc. Natl. Acad. Sci. USA*. **95**: 3770–3775.

Deneke J, Ziegelin G, Lurz R, Lanka E. (2000) The protelomerase of temperate *Escherichia coli* phage N15 has cleaving-joining activity. *Proc. Natl. Acad. Sci. USA*. **97**: 7721–7726.

Deneke J, Ziegelin G, Lurz R, Lanka E. (2002) Phage N15 telomere resolution. Target requirements for recognition and processing by the protelomerase. *J. Biol. Chem.* **277**: 10410–10419.

Domingo E, Holland JJ. (1997) RNA virus mutations and fitness for survival. *Annu. Rev. Microbiol.* **51**: 151–178.

Doulatov S, Hodes A, Dai L, Mandhana N, Liu M, Deora R, Simons RW, Zimmerly S, Miller JF. (2004) Tropism switching in *Bordetella* bacteriophage defines a family of diversity-generating retroelements. *Nature*. **431**: 476–481.

Eickbush TH. (1997) Telomerase and retrotransposons: which came first? *Science*. **277**: 911–912.

Eickbush TH, Jamburuthugoda VK. (2008) The diversity of retrotransposons and the properties of their reverse transcriptases. *Virus Res.* **134**: 221–234.

Eickbush TH, Malik H. (2002) Origins and evolution of retrotransposons. In: Craig NL, Craigie R, Gellert M, Lambowitz AM,eds., *Mobile DNA II*. Washington DC: ASM Press, pp. 1111–1144.

Evgen'ev MB, Arkhipova IR. (2005) Penelope-like elements–a new class of retroelements: distribution, function and possible evolutionary significance. *Cytogenet. Genome Res.* **110**: 510–521.

Fang G, Cech TR. (1993) The β subunit of *Oxytricha* telomere-binding protein promotes G-quartet formation by telomeric DNA. *Cell*. **74**: 875–885.

Frydrychová R, Grossmann P, Trubac P, Vítková M, Marec F. (2004) Phylogenetic distribution of TTAGG telomeric repeats in insects. *Genome*. **47**: 163–178.

Fujiwara H, Osanai M, Matsumoto T, Kojima KK. (2005) Telomere-specific non-LTR retrotransposons and telomere maintenance in the silkworm, *Bombyx mori*. *Chromosome Res.* **13**: 455–467.

Fuller AM, Cook EG, Kelley KJ, Pardue ML. (2010) Gag proteins of *Drosophila* telomeric retrotransposons: collaborative targeting to chromosome ends. *Genetics*. **184**: 629–636.

Gafner J, Philippsen P. (1980) The yeast transposon Ty1 generates duplications of target DNA on insertion. *Nature*. **286**: 414–418.

George JA, DeBaryshe PG, Traverse KL, Celniker SE, Pardue ML. (2006) Genomic organization of the *Drosophila* telomere retrotransposable elements. *Genome Res.* **16**: 1231–1240.

Gladyshev EA, Arkhipova IR. (2007) Telomere-associated endonuclease-deficient Penelope-like retroelements in diverse eukaryotes. *Proc. Natl. Acad. Sci. USA.* **104**: 9352–9357.

Gladyshev EA, Arkhipova IR. (2011) A widespread class of reverse transcriptase-related cellular genes. *Proc. Natl. Acad. Sci. USA.* **51**: 20311–20316.

Guo H, Tse LV, Barbalat R, Sivaamnuaiphorn S, Xu M, Doulatov S, Miller JF. (2008) Diversity-generating retroelement homing regenerates target sequences for repeated rounds of codon rewriting and protein diversification. *Mol. Cell.* **31**: 813–823.

Higashiyama T, Noutoshi Y, Fujie M, Yamada T. (1997) Zepp, a LINE-like retrotransposon accumulated in the *Chlorella* telomeric region. *EMBO J.* **16**: 3715–3723.

Hou G, Le Blancq S E.Y., Zhu H, Lee M. (1995) Structure of a frequently rearranged rRNA-encoding chromosome in *Giardia lamblia. Nucl. Acids Res.* **23**: 3310–3317.

Holt RA, Subramanian GM, Halpern A, Sutton GG, Charlab R, et al. (2002) The genome sequence of the malaria mosquito *Anopheles gambiae. Science.* **298**: 129–149.

Huson DH, Bryant D. (2006) Application of phylogenetic networks in evolutionary studies. *Mol. Biol. Evol.* **23**: 254–267.

Ichiyanagi K, Beauregard A, Lawrence S, Smith D, Cousineau B, Belfort M. (2002) Retrotransposition of the Ll.LtrB group II intron proceeds predominantly via reverse splicing into DNA targets. *Mol. Microbiol.* **46**: 1259–1272.

Ichiyanagi K, Beauregard A, Belfort M. (2003) A bacterial group II intron favors retrotransposition into plasmid targets. *Proc. Natl. Acad. Sci. USA.* **100**: 15742–15747.

Jacobs SA, Podell ER, Cech TR. (2006) Crystal structure of the essential N-terminal domain of telomerase reverse transcriptase. *Nat. Struct. Mol. Biol.* **13**: 218–225.

Kahn T, Savitsky M, Georgiev P. (2000) Attachment of HeT-A sequences to chromosomal termini in *Drosophila melanogaster* may occur by different mechanisms. *Mol. Cell. Biol.* **20**: 7634–7642.

Kamnert I, López CC, Rosén M, Edström JE. (1997) Telomeres terminating with long complex tandem repeats. *Hereditas.* **127**: 175–180.

Kajikawa M, Okada N. (2002) LINEs mobilize SINEs in the eel through a shared 3' sequence. *Cell.* **111**: 433–444.

Kapitonov VV, Jurka J. (2003) The esterase and PHD domains in CR1-like non-LTR retrotransposons. *Mol. Biol. Evol.* **20**: 38–46.

Kistler HC, Benny U, Powell WA. (1997) Linear mitochondrial plasmids of *Fusarium oxysporum* contain genes with sequence similarity to genes encoding a reverse transcriptase from *Neurospora* spp. *Appl. Environ. Microbiol.* **63**: 3311–3313.

Kojima KK, Fujiwara H. (2005) An extraordinary retrotransposon family encoding dual endonucleases. *Genome Res.* **15**: 1106–1117.

Kojima KK, Kanehisa M. (2008) Systematic survey for novel types of prokaryotic retroelements based on gene neighborhood and protein architecture. *Mol. Biol. Evol.* **25**: 1395–1404.

Koonin EV, Wolf YI, Nagasaki K, Dolja VV. (2008) The Big Bang of picorna-like virus evolution antedates the radiation of eukaryotic supergroups. *Nat. Rev. Microbiol.* **6**: 925–939.

Kuiper MT, Lambowitz AM. (1988) A novel reverse transcriptase activity associated with mitochondrial plasmids of *Neurospora. Cell.* **55**: 693–704.

Lambowitz AM, Zimmerly S. (2004) Mobile group II introns. *Annu. Rev. Genet.* **38**: 1–35.

Lambowitz AM, Zimmerly S. (2010) Group II introns: mobile ribozymes that invade DNA. *Cold Spring Harb. Perspect. Biol.* DOI: 10.1101/cshperspect.a003616

Lampson BC, Sun J, Hsu M-Y., Vallejo-Ramirez J, Inouye S, Inouye M. (1989) Reverse transcriptase in a clinical strain of *E. coli*: its requirements for production of branched RNA-linked msDNA. *Science*. **243**: 1033–1038.

Lampson BC, Inouye M, Inouye S. (2005) Retrons, msDNA, and the bacterial genome. *Cytogenet. Genome Res.* **110**: 491–499.

de Lange T. (2004) T-loops and the origin of telomeres. *Nat. Rev. Mol. Cell. Biol.* **5**: 323–329.

Levis R, Dunsmuir P, Rubin GM. (1980) Terminal repeats of the *Drosophila* transposable element copia: nucleotide sequence and genomic organization. *Cell*. **21**: 581–588.

Levis RW, Ganesan R, Houtchens K, Tolar LA, Sheen FM. (1993) Transposons in place of telomeric repeats at a *Drosophila* telomere. *Cell*. **75**: 1083–1093.

Lim D, Maas WK. (1989) Reverse transcriptase-dependent synthesis of a covalently linked, branched DNA-RNA compound in *E. coli* B. *Cell*. **56**: 891–904.

Lingner J, Hughes TR, Shevchenko A, Mann M, Lundblad V, Cech TR. (1997) Reverse transcriptase motifs in the catalytic subunit of telomerase. *Science*. **276**: 561–567.

Luan DD, Korman MH, Jakubczak JL, Eickbush TH. (1993) Reverse transcription of R2Bm RNA is primed by a nick at the chromosomal target site: a mechanism for non-LTR retrotransposition. *Cell*. **72**: 595–605.

Lundblad V (2002) Telomere maintenance without telomerase. *Oncogene* **21**: 522–531.

Lyozin GT, Makarova KS, Velikodvorskaja VV, Zelentsova HS, Khechumian RR, Kidwell MG, Koonin EV, Evgen'ev MB. (2001) The structure and evolution of Penelope in the virilis species group of *Drosophila*: an ancient lineage of retroelements. *J. Mol. Evol.* **52**: 445–456.

Maizels N, Weiner AM. (1993) The genomic tag hypothesis: modern viruses as molecular fossils of ancient strategies for genomic replication. Pp. 577–602 in *The RNA World*, 2nd ed. R.F. Gesteland, J.F. Atkins,eds. Cold Spring Harbor Press, Cold Spring Harbor, NY.

Maizels N, Weiner AM. (1993) The genomic tag hypothesis: what molecular fossils tell us about the evolution of tRNA. Pp. 79–111 in *The RNA World*, 2nd ed. R.F. Gestel J.F. Atkins, eds. Cold Spring Harbor Press, Cold Spring Harbor, NY.

Malik HS, Burke WD, Eickbush TH. (1999) The age and evolution of non-LTR retrotransposable elements. *Mol. Biol. Evol.* **16**: 793–805.

Malik HS, Burke WD, Eickbush TH. (2000a) Putative telomerase catalytic subunits from *Giardia lamblia* and *Caenorhabditis elegans*. *Gene*. **251**: 101–108.

Malik HS, Henikoff S, Eickbush TH. (2000b) Poised for contagion: evolutionary origins of the infectious abilities of invertebrate retroviruses. *Genome Res.* **10**: 1307–1318.

Malik HS, Eickbush TH. (2001) Phylogenetic analysis of ribonuclease H domains suggests a late, chimeric origin of LTR retrotransposable elements and retroviruses. *Genome Res.* **11**: 1187–1197.

Mason JM, Frydrychova RC, Biessmann H. (2008) *Drosophila* telomeres: an exception providing new insights. *Bioessays*. **30**: 25–37.

Matsumoto T, Takahashi H, Fujiwara H. (2004) Targeted nuclear import of open reading frame 1 is required for *in vivo* retrotransposition of a telomere-specific non-long terminal repeat retrotransposon, SART1. *Mol. Cell. Biol.* **24**: 105–122.

Maxwell PH, Coombes C, Kenny AE, Lawler JF, Boeke JD, Curcio MJ. (2004) Ty1 mobilizes subtelomeric Y′ elements in telomerase-negative *Saccharomyces cerevisiae* survivors. *Mol. Cell. Biol.* **24**: 9887–9898.

Maxwell PH, Curcio MJ. (2008) Incorporation of Y′-Ty1 cDNA destabilizes telomeres in *Saccharomyces cerevisiae* telomerase-negative mutants. *Genetics*. **179**: 2313–2317.

McEachern MJ, Krauskopf A, Blackburn EH. (2000) Telomeres and their control. *Annu Rev Genet*. **34**: 331–358.

Medhekar B, Miller JF. (2007) Diversity-generating retroelements. *Curr. Opin. Microbiol.* **10**: 388–395.

Morin GB, Cech TR. (1986) The telomeres of the linear mitochondrial DNA of *Tetrahymena thermophila* consist of 53 bp tandem repeats. *Cell*. **46**: 873–883.

Morin GB, Cech TR. (1988) Mitochondrial telomeres: surprising diversity of repeated telomeric DNA sequences among six species of *Tetrahymena*. *Cell*. **52**: 367–374.

Morrish TA, Gilbert N, Myers JS, Vincent BJ, Stamato TD, Taccioli GE, Batzer MA, Moran JV. (2002) DNA repair mediated by endonuclease-independent LINE-1 retrotransposition. *Nat. Genet.* **31**: 159–165.

Morrish TA, Garcia-Perez JL, Stamato TD, Taccioli GE, Sekiguchi J, Moran JV. (2007) Endonuclease-independent LINE-1 retrotransposition at mammalian telomeres. *Nature*. **446**: 208–212.

Nakamura TM, Morin GB, Chapman KB, Weinrich SL, Andrews WH, Lingner J, Harley CB, Cech TR. (1997) Telomerase catalytic subunit homologs from fission yeast and human. *Science*. **277**: 955–959.

Nakamura TM, Cech TR. (1998) Reversing time: origin of telomerase. *Cell*. **92**: 587–590.

Neidle S, Parkinson GN. (2003) The structure of telomeric DNA. *Curr. Opin. Struct. Biol.* **13**: 275–283.

Nene V, Wortman JR, Lawson D, Haas B, Kodira C, et al. (2007) Genome sequence of *Aedes aegypti*, a major arbovirus vector. *Science*. **316**: 1718–1723.

Nosek J, Tomáska L, Fukuhara H, Suyama Y, Kovác L (1998) Linear mitochondrial genomes: 30 years down the line. *Trends Genet*. **14**: 184–188.

Noutoshi Y, Arai R, Fujie M, Yamada T. (1998) Structure of the *Chlorella* Zepp retrotransposon: nested Zepp clusters in the genome. *Mol. Gen. Genet.* **259**: 256–263.

Oganesian L, Bryan TM. (2007) Physiological relevance of telomeric G-quadruplex formation: a potential drug target. *Bioessays*. **29**: 155–165.

Okazaki S, Tsuchida K, Maekawa H, Ishikawa H, Fujiwara H. (1993) Identification of a pentanucleotide telomeric sequence, $(TTAGG)_n$, in the silkworm *Bombyx mori* and in other insects. *Mol. Cell. Biol.* **13**: 1424–1432.

Osanai M, Kojima KK, Futahashi R, Yaguchi S, Fujiwara H. (2006) Identification and characterization of the telomerase reverse transcriptase of *Bombyx mori* (silkworm) and *Tribolium castaneum* (flour beetle). *Gene*. **376**: 281–289.

Ostertag EM, Kazazian HH Jr. (2001) Twin priming: a proposed mechanism for the creation of inversions in L1 retrotransposition. *Genome Res*. **11**: 2059–2065.

Pardue ML, Danilevskaya ON, Traverse KL, Lowenhaupt K. (1997) Evolutionary links between telomeres and transposable elements. *Genetica*. **100**: 73–84.

Pardue ML, DeBaryshe PG. (2003) Retrotransposons provide an evolutionarily robust non-telomerase mechanism to maintain telomeres. *Annu. Rev. Genet.* **37**: 485–511.

Pardue ML, Rashkova S, Casacuberta E, DeBaryshe PG, George JA, Traverse KL. (2005) Two retrotransposons maintain telomeres in *Drosophila*. *Chromosome Res*. **13**: 443–453.

Pardue ML, Debaryshe PG. (2008) *Drosophila* telomeres: a variation on the telomerase theme. *Fly (Austin)*. **2** (3).

Poulter RT, Goodwin TJ. (2005) DIRS-1 and the other tyrosine recombinase retrotransposons. *Cytogenet. Genome Res.* **110**: 575–588.

Rashkova S, Karam SE, Pardue ML. (2002a) Element-specific localization of *Drosophila* retrotransposon Gag proteins occurs in both nucleus and cytoplasm. *Proc. Natl. Acad. Sci. USA*. **99**: 3621–3626.

Rashkova S, Karam SE, Kellum R, Pardue ML. (2002b) Gag proteins of the two *Drosophila* telomeric retrotransposons are targeted to chromosome ends. *J. Cell. Biol.* **159**: 397–402.

Ricchetti M, Buc H. (1996) A reiterative mode of DNA synthesis adopted by HIV-1 reverse transcriptase after a misincorporation. *Biochemistry*. **35**: 14970–14983.

Robertson HM, Gordon KH. (2006) Canonical TTAGG-repeat telomeres and telomerase in the honey bee, *Apis mellifera*. *Genome Res.* **16**: 1345–1351.

Sasaki T, Fujiwara H. (2000) Detection and distribution patterns of telomerase activity in insects. *Eur. J. Biochem.* **267**: 3025–3031.

Sen D, Gilbert W. (1992) Guanine quartet structures. *Methods Enzymol.* **211**: 191–199.

Simon DM, Zimmerly S. (2008) A diversity of uncharacterized reverse transcriptases in bacteria. *Nucl. Acids Res.* **36**: 7219–7229.

Simpson EB, Ross SL, Marchetti SE, Kennell JC. (2004) Relaxed primer specificity associated with reverse transcriptases encoded by the pFOXC retroplasmids of *Fusarium oxysporum*. *Eukaryot. Cell.* **3**: 1589–1600.

Scholes DT, Kenny AE, Gamache ER, Mou Z, Curcio MJ. (2003) Activation of a LTR-retrotransposon by telomere erosion. *Proc. Natl. Acad. Sci. USA*. **100**: 15736–15741.

Temin HM, Mizutani S. (1970) RNA-dependent DNA polymerase in virions of Rous sarcoma virus. *Nature*. **226**: 1211–1213.

Temin HM. (1985) Reverse transcription in the eukaryotic genome: retroviruses, pararetroviruses, retrotransposons, and retrotranscripts. *Mol. Biol. Evol.* **2**: 455–468.

Toro N, Jiménez-Zurdo JI, García-Rodríguez FM. (2007) Bacterial group II introns: not just splicing. *FEMS Microbiol. Rev.* **31**: 342–358.

Traverse KL, George JA, Debaryshe PG, Pardue ML. (2010) Evolution of species-specific promoter-associated mechanisms for protecting chromosome ends by *Drosophila* Het-A telomeric transposons. *Proc. Natl. Acad. Sci. USA*. **107**: 5064–5069.

Van Roey P, Derbyshire V. (2005) GIY-YIG endonucleases—beads on a string. In: *Homing Endonucleases and Inteins*, Belfort M, Derbyshire V, Stoddard BL, Wood DW,eds. Springer, New York pp. 67–83.

Villasante A, Abad JP, Planelló R, Méndez-Lago M, Celniker SE, de Pablos B. (2007) *Drosophila* telomeric retrotransposons derived from an ancestral element that was recruited to replace telomerase. *Genome Res.* **17**: 1909–1918.

Villasante A, de Pablos B, Méndez-Lago M, Abad JP. (2008) Telomere maintenance in *Drosophila*: rapid transposon evolution at chromosome ends. *Cell Cycle*. **7**: 2134–2138.

Vítková M, Král J, Traut W, Zrzavý J, Marec F. (2005) The evolutionary origin of insect telomeric repeats, (TTAGG)*n*. *Chromosome Res.* **13**: 145–156.

Volff JN, Altenbuchner J. (2000) A new beginning with new ends: linearisation of circular chromosomes during bacterial evolution. *FEMS Microbiol. Lett.* **186**: 143–150.

Walther TC, Kennell JC. (1999) Linear mitochondrial plasmids of *F. oxysporum* are novel, telomere-like retroelements. *Mol. Cell.* **4**: 229–238.

Wang H, Lambowitz AM. (1993) The Mauriceville plasmid reverse transcriptase can initiate cDNA synthesis *de novo* and may be related to reverse transcriptase and DNA polymerase progenitor. *Cell.* **75**: 1071–1081.

Weiner AM, Maizels N. (1987) 3′ terminal tRNA-like structures tag genomic RNA molecules for replication: implications for the origin of protein synthesis. *Proc. Natl. Acad. Sci. USA.* **84**: 7383–7387.

Xie W, Gai X, Zhu Y, Zappulla DC, Sternglanz R, Voytas DF. (2001) Targeting of the yeast Ty5 retrotransposon to silent chromatin is mediated by interactions between integrase and Sir4p. *Mol. Cell. Biol.* **21**: 6606–6614.

Xiong Y, Eickbush TH. (1988) The site-specific ribosomal DNA insertion element R1Bm belongs to a class of non-long-terminal-repeat retrotransposons. *Mol. Cell. Biol.* **8**: 114–123.

Yamamoto Y, Fujimoto Y, Arai R, Fujie M, Usami S, Yamada T. (2003) Retrotransposon-mediated restoration of *Chlorella* telomeres: accumulation of Zepp retrotransposons at termini of newly formed minichromosomes. *Nucl. Acids Res.* **31**: 4646–4653.

Yamamoto Y, Noutoshi Y, Fujie M, Usami S, Yamada T. (2000) Analysis of double-strand-break repair by *Chlorella* retrotransposon Zepp. *Nucl. Acids Symp. Ser.* **44**: 101–102.

Zakian VA. (1995) Telomeres: beginning to understand the end. *Science.* **270**: 1601–1607.

Zhang ML, Tong XJ, Fu XH, Zhou BO, Wang J, Liao XH, Li QJ, Shen N, Ding J, Zhou JQ. (2010) Yeast telomerase subunit Est1p has guanine quadruplex-promoting activity that is required for telomere elongation. *Nat. Struct. Mol. Biol.* **17**: 202–209.

Zimmerly S, Guo H, Perlman PS, Lambowitz AM. (1995) Group II intron mobility occurs by target DNA-primed reverse transcription. *Cell.* **82**: 545–554.

Zou S, Wright DA, Voytas DF. (1995) The *Saccharomyces* Ty5 retrotransposon family is associated with origins of DNA replication at the telomeres and the silent mating locus HMR. *Proc. Natl. Acad. Sci. USA.* **92**: 920–924.

Zou S, Ke N, Kim JM, Voytas DF. (1996) The *Saccharomyces* retrotransposon Ty5 integrates preferentially into regions of silent chromatin at the telomeres and mating loci. *Genes Dev.* **10**: 634–645.

Zou S, Voytas DF. (1997) Silent chromatin determines target preference of the *Saccharomyces* retrotransposon Ty5. *Proc. Natl. Acad. Sci. USA.* **94**: 7412–7416.

INDEX

Telomerases: Chemistry, Biology, and Clinical Applications, First Edition.
Edited by Neal F. Lue and Chantal Autexier.